深入浅出 Spring Security

王松 著

清华大学出版社
北京

内 容 简 介

Spring Security 是 Java 企业级开发中常用的安全管理框架,也能完美支持 OAuth2。同时,Spring Security 作为 Spring 家族的一员,与 Spring Boot、Spring Cloud 等框架整合使用也非常方便。

本书分为 15 章,讲解 Spring Security 框架、认证、认证流程分析、过滤器链分析、密码加密、RememberMe、会话管理、HttpFirewall、漏洞保护、HTTP 认证、跨域问题、异常处理、权限管理、权限模型、OAuth2 等内容。本书致力于让读者在学会 Spring Security 用法的同时,也能通过阅读源码来理解它的实现原理。

本书适合具有 Spring Boot 基础的读者、Java 企业应用开发工程师,也适合作为高等院校和培训机构计算机相关专业师生的教学参考书。

本书封面贴有清华大学出版社防伪标签,无标签者不得销售。
版权所有,侵权必究。举报: 010-62782989, beiqinquan@tup.tsinghua.edu.cn。

图书在版编目(CIP)数据

深入浅出 Spring Security / 王松著.—北京: 清华大学出版社,2021.1(2024.3重印)
ISBN 978-7-302-57276-3

Ⅰ.①深… Ⅱ.①王… Ⅲ.①JAVA 语言—程序设计 Ⅳ.①TP312.8

中国版本图书馆 CIP 数据核字(2021)第 005021 号

责任编辑: 夏毓彦
封面设计: 王 翔
责任校对: 闫秀华
责任印制: 丛怀宇

出版发行: 清华大学出版社
网　址: https://www.tup.com.cn, https://www.wqxuetang.com
地　址: 北京清华大学学研大厦 A 座　　邮　编: 100084
社 总 机: 010-83470000　　邮　购: 010-62786544
投稿与读者服务: 010-62776969, c-service@tup.tsinghua.edu.cn
质 量 反 馈: 010-62772015, zhiliang@tup.tsinghua.edu.cn

印 装 者: 三河市人民印务有限公司
经　销: 全国新华书店
开　本: 190mm×260mm　　印　张: 26.25　　字　数: 672 千字
版　次: 2021 年 3 月第 1 版　　印　次: 2024 年 3 月第 4 次印刷
定　价: 99.00 元

产品编号: 089794-01

前　言

安全管理是 Java 应用开发中无法避免的问题，目前主流的安全管理框架就是 Spring Security 和 Shiro，其中 Shiro 一直以使用简单和轻量级著称。然而，随着 Spring Boot 和微服务的流行，Spring Security 受到越来越多开发者的重视，因为 Spring Security 在和 Spring Boot 整合时具有先天优势。

目前市面上缺少系统介绍 Spring Security 的书籍，网上的博客内容又比较零散，这为很多初次接触 Spring Security 的 Java 工程师学习这门技术带来诸多不便。

笔者最早于个人博客上连载 Spring Security 系列教程，连载期间有不少读者加笔者微信讨论 Spring Security 的相关技术点，让笔者感受到读者对 Spring Security 的热情，也因此萌生了写一本技术图书来系统介绍 Spring Security 的想法。在朋友和家人的鼓励之下，这一想法逐步付诸实践，最终完成大家现在看到的这本《深入浅出 Spring Security》。

本书以 Spring Security 5.3.4 为基础，详细介绍 Spring Security 的基本用法以及相关原理。得益于 Spring Boot 中的自动化配置，Spring Security 上手非常容易，然而这种自动化配置，也让很多初次接触 Spring Security 的开发者"知其然，而不知其所以然"，仅限于会用，一旦出了漏洞，或者想要定制功能时，就会不知所措。因此，在写作本书过程中，除了基本功能的 Demo 演示外，还对 Spring Security 的相关源码做了深入分析，以便读者"知其然，更知其所以然"。

学习 Spring Security 不仅仅是学习安全管理框架，也是一个学习各种网络攻击与防御策略的过程，Spring Security 对很多常见网络攻击，如计时攻击、CSRF、XSS 等，都提供了相应的防御策略，因此，我们在学习 Spring Security 时，也可以顺便研究一下这些常见的网络攻击，以便设计出更加安全健壮的权限管理系统。

本书分为四部分：

第一部分：第 1 章，这一部分总体介绍 Spring Security 架构，方便读者从整体上把握 Spring Security 的功能。

第二部分：第 2～12 章，这一部分主要介绍 Spring Security 中的认证功能，以及由此衍生出来的会话管理、HTTP 防火墙、跨域管理等。

第三部分：第 13～14 章，这一部分主要介绍 Spring Security 中的授权功能，以及常见的权限模型 ACL 和 RBAC。

第四部分：第 15 章，这一部分主要介绍 OAuth2 协议在 Spring Security 框架中的落地。

示例代码约定

为了减少代码冗余和本书篇幅，书中的所有示例代码片段都省略了 package 和 import 部分，像下面这样：

```java
@Configuration
public class SecurityConfig extends WebSecurityConfigurerAdapter {
    @Override
    protected void configure(HttpSecurity http) throws Exception {
        http.authorizeRequests()
                .anyRequest().authenticated()
                .and()
                .formLogin()
                .loginPage("/login.html")
                .loginProcessingUrl("/doLogin")
                .defaultSuccessUrl("/index ")
                .failureUrl("/login.html")
                .usernameParameter("uname")
                .passwordParameter("passwd")
                .permitAll()
                .and()
                .csrf().disable();
    }
}
```

有时候为了向读者演示代码的运行效果，一个案例可能会被反复修改多次，那么在后面展示代码时，将不再列出不变的部分，仅仅列出发生变化的代码片段，像下面这样：

```java
@Autowired
TokenStore tokenStore;
@Autowired
JwtAccessTokenConverter jwtAccessTokenConverter;
@Bean
AuthorizationServerTokenServices tokenServices() {
    DefaultTokenServices services = new DefaultTokenServices();
    services.setClientDetailsService(clientDetailsService);
    services.setSupportRefreshToken(true);
    services.setTokenStore(tokenStore);
    TokenEnhancerChain tokenEnhancerChain = new TokenEnhancerChain();
    tokenEnhancerChain
            .setTokenEnhancers(Arrays.asList(jwtAccessTokenConverter));
    services.setTokenEnhancer(tokenEnhancerChain);
    return services;
}
//省略其他
```

正常情况下，这样的代码片段并不会影响大家理解本书内容。如果读者想要看到完整的代码片段，可以下载本书提供的示例代码进行对照理解。

源码省略约定

在分析 Spring Security 源码时，为了简化源码和篇幅以便于读者理解，源码中的日志输出、注释以及一些无关紧要的代码会被移除掉，像下面这样：

```java
@ConfigurationProperties(prefix = "spring.security")
public class SecurityProperties {
    private User user = new User();
    public User getUser() {
        return this.user;
    }
    public static class User {
        private String name = "user";
        private String password = UUID.randomUUID().toString();
        private List<String> roles = new ArrayList<>();
        //省略 getter/setter
    }
}
```

如果读者觉得这样阅读"不过瘾"，也可以下载 Spring Security 源码对照理解。

读者定位

阅读本书需要有一定的 Spring Boot 基础，对于无 Spring Boot 基础的读者，可以先学习 Spring Boot 然后再来阅读本书。学习 Spring Boot，可以参考笔者编写的图书《Spring Boot+Vue 全栈开发实战》或者笔者的教程：http://springboot.javaboy.org。

源码获取

本书所有的示例代码均存放在 GitHub 上，地址如下：
https://github.com/lenve/spring-security-book-samples
所有工程均为标准的 Maven 工程，可以用 IntelliJ IDEA 或者 Eclipse 打开。

纠错与勘误

如果读者在阅读本书时发现错误，可以将错误提交到 https://github.com/lenve/spring-security-book-samples/issues，笔者将错误内容汇总后同步发布在 http://www.javaboy.org/spring-security-book 以及微信公众号"江南一点雨"。修正后的内容将在后续重印的书中得到体现。

交流社区

学无止境，笔者将继续对 Spring Security 的发展保持关注。关于 Spring Security 的最新变化，笔者都将发布在微信公众号"江南一点雨"上，读者关注微信公众号后，也可以进入本书微信交流群进行交流。

王松

2021 年 1 月

目 录

第 1 章 Spring Security 架构概览 1

1.1 Spring Security 简介 1
1.2 Spring Security 核心功能 2
    1.2.1 认证 3
    1.2.2 授权 3
    1.2.3 其他 3
1.3 Spring Security 整体架构 4
    1.3.1 认证和授权 4
    1.3.2 Web 安全 6
    1.3.3 登录数据保存 9
1.4 小结 9

第 2 章 Spring Security 认证 10

2.1 Spring Security 基本认证 10
    2.1.1 快速入门 10
    2.1.2 流程分析 11
    2.1.3 原理分析 12
2.2 登录表单配置 19
    2.2.1 快速入门 19
    2.2.2 配置细节 23
2.3 登录用户数据获取 39
    2.3.1 从 SecurityContextHolder 中获取 41
    2.3.2 从当前请求对象中获取 59
2.4 用户定义 64
    2.4.1 基于内存 64
    2.4.2 基于 JdbcUserDetailsManager 65

2.4.3 基于 MyBatis ... 68
2.4.4 基于 Spring Data JPA ... 74
2.5 小结 ... 77

第 3 章 认证流程分析 ... 78

3.1 登录流程分析 ... 78
3.1.1 AuthenticationManager ... 78
3.1.2 AuthenticationProvider ... 79
3.1.3 ProviderManager ... 86
3.1.4 AbstractAuthenticationProcessingFilter ... 89
3.2 配置多个数据源 ... 94
3.3 添加登录验证码 ... 95
3.4 小结 ... 99

第 4 章 过滤器链分析 ... 100

4.1 初始化流程分析 ... 100
4.1.1 ObjectPostProcessor ... 101
4.1.2 SecurityFilterChain ... 102
4.1.3 SecurityBuilder ... 103
4.1.4 FilterChainProxy ... 117
4.1.5 SecurityConfigurer ... 120
4.1.6 初始化流程分析 ... 128
4.2 ObjectPostProcessor 使用 ... 136
4.3 多种用户定义方式 ... 137
4.4 定义多个过滤器链 ... 141
4.5 静态资源过滤 ... 144
4.6 使用 JSON 格式登录 ... 146
4.7 添加登录验证码 ... 150
4.8 小结 ... 152

第 5 章 密码加密 ... 153

5.1 密码为什么要加密 ... 153
5.2 密码加密方案进化史 ... 154

目录 | VII

- 5.3 PasswordEncoder 详解 154
  - 5.3.1 PasswordEncoder 常见实现类 155
  - 5.3.2 DelegatingPasswordEncoder 156
- 5.4 实战 159
- 5.5 加密方案自动升级 161
- 5.6 是谁的 PasswordEncoder 166
- 5.7 小结 168

第 6 章 RememberMe 169

- 6.1 RememberMe 简介 169
- 6.2 RememberMe 基本用法 170
- 6.3 持久化令牌 172
- 6.4 二次校验 174
- 6.5 原理分析 176
- 6.6 小结 189

第 7 章 会话管理 190

- 7.1 会话简介 190
- 7.2 会话并发管理 191
  - 7.2.1 实战 191
  - 7.2.2 原理分析 194
- 7.3 会话固定攻击与防御 206
  - 7.3.1 什么是会话固定攻击 206
  - 7.3.2 会话固定攻击防御策略 207
- 7.4 Session 共享 208
  - 7.4.1 集群会话方案 208
  - 7.4.2 实战 210
- 7.5 小结 212

第 8 章 HttpFirewall 213

- 8.1 HttpFirewall 简介 213
- 8.2 HttpFirewall 严格模式 215
  - 8.2.1 rejectForbiddenHttpMethod 216

8.2.2　rejectedBlacklistedUrls 217
　　8.2.3　rejectedUntrustedHosts 218
　　8.2.4　isNormalized 219
　　8.2.5　containsOnlyPrintableAsciiCharacters 220
8.3　HttpFirewall 普通模式 220
8.4　小结 221

第 9 章　漏洞保护 222

9.1　CSRF 攻击与防御 222
　　9.1.1　CSRF 简介 222
　　9.1.2　CSRF 攻击演示 223
　　9.1.3　CSRF 防御 224
　　9.1.4　源码分析 231
9.2　HTTP 响应头处理 237
　　9.2.1　缓存控制 239
　　9.2.2　X-Content-Type-Options 240
　　9.2.3　Strict-Transport-Security 241
　　9.2.4　X-Frame-Options 244
　　9.2.5　X-XSS-Protection 245
　　9.2.6　Content-Security-Policy 246
　　9.2.7　Referrer-Policy 248
　　9.2.8　Feature-Policy 249
　　9.2.9　Clear-Site-Data 249
9.3　HTTP 通信安全 250
　　9.3.1　使用 HTTPS 250
　　9.3.2　代理服务器配置 253
9.4　小结 254

第 10 章　HTTP 认证 255

10.1　HTTP Basic authentication 255
　　10.1.1　简介 255
　　10.1.2　具体用法 257
　　10.1.3　源码分析 257
10.2　HTTP Digest authentication 260

			10.2.1	简介	260

			10.2.2	具体用法	261

			10.2.3	源码分析	263

10.3 小结 268

第 11 章 跨域问题 269

11.1 什么是 CORS 269

11.2 Spring 处理方案 270

    11.2.1 @CrossOrigin 271

    11.2.2 addCorsMappings 272

    11.2.3 CorsFilter 273

11.3 Spring Security 处理方案 274

    11.3.1 特殊处理 OPTIONS 请求 275

    11.3.2 继续使用 CorsFilter 275

    11.3.3 专业解决方案 276

11.4 小结 279

第 12 章 异常处理 280

12.1 Spring Security 异常体系 280

12.2 ExceptionTranslationFilter 原理分析 281

12.3 自定义异常配置 287

12.4 小结 290

第 13 章 权限管理 291

13.1 什么是权限管理 291

13.2 Spring Security 权限管理策略 292

13.3 核心概念 292

    13.3.1 角色与权限 292

    13.3.2 角色继承 294

    13.3.3 两种处理器 295

    13.3.4 前置处理器 296

    13.3.5 后置处理器 299

    13.3.6 权限元数据 300

13.3.7　权限表达式 303
13.4　基于 URL 地址的权限管理 305
　　13.4.1　基本用法 306
　　13.4.2　角色继承 308
　　13.4.3　自定义表达式 309
　　13.4.4　原理剖析 310
　　13.4.5　动态管理权限规则 316
13.5　基于方法的权限管理 325
　　13.5.1　注解介绍 325
　　13.5.2　基本用法 326
　　13.5.3　原理剖析 331
13.6　小结 338

第 14 章　权限模型 339

14.1　常见的权限模型 339
14.2　ACL 340
　　14.2.1　ACL 权限模型介绍 340
　　14.2.2　ACL 核心概念介绍 341
　　14.2.3　ACL 数据库分析 343
　　14.2.4　实战 345
14.3　RBAC 354
　　14.3.1　RBAC 权限模型介绍 354
　　14.3.2　RBAC 权限模型分类 355
　　14.3.3　RBAC 小结 357
14.4　小结 357

第 15 章　OAuth2 358

15.1　OAuth2 简介 358
15.2　OAuth2 四种授权模式 359
　　15.2.1　授权码模式 360
　　15.2.2　简化模式 361
　　15.2.3　密码模式 363
　　15.2.4　客户端模式 363
15.3　Spring Security OAuth2 364

- 15.4 GitHub 授权登录 ... 365
  - 15.4.1 准备工作 ... 365
  - 15.4.2 项目开发 ... 367
  - 15.4.3 测试 ... 368
  - 15.4.4 原理分析 ... 369
  - 15.4.5 自定义配置 ... 375
- 15.5 授权服务器与资源服务器 ... 379
  - 15.5.1 项目规划 ... 379
  - 15.5.2 项目搭建 ... 380
  - 15.5.3 测试 ... 391
  - 15.5.4 原理分析 ... 393
  - 15.5.5 自定义请求 ... 396
- 15.6 使用 Redis ... 397
- 15.7 客户端信息存入数据库 ... 399
- 15.8 使用 JWT ... 401
  - 15.8.1 JWT ... 401
  - 15.8.2 JWT 数据格式 ... 402
  - 15.8.3 OAuth2 中使用 JWT ... 403
- 15.9 小结 ... 406

第 1 章

Spring Security 架构概览

Spring Security 虽然历史悠久，但是从来没有像今天这样受到开发者这么多的关注。究其原因，还是沾了微服务的光。作为 Spring 家族中的一员，在和 Spring 家族中的其他产品如 Spring Boot、Spring Cloud 等进行整合时，Spring Security 拥有众多同类型框架无可比拟的优势。本章我们就先从整体上了解一下 Spring Security 及其工作原理。

本章涉及的主要知识点有：

- Spring Security 简介。
- Spring Security 整体架构。

1.1 Spring Security 简介

Java 企业级开发生态丰富，无论你想做哪方面的功能，都有众多的框架和工具可供选择，以至于 SUN 公司在早些年不得不制定了很多规范，这些规范在今天依然影响着我们的开发，安全领域也是如此。然而，不同于其他领域，在 Java 企业级开发中，安全管理方面的框架非常少，一般来说，主要有三种方案：

- Shiro
- Spring Security
- 开发者自己实现

Shiro 本身是一个老牌的安全管理框架，有着众多的优点，例如轻量、简单、易于集成、可以在 JavaSE 环境中使用等。不过，在微服务时代，Shiro 就显得力不从心了，在微服务面前，它无法充分展示自己的优势。

也有开发者选择自己实现安全管理，据笔者所知，这一部分人不在少数。但是一个系统的安全，不仅仅是登录和权限控制这么简单，我们还要考虑各种各样可能存在的网络攻击以及防御策略，从这个角度来说，开发者自己实现安全管理也并非是一件容易的事情，只有大公司才有足够的人力物力去支持这件事情。

Spring Security 作为 Spring 家族的一员，在和 Spring 家族的其他成员如 Spring Boot、Spring Cloud 等进行整合时，具有其他框架无可比拟的优势，同时对 OAuth2 有着良好的支持，再加上 Spring Cloud 对 Spring Security 的不断加持（如推出 Spring Cloud Security），让 Spring Security 不知不觉中成为微服务项目的首选安全管理方案。

陈年旧事

Spring Security 最早叫 Acegi Security，这个名称并不是说它和 Spring 就没有关系，它依然是为 Spring 框架提供安全支持的。Acegi Security 基于 Spring，可以帮助我们为项目建立丰富的角色与权限管理系统。Acegi Security 虽然好用，但是最为人诟病的则是它臃肿烦琐的配置，这一问题最终也遗传给了 Spring Security。

Acegi Security 最终被并入 Spring Security 项目中，并于 2008 年 4 月发布了改名后的第一个版本 Spring Security 2.0.0，截止本书写作时，Spring Security 的最新版本已经到了 5.3.4。

和 Shiro 相比，Spring Security 重量级并且配置烦琐，直至今天，依然有人以此为理由而拒绝了解 Spring Security。其实，自从 Spring Boot 推出后，就彻底颠覆了传统了 JavaEE 开发，自动化配置让许多事情变得非常容易，包括 Spring Security 的配置。在一个 Spring Boot 项目中，我们甚至只需要引入一个依赖，不需要任何额外配置，项目的所有接口就会被自动保护起来了。在 Spring Cloud 中，很多涉及安全管理的问题，也是一个 Spring Security 依赖两行配置就能搞定，在和 Spring 家族的产品一起使用时，Spring Security 的优势就非常明显了。

因此，在微服务时代，我们不需要纠结要不要学习 Spring Security，我们要考虑的是如何快速掌握 Spring Security，并且能够使用 Spring Security 实现我们微服务的安全管理。

1.2　Spring Security 核心功能

对于一个安全管理框架而言，无论是 Shiro 还是 Spring Security，最核心的功能，无非就是如下两方面：

- 认证
- 授权

通俗点说，认证就是身份验证（你是谁？），授权就是访问控制（你可以做什么？）。

1.2.1 认证

Spring Security 支持多种不同的认证方式，这些认证方式有的是 Spring Security 自己提供的认证功能，有的是第三方标准组织制订的。Spring Security 集成的主流认证机制主要有如下几种：

- 表单认证。
- OAuth2.0 认证。
- SAML2.0 认证。
- CAS 认证。
- RememberMe 自动认证。
- JAAS 认证。
- OpenID 去中心化认证。
- Pre-Authentication Scenarios 认证。
- X509 认证。
- HTTP Basic 认证。
- HTTP Digest 认证。

作为一个开放的平台，Spring Security 提供的认证机制不仅仅包括上面这些，我们还可以通过引入第三方依赖来支持更多的认证方式，同时，如果这些认证方式无法满足我们的需求，我们也可以自定义认证逻辑，特别是当我们和一些"老破旧"的系统进行集成时，自定义认证逻辑就显得非常重要了。

1.2.2 授权

无论采用了上面哪种认证方式，都不影响在 Spring Security 中使用授权功能。Spring Security 支持基于 URL 的请求授权、支持方法访问授权、支持 SpEL 访问控制、支持域对象安全（ACL），同时也支持动态权限配置、支持 RBAC 权限模型等，总之，我们常见的权限管理需求，Spring Security 基本上都是支持的。

1.2.3 其他

在认证和授权这两个核心功能之外，Spring Security 还提供了很多安全管理的"周边功能"，这也是一个非常重要的特色。

大部分 Java 工程师都不是专业的 Web 安全工程师，自己开发的安全管理框架可能会存在大大小小的安全漏洞。而 Spring Security 的强大之处在于，即使你不了解很多网络攻击，只要使用了 Spring Security，它会帮助我们自动防御很多网络攻击，例如 CSRF 攻击、会话固定攻击等，同时 Spring Security 还提供了 HTTP 防火墙来拦截大量的非法请求。由此可见，研究

Spring Security，也是研究常见的网络攻击以及防御策略。

对于大部分的 Java 项目而言，无论是从经济性还是安全性来考虑，使用 Spring Security 无疑是最佳方案。

1.3 Spring Security 整体架构

在具体学习 Spring Security 各种用法之前，我们先介绍一下 Spring Security 中常见的概念，以及认证、授权思路，方便读者从整体上把握 Spring Security 架构，这里涉及的所有组件，在后面的章节中还会做详细介绍。

1.3.1 认证和授权

1.3.1.1 认证

在 Spring Security 的架构设计中，认证（Authentication）和授权（Authorization）是分开的，在本书后面的章节中读者可以看到，无论使用什么样的认证方式，都不会影响授权，这是两个独立的存在，这种独立带来的好处之一，就是 Spring Security 可以非常方便地整合一些外部的认证方案。

在 Spring Security 中，用户的认证信息主要由 Authentication 的实现类来保存，Authentication 接口定义如下：

```
public interface Authentication extends Principal, Serializable {
    Collection<? extends GrantedAuthority> getAuthorities();
    Object getCredentials();
    Object getDetails();
    Object getPrincipal();
    boolean isAuthenticated();
    void setAuthenticated(boolean isAuthenticated);
}
```

这里接口中定义的方法如下：

- getAuthorities 方法：用来获取用户的权限。
- getCredentials 方法：用来获取用户凭证，一般来说就是密码。
- getDetails 方法：用来获取用户携带的详细信息，可能是当前请求之类等。
- getPrincipal 方法：用来获取当前用户，例如是一个用户名或者一个用户对象。
- isAuthenticated：当前用户是否认证成功。

当用户使用用户名/密码登录或使用 Remember-me 登录时，都会对应一个不同的 Authentication 实例。

Spring Security 中的认证工作主要由 AuthenticationManager 接口来负责，下面来看一下该接口的定义：

```
public interface AuthenticationManager {
    Authentication authenticate(Authentication authentication)
        throws AuthenticationException;
}
```

AuthenticationManager 只有一个 authenticate 方法可以用来做认证，该方法有三个不同的返回值：

- 返回 Authentication，表示认证成功。
- 抛出 AuthenticationException 异常，表示用户输入了无效的凭证。
- 返回 null，表示不能断定。

AuthenticationManager 最主要的实现类是 ProviderManager，ProviderManager 管理了众多的 AuthenticationProvider 实例，AuthenticationProvider 有点类似于 AuthenticationManager，但是它多了一个 supports 方法用来判断是否支持给定的 Authentication 类型。

```
public interface AuthenticationProvider {
    Authentication authenticate(Authentication authentication)
        throws AuthenticationException;
    boolean supports(Class<?> authentication);
}
```

由于 Authentication 拥有众多不同的实现类，这些不同的实现类又由不同的 AuthenticationProvider 来处理，所以 AuthenticationProvider 会有一个 supports 方法，用来判断当前的 Authentication Provider 是否支持对应的 Authentication。

在一次完整的认证流程中，可能会同时存在多个 AuthenticationProvider（例如，项目同时支持 form 表单登录和短信验证码登录），多个 AuthenticationProvider 统一由 ProviderManager 来管理。同时，ProviderManager 具有一个可选的 parent，如果所有的 AuthenticationProvider 都认证失败，那么就会调用 parent 进行认证。parent 相当于一个备用认证方式，即各个 AuthenticationProvider 都无法处理认证问题的时候，就由 parent 出场收拾残局。

1.3.1.2 授权

当完成认证后，接下来就是授权了。在 Spring Security 的授权体系中，有两个关键接口：

- AccessDecisionManager
- AccessDecisionVoter

AccessDecisionVoter 是一个投票器，投票器会检查用户是否具备应有的角色，进而投出赞成、反对或者弃权票；AccessDecisionManager 则是一个决策器，来决定此次访问是否被允许。AccessDecisionVoter 和 AccessDecisionManager 都有众多的实现类，在 AccessDecisionManager 中会挨个遍历 AccessDecisionVoter，进而决定是否允许用户访问，因而 AccessDecisionVoter 和 AccessDecisionManager 两者的关系类似于 AuthenticationProvider 和 ProviderManager 的关系。

在 Spring Security 中，用户请求一个资源（通常是一个网络接口或者一个 Java 方法）所需要的角色会被封装成一个 ConfigAttribute 对象，在 ConfigAttribute 中只有一个 getAttribute 方法，该方法返回一个 String 字符串，就是角色的名称。一般来说，角色名称都带有一个 ROLE_ 前缀，投票器 AccessDecisionVoter 所做的事情，其实就是比较用户所具备的角色和请求某个资源所需的 ConfigAttribute 之间的关系。

1.3.2 Web 安全

在 Spring Security 中，认证、授权等功能都是基于过滤器来完成的。表 1-1 列出了 Spring Security 中常见的过滤器，注意这里说的是否默认加载是指引入 Spring Security 依赖之后，开发者不做任何配置时，会自动加载的过滤器。

表 1-1 Spring Security 中的过滤器

过滤器	过滤器作用	是否默认加载
ChannelProcessingFilter	过滤请求协议，如 HTTPS 和 HTTP	NO
WebAsyncManagerIntegrationFilter	将 WebAsyncManager 与 Spring Security 上下文进行集成	YES
SecurityContextPersistenceFilter	在处理请求之前，将安全信息加载到 SecurityContextHolder 中以方便后续使用。请求结束后，再擦除 SecurityContextHolder 中的信息	YES
HeaderWriterFilter	头信息加入到响应中	YES
CorsFilter	处理跨域问题	NO
CsrfFilter	处理 CSRF 攻击	YES
LogoutFilter	处理注销登录	YES
OAuth2AuthorizationRequestRedirectFilter	处理 OAuth2 认证重定向	NO
Saml2WebSsoAuthenticationRequestFilter	处理 SAML 认证	NO
X509AuthenticationFilter	处理 X509 认证	NO
AbstractPreAuthenticatedProcessingFilter	处理预认证问题	NO
CasAuthenticationFilter	处理 CAS 单点登录	NO
OAuth2LoginAuthenticationFilter	处理 OAuth2 认证	NO
Saml2WebSsoAuthenticationFilter	处理 SAML 认证	NO
UsernamePasswordAuthenticationFilter	处理表单登录	YES
OpenIDAuthenticationFilter	处理 OpenID 认证	NO
DefaultLoginPageGeneratingFilter	配置默认登录页面	YES
DefaultLogoutPageGeneratingFilter	配置默认注销页面	YES
ConcurrentSessionFilter	处理 Session 有效期	NO
DigestAuthenticationFilter	处理 HTTP 摘要认证	NO
BearerTokenAuthenticationFilter	处理 OAuth2 认证时的 Access Token	NO
BasicAuthenticationFilter	处理 HttpBasic 登录	YES
RequestCacheAwareFilter	处理请求缓存	YES

(续表)

过滤器	过滤器作用	是否默认加载
SecurityContextHolderAwareRequestFilter	包装原始请求	YES
JaasApiIntegrationFilter	处理 JAAS 认证	NO
RememberMeAuthenticationFilter	处理 RememberMe 登录	NO
AnonymousAuthenticationFilter	配置匿名认证	YES
OAuth2AuthorizationCodeGrantFilter	处理 OAuth2 认证中的授权码	NO
SessionManagementFilter	处理 Session 并发问题	YES
ExceptionTranslationFilter	处理异常认证/授权中的情况	YES
FilterSecurityInterceptor	处理授权	YES
SwitchUserFilter	处理账户切换	NO

开发者所见到的 Spring Security 提供的功能，都是通过这些过滤器来实现的，这些过滤器按照既定的优先级排列，最终形成一个过滤器链。开发者也可以自定义过滤器，并通过@Order 注解去调整自定义过滤器在过滤器链中的位置。

需要注意的是，默认过滤器并不是直接放在 Web 项目的原生过滤器链中，而是通过一个 FilterChainProxy 来统一管理。Spring Security 中的过滤器链通过 FilterChainProxy 嵌入到 Web 项目的原生过滤器链中，如图 1-1 所示。

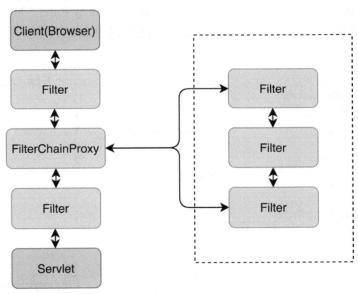

图 1-1　过滤器链通过 FilterChainProxy 出现在 Web 容器中

在 Spring Security 中，这样的过滤器链不仅仅只有一个，可能会有多个，如图 1-2 所示。当存在多个过滤器链时，多个过滤器链之间要指定优先级，当请求到达后，会从 FilterChainProxy 进行分发，先和哪个过滤器链匹配上，就用哪个过滤器链进行处理。当系统中存在多个不同的认证体系时，那么使用多个过滤器链就非常有效。

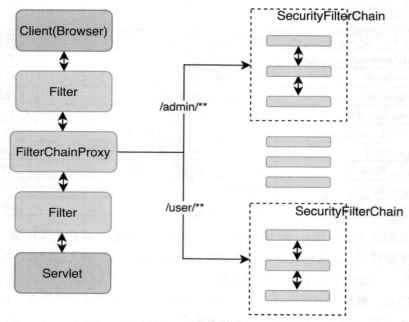

图 1-2　存在多个过滤器链时通过优先级进行匹配

FilterChainProxy 作为一个顶层管理者，将统一管理 Security Filter。FilterChainProxy 本身将通过 Spring 框架提供的 DelegatingFilterProxy 整合到原生过滤器链中，所以图 1-2 还可以做进一步的优化，如图 1-3 所示。

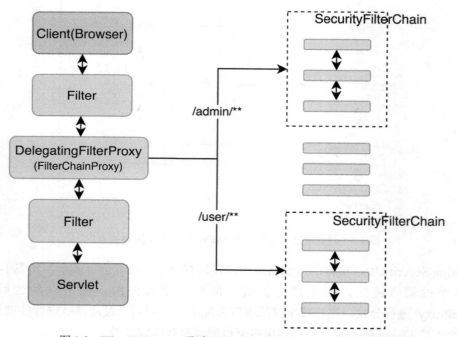

图 1-3　FilterChainProxy 通过 DelegatingFilterProxy 整合进 Web Filter

1.3.3 登录数据保存

如果不使用 Spring Security 这一类的安全管理框架，大部分的开发者可能会将登录用户数据保存在 Session 中，事实上，Spring Security 也是这么做的。但是，为了使用方便，Spring Security 在此基础上还做了一些改进，其中最主要的一个变化就是线程绑定。

当用户登录成功后，Spring Security 会将登录成功的用户信息保存到 SecurityContextHolder 中。SecurityContextHolder 中的数据保存默认是通过 ThreadLocal 来实现的，使用 ThreadLocal 创建的变量只能被当前线程访问，不能被其他线程访问和修改，也就是用户数据和请求线程绑定在一起。当登录请求处理完毕后，Spring Security 会将 SecurityContextHolder 中的数据拿出来保存到 Session 中，同时将 SecurityContextHolder 中的数据清空。以后每当有请求到来时，Spring Security 就会先从 Session 中取出用户登录数据，保存到 SecurityContextHolder 中，方便在该请求的后续处理过程中使用，同时在请求结束时将 SecurityContextHolder 中的数据拿出来保存到 Session 中，然后将 SecurityContextHolder 中的数据清空。

这一策略非常方便用户在 Controller 或者 Service 层获取当前登录用户数据，但是带来的另外一个问题就是，在子线程中想要获取用户登录数据就比较麻烦。Spring Security 对此也提供了相应的解决方案，如果开发者使用@Async 注解来开启异步任务的话，那么只需要添加如下配置，使用 Spring Security 提供的异步任务代理，就可以在异步任务中从 SecurityContextHolder 里边获取当前登录用户的信息：

```
@Configuration
public class ApplicationConfiguration extends AsyncConfigurerSupport {
    @Override
    public Executor getAsyncExecutor() {
        return new DelegatingSecurityContextExecutorService(
                            Executors.newFixedThreadPool(5));
    }
}
```

1.4 小 结

本章主要介绍了 Spring Security 的基本原理与整体架构，方便读者从整体上把握 Spring Security 认证、授权的实现原理。在接下来的章节中，我们将继续详细介绍 Spring Security 认证与授权的每一个实现细节。

第 2 章

Spring Security 认证

对于安全管理框架而言，认证功能可以说是一切的起点，所以我们要研究 Spring Security，就要从最基本的认证开始。在 Spring Security 中，对认证功能做了大量的封装，以至于开发者只需要稍微配置一下就能使用认证功能，然而要深刻理解其源码却并非易事。本章将带领读者从最基本的用法开始讲解，最终再扩展到对源码的理解。

本章涉及的主要知识点有：

- Spring Security 基本认证。
- 登录表单配置。
- 登录用户数据获取。
- 用户的四种定义方式。

2.1 Spring Security 基本认证

2.1.1 快速入门

在 Spring Boot 项目中使用 Spring Security 非常方便，创建一个新的 Spring Boot 项目，我们只需要引入 Web 和 Spring Security 依赖即可，具体代码如下：

```
<dependency>
    <groupId>org.springframework.boot</groupId>
    <artifactId>spring-boot-starter-security</artifactId>
</dependency>
<dependency>
```

```
    <groupId>org.springframework.boot</groupId>
    <artifactId>spring-boot-starter-web</artifactId>
</dependency>
```

然后我们在项目中提供一个用于测试的/hello 接口，代码如下：

```
@RestController
public class HelloController {
    @GetMapping("/hello")
    public String hello() {
        return "hello spring security";
    }
}
```

接下来启动项目，/hello 接口就已经被自动保护起来了。当用户访问/hello 接口时，会自动跳转到登录页面，如图 2-1 所示，用户登录成功后，才能访问到/hello 接口。

图 2-1　Spring Security 默认登录页面

默认的登录用户名是 user，登录密码则是一个随机生成的 UUID 字符串，在项目启动日志中可以看到登录密码（这也意味着项目每次启动时，密码都会发生变化）：

```
Using generated security password: 8ef9c800-17cf-47a3-9984-8ff936db6dd8
```

输入默认的用户名和密码，就可以成功登录了。这就是 Spring Security 的强大之处，只需要引入一个依赖，所有的接口就会被自动保护起来。

2.1.2　流程分析

通过一个简单的流程图来看一下上面案例中的请求流程，如图 2-2 所示。

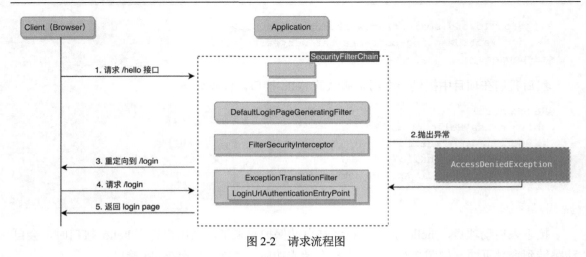

图 2-2 请求流程图

流程图比较清晰地说明了整个请求过程：

（1）客户端（浏览器）发起请求去访问/hello 接口，这个接口默认是需要认证之后才能访问的。

（2）这个请求会走一遍 Spring Security 中的过滤器链，在最后的 FilterSecurityInterceptor 过滤器中被拦截下来，因为系统发现用户未认证。请求拦截下来之后，接下来会抛出 AccessDeniedException 异常。

（3）抛出的 AccessDeniedException 异常在 ExceptionTranslationFilter 过滤器中被捕获，ExceptionTranslationFilter 过滤器通过调用 LoginUrlAuthenticationEntryPoint#commence 方法给客户端返回 302，要求客户端重定向到/login 页面。

（4）客户端发送/login 请求。

（5）/login 请求被 DefaultLoginPageGeneratingFilter 过滤器拦截下来，并在该过滤器中返回登录页面。所以当用户访问/hello 接口时会首先看到登录页面。

在整个过程中，相当于客户端一共发送了两个请求，第一个请求是/hello，服务端收到之后，返回 302，要求客户端重定向到/login，于是客户端又发送了/login 请求。

读者现在去理解上面这一个流程图可能还有些困难，等阅读完本章后面的内容之后，再回过头来看这个流程图，应该就会比较清晰了。

2.1.3 原理分析

在 2.1.1 小节中，虽然开发者只是引入了一个依赖，代码不多，但是 Spring Boot 背后却默默做了很多事情：

- 开启 Spring Security 自动化配置，开启后，会自动创建一个名为 springSecurityFilterChain 的过滤器，并注入到 Spring 容器中，这个过滤器将负责所有的安全管理，包括用户的认证、授权、重定向到登录页面等（springSecurityFilterChain 实际上代理了 Spring Security

中的过滤器链）。
- 创建一个 UserDetailsService 实例，UserDetailsService 负责提供用户数据，默认的用户数据是基于内存的用户，用户名为 user，密码则是随机生成的 UUID 字符串。
- 给用户生成一个默认的登录页面。
- 开启 CSRF 攻击防御。
- 开启会话固定攻击防御。
- 集成 X-XSS-Protection。
- 集成 X-Frame-Options 以防止单击劫持。

这里涉及的细节还是非常多的，登录的细节我们会在后面的章节继续详细介绍，这里主要分析一下默认用户的生成以及默认登录页面的生成。

2.1.3.1 默认用户生成

Spring Security 中定义了 UserDetails 接口来规范开发者自定义的用户对象，这样方便一些旧系统、用户表已经固定的系统集成到 Spring Security 认证体系中。

UserDetails 接口定义如下：

```
public interface UserDetails extends Serializable {
    Collection<? extends GrantedAuthority> getAuthorities();
    String getPassword();
    String getUsername();
    boolean isAccountNonExpired();
    boolean isAccountNonLocked();
    boolean isCredentialsNonExpired();
    boolean isEnabled();
}
```

该接口中一共定义了 7 个方法：

（1）getAuthorities 方法：返回当前账户所具备的权限。
（2）getPassword 方法：返回当前账户的密码。
（3）getUsername 方法：返回当前账户的用户名。
（4）isAccountNonExpired 方法：返回当前账户是否未过期。
（5）isAccountNonLocked 方法：返回当前账户是否未锁定。
（6）isCredentialsNonExpired 方法：返回当前账户凭证（如密码）是否未过期。
（7）isEnabled 方法：返回当前账户是否可用。

这是用户对象的定义，而负责提供用户数据源的接口是 UserDetailsService，UserDetailsService 中只有一个查询用户的方法，代码如下：

```
public interface UserDetailsService {
    UserDetails loadUserByUsername(String username)
            throws UsernameNotFoundException;
}
```

loadUserByUsername 有一个参数是 username，这是用户在认证时传入的用户名，最常见的就是用户在登录表单中输入的用户名（实际开发时还可能存在其他情况，例如使用 CAS 单点登录时，username 并非表单输入的用户名，而是 CAS Server 认证成功后回调的用户名参数），开发者在这里拿到用户名之后，再去数据库中查询用户，最终返回一个 UserDetails 实例。

在实际项目中，一般需要开发者自定义 UserDetailsService 的实现。如果开发者没有自定义 UserDetailsService 的实现，Spring Security 也为 UserDetailsService 提供了默认实现，如图 2-3 所示。

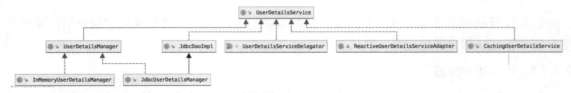

图 2-3　UserDetailsService 的默认实现类

- UserDetailsManager 在 UserDetailsService 的基础上，继续定义了添加用户、更新用户、删除用户、修改密码以及判断用户是否存在共 5 种方法。
- JdbcDaoImpl 在 UserDetailsService 的基础上，通过 spring-jdbc 实现了从数据库中查询用户的方法。
- InMemoryUserDetailsManager 实现了 UserDetailsManager 中关于用户的增删改查方法，不过都是基于内存的操作，数据并没有持久化。
- JdbcUserDetailsManager 继承自 JdbcDaoImpl 同时又实现了 UserDetailsManager 接口，因此可以通过 JdbcUserDetailsManager 实现对用户的增删改查操作，这些操作都会持久化到数据库中。不过 JdbcUserDetailsManager 有一个局限性，就是操作数据库中用户的 SQL 都是提前写好的，不够灵活，因此在实际开发中 JdbcUserDetailsManager 使用并不多。
- CachingUserDetailsService 的特点是会将 UserDetailsService 缓存起来。
- UserDetailsServiceDelegator 则是提供了 UserDetailsService 的懒加载功能。
- ReactiveUserDetailsServiceAdapter 是 webflux-web-security 模块定义的 UserDetailsService 实现。

当我们使用 Spring Security 时，如果仅仅只是引入一个 Spring Security 依赖，则默认使用的用户就是由 InMemoryUserDetailsManager 提供的。

大家知道，Spring Boot 之所以能够做到零配置使用 Spring Security，就是因为它提供了众多的自动化配置类。其中，针对 UserDetailsService 的自动化配置类是 UserDetailsServiceAutoConfiguration，这个类的源码并不长，我们一起来看一下：

```
@Configuration(proxyBeanMethods = false)
@ConditionalOnClass(AuthenticationManager.class)
@ConditionalOnBean(ObjectPostProcessor.class)
@ConditionalOnMissingBean(
        value = { AuthenticationManager.class,
            AuthenticationProvider.class,
```

```java
                    UserDetailsService.class },
        type = { "org.springframework.security.oauth2.jwt.JwtDecoder",
                 "org.springframework.security.oauth2.server.resource.
                         introspection.OpaqueTokenIntrospector" })
public class UserDetailsServiceAutoConfiguration {
    private static final String NOOP_PASSWORD_PREFIX = "{noop}";
    private static final Pattern PASSWORD_ALGORITHM_PATTERN
                        = Pattern.compile("^\\{.+}.*$");
    @Bean
    @ConditionalOnMissingBean(
        type = "org.springframework.security.oauth2.client.
                        registration.ClientRegistrationRepository")
    @Lazy
    public InMemoryUserDetailsManager inMemoryUserDetailsManager(
        SecurityProperties properties,
        ObjectProvider<PasswordEncoder> passwordEncoder) {
        SecurityProperties.User user = properties.getUser();
        List<String> roles = user.getRoles();
        return new InMemoryUserDetailsManager(
            User.withUsername(user.getName())
                .password(getOrDeducePassword(user,
                        passwordEncoder.getIfAvailable()))
                .roles(StringUtils.toStringArray(roles)).build());
    }
    private String getOrDeducePassword(
                SecurityProperties.User user, PasswordEncoder encoder) {
        String password = user.getPassword();
        if (user.isPasswordGenerated()) {
            logger.info(String
                .format("%n%nUsing generated security password: %s%n",
                                        user.getPassword()));
        }
        if (encoder != null
            || PASSWORD_ALGORITHM_PATTERN.matcher(password).matches()) {
            return password;
        }
        return NOOP_PASSWORD_PREFIX + password;
    }
}
```

从上述代码中可以看到，有两个比较重要的条件促使系统自动提供一个 InMemoryUserDetailsManager 的实例：

（1）当前 classpath 下存在 AuthenticationManager 类。

（2）当前项目中，系统没有提供 AuthenticationManager、AuthenticationProvider、UserDetailsService 以及 ClientRegistrationRepository 实例。

默认情况下，上面的条件都会满足，此时 Spring Security 会提供一个 InMemoryUser

DetailsManager 实例。从 inMemoryUserDetailsManager 方法中可以看到，用户数据源自 SecurityProperties#getUser 方法：

```
@ConfigurationProperties(prefix = "spring.security")
public class SecurityProperties {
    private User user = new User();
    public User getUser() {
        return this.user;
    }
    public static class User {
        private String name = "user";
        private String password = UUID.randomUUID().toString();
        private List<String> roles = new ArrayList<>();
        //省略 getter/setter
    }
}
```

从 SecurityProperties.User 类中，我们就可以看到默认的用户名是 user，默认的密码是一个 UUID 字符串。

再回到 inMemoryUserDetailsManager 方法中，构造 InMemoryUserDetailsManager 实例时需要一个 User 对象。这里的 User 对象不是 SecurityProperties.User，而是 org.springframework.security.core.userdetails.User，这是 Spring Security 提供的一个实现了 UserDetails 接口的用户类，该类提供了相应的静态方法，用来构造一个默认的 User 实例。同时，默认的用户密码还在 getOrDeducePassword 方法中进行了二次处理，由于默认的 encoder 为 null，所以密码的二次处理只是给密码加了一个前缀{noop}，表示密码是明文存储的（关于{noop}将在第 5 章密码加密中做详细介绍）。

经过以上的源码梳理，相信大家已经明白了 Spring Security 默认的用户名/密码是来自哪里了！

另外，当看了 SecurityProperties 的源码后，只要对 Spring Boot 中 properties 属性的加载机制有一点了解，就会明白，只要我们在项目的 application.properties 配置文件中添加如下配置，就能定制 SecurityProperties.User 类中各属性的值：

```
spring.security.user.name=javaboy
spring.security.user.password=123
spring.security.user.roles=admin,user
```

配置完成后，重启项目，此时登录的用户名就是 javaboy，登录密码就是 123，登录成功后用户具备 admin 和 user 两个角色。

2.1.3.2 默认页面生成

在 2.1.1 小节的案例中，一共存在两个默认页面，一个就是图 2-1 所示的登录页面，另外一个则是注销登录页面。当用户登录成功之后，在浏览器中输入 http://localhost:8080/logout 就可以看到注销登录页面，如图 2-4 所示。

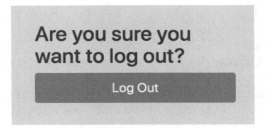

图 2-4 注销登录页面

那么这两个页面是从哪里来的呢？这里剖析一下。

在 1.3.2 小节中，我们介绍了 Spring Security 中常见的过滤器，在这些常见的过滤器中就包含两个和页面相关的过滤器：DefaultLoginPageGeneratingFilter 和 DefaultLogoutPageGeneratingFilter。

通过过滤器的名字就可以分辨出 DefaultLoginPageGeneratingFilter 过滤器用来生成默认的登录页面，DefaultLogoutPageGeneratingFilter 过滤器则用来生成默认的注销页面。

先来看 DefaultLoginPageGeneratingFilter。作为 Spring Security 过滤器链中的一员，在第一次请求/hello 接口的时候，就会经过 DefaultLoginPageGeneratingFilter 过滤器，但是由于/hello 接口和登录无关，因此 DefaultLoginPageGeneratingFilter 过滤器并未干涉/hello 接口。等到第二次重定向到/login 页面的时候，这个时候就和 DefaultLoginPageGeneratingFilter 有关系了，此时请求就会在 DefaultLoginPageGeneratingFilter 中进行处理，生成登录页面返回给客户端。

我们来看一下 DefaultLoginPageGeneratingFilter 的源码，源码比较长，这里仅列出核心部分：

```java
public class DefaultLoginPageGeneratingFilter extends GenericFilterBean {
    public void doFilter(ServletRequest req,
                         ServletResponse res, FilterChain chain)
                throws IOException, ServletException {
        HttpServletRequest request = (HttpServletRequest) req;
        HttpServletResponse response = (HttpServletResponse) res;
        boolean loginError = isErrorPage(request);
        boolean logoutSuccess = isLogoutSuccess(request);
        if (isLoginUrlRequest(request) || loginError || logoutSuccess) {
            String loginPageHtml =
                generateLoginPageHtml(request, loginError, logoutSuccess);
            response.setContentType("text/html;charset=UTF-8");
            response.setContentLength(loginPageHtml
                        .getBytes(StandardCharsets.UTF_8).length);
            response.getWriter().write(loginPageHtml);
            return;
        }
        chain.doFilter(request, response);
    }
    private String generateLoginPageHtml(HttpServletRequest request,
                         boolean loginError,
```

```
                                    boolean logoutSuccess) {
    String errorMsg = "Invalid credentials";
    if (loginError) {
        HttpSession session = request.getSession(false);
        if (session != null) {
            AuthenticationException ex = (AuthenticationException) session
                .getAttribute(WebAttributes.AUTHENTICATION_EXCEPTION);
            errorMsg = ex != null ? ex.getMessage() : "Invalid credentials";
        }
    }
    StringBuilder sb = new StringBuilder();
    String contextPath = request.getContextPath();
    if (this.formLoginEnabled) {
        sb.append("");
    }
    if (openIdEnabled) {
        sb.append("");
    }
    if (oauth2LoginEnabled) {
        sb.append("");
    }
    if (this.saml2LoginEnabled) {
        sb.append("");
    }
    return sb.toString();
    }
}
```

DefaultLoginPageGeneratingFilter 的源码执行流程还是非常清晰的，我们梳理一下：

（1）在 doFilter 方法中，首先判断出当前请求是否为登录出错请求、注销成功请求或者登录请求。如果是这三种请求中的任意一个，就会在 DefaultLoginPageGeneratingFilter 过滤器中生成登录页面并返回，否则请求继续往下走，执行下一个过滤器（这就是一开始的/hello 请求为什么没有被 DefaultLoginPageGeneratingFilter 拦截下来的原因）。

（2）如果当前请求是登录出错请求、注销成功请求或者登录请求中的任意一个，就会调用 generateLoginPageHtml 方法去生成登录页面。在该方法中，如果有异常信息就把异常信息取出来一同返回给前端，然后根据不同的登录场景，生成不同的登录页面。生成过程其实就是字符串拼接，拼接出不同的登录表单（由于源码太长，上面没有贴出来具体的字符串拼接源码，读者可以自行查看 DefaultLoginPageGeneratingFilter 类的源码）。

（3）登录页面生成后，接下来通过 HttpServletResponse 将登录页面写回到前端，然后调用 return 方法跳出过滤器链。

这就是 DefaultLoginPageGeneratingFilter 的工作过程。这里重点搞明白为什么/hello 请求没有被拦截，而/login 请求却被拦截了，其他都很好懂。

理解了 DefaultLoginPageGeneratingFilter，再来看 DefaultLogoutPageGeneratingFilter 就更

容易了，DefaultLogoutPageGeneratingFilter 部分核心源码如下：

```java
public class DefaultLogoutPageGeneratingFilter extends OncePerRequestFilter {
    private RequestMatcher matcher = new AntPathRequestMatcher("/logout", "GET");
    @Override
    protected void doFilterInternal(HttpServletRequest request,
            HttpServletResponse response, FilterChain filterChain)
            throws ServletException, IOException {
        if (this.matcher.matches(request)) {
            renderLogout(request, response);
        } else {
            filterChain.doFilter(request, response);
        }
    }
    private void renderLogout(HttpServletRequest request,
                              HttpServletResponse response)
                    throws IOException {
        String page = "";
        response.setContentType("text/html;charset=UTF-8");
        response.getWriter().write(page);
    }
}
```

从上述源码中可以看出，请求到来之后，会先判断是否是注销请求/logout，如果是/logout请求，则渲染一个注销请求的页面返回给客户端，渲染过程和前面登录页面的渲染过程类似，也是字符串拼接（这里省略了字符串拼接，读者可以参考 DefaultLogoutPageGeneratingFilter 的源码）；否则请求继续往下走，执行下一个过滤器。

通过前面的分析，相信大家对这个简单的案例已经有所了解，看似只是加了一个依赖，但实际上 Spring Security 和 Spring Boot 在背后都默默做了很多事情，当然还有很多没有介绍到的，我们将在后面的章节中和大家一起继续深究。

2.2 登录表单配置

2.2.1 快速入门

理解了入门案例之后，接下来我们再来看一下登录表单的详细配置。

首先创建一个新的 Spring Boot 项目，引入 Web 和 Spring Security 依赖，代码如下：

```xml
<dependency>
    <groupId>org.springframework.boot</groupId>
    <artifactId>spring-boot-starter-security</artifactId>
</dependency>
<dependency>
```

```xml
        <groupId>org.springframework.boot</groupId>
        <artifactId>spring-boot-starter-web</artifactId>
</dependency>
```

项目创建好之后，为了方便测试，需要在 application.properties 中添加如下配置，将登录用户名和密码固定下来：

```
spring.security.user.name=javaboy
spring.security.user.password=123
```

接下来，我们在 resources/static 目录下创建一个 login.html 页面，这个是我们自定义的登录页面：

```html
<!DOCTYPE html>
<html lang="en">
<head>
    <meta charset="UTF-8">
    <title>登录</title>
    <link
        href="//maxcdn.bootstrapcdn.com/bootstrap/4.1.1/css/bootstrap.min.css"
        rel="stylesheet" id="bootstrap-css">
    <script
        src="//maxcdn.bootstrapcdn.com/bootstrap/4.1.1/js/bootstrap.min.js">
    </script>
    <script
        src="//cdnjs.cloudflare.com/ajax/libs/jquery/3.2.1/jquery.min.js">
    </script>
</head>
<style>
    #login .container #login-row #login-column #login-box {
        border: 1px solid #9C9C9C;
        background-color: #EAEAEA;
    }
</style>
<body>
<div id="login">
    <div class="container">
        <div id="login-row"
                class="row justify-content-center align-items-center">
            <div id="login-column" class="col-md-6">
                <div id="login-box" class="col-md-12">
                    <form id="login-form" class="form"
                                        action="/doLogin" method="post">
                        <h3 class="text-center text-info">登录</h3>
                        <div class="form-group">
                            <label for="username"
                                    class="text-info">用户名:</label><br>
                            <input type="text" name="uname"
                                    id="username" class="form-control">
```

```html
                </div>
                <div class="form-group">
                    <label for="password"
                                class="text-info">密码:</label><br>
                    <input type="text" name="passwd"
                                id="password" class="form-control">
                </div>
                <div class="form-group">
                    <input type="submit" name="submit"
                                class="btn btn-info btn-md" value="登录">
                </div>
            </form>
        </div>
    </div>
</div>
</body>
```

这个 logint.html 中的核心内容就是一个登录表单，登录表单中有三个需要注意的地方：

（1）form 的 action，这里给出的是/doLogin，表示表单要提交到/doLogin 接口上。

（2）用户名输入框的 name 属性值为 uname，当然这个值是可以自定义的，这里采用了 uname。

（3）密码输入框的 name 属性值为 passwd，passwd 也是可以自定义的。

login.html 定义好之后，接下来定义两个测试接口，作为受保护的资源。当用户登录成功后，就可以访问到受保护的资源。接口定义如下：

```java
@RestController
public class LoginController {
    @RequestMapping("/index")
    public String index() {
        return "login success";
    }
    @RequestMapping("/hello")
    public String hello() {
        return "hello spring security";
    }
}
```

最后再提供一个 Spring Security 的配置类：

```java
@Configuration
public class SecurityConfig extends WebSecurityConfigurerAdapter {
    @Override
    protected void configure(HttpSecurity http) throws Exception {
        http.authorizeRequests()
                .anyRequest().authenticated()
```

```
            .and()
            .formLogin()
            .loginPage("/login.html")
            .loginProcessingUrl("/doLogin")
            .defaultSuccessUrl("/index ")
            .failureUrl("/login.html")
            .usernameParameter("uname")
            .passwordParameter("passwd")
            .permitAll()
            .and()
            .csrf().disable();
    }
}
```

在 Spring Security 中，如果我们需要自定义配置，基本上都是继承自 WebSecurityConfigurerAdapter 来实现的，当然 WebSecurityConfigurerAdapter 本身的配置还是比较复杂，同时也是比较丰富的，这里先不做过多的展开，仅就结合上面的代码来解释，在下一个小节中我们将会对这里的配置再做更加详细的介绍。

（1）首先 configure 方法中是一个链式配置，当然也可以不用链式配置，每一个属性配置完毕后再从 http.重新开始写起。

（2）authorizeRequests()方法表示开启权限配置（该方法的含义其实比较复杂，我们在 13.4.4 小节还会再次介绍该方法），.anyRequest().authenticated()表示所有的请求都要认证之后才能访问。

（3）有的读者会对 and()方法表示疑惑，and()方法会返回 HttpSecurityBuilder 对象的一个子类（实际上就是 HttpSecurity），所以 and()方法相当于又回到 HttpSecurity 实例，重新开启新一轮的配置。如果觉得 and() 方法很难理解，也可以不用 and() 方法，在.anyRequest().authenticated()配置完成后直接用分号（;）结束，然后通过 http.formLogin()继续配置表单登录。

（4）formLogin() 表示开启表单登录配置，loginPage 用来配置登录页面地址；loginProcessingUrl 用来配置登录接口地址；defaultSuccessUrl 表示登录成功后的跳转地址；failureUrl 表示登录失败后的跳转地址；usernameParameter 表示登录用户名的参数名称；passwordParameter 表示登录密码的参数名称；permitAll 表示跟登录相关的页面和接口不做拦截，直接通过。需要注意的是，loginProcessingUrl、usernameParameter、passwordParameter 需要和 login.html 中登录表单的配置一致。

（5）最后的 csrf().disable()表示禁用 CSRF 防御功能，Spring Security 自带了 CSRF 防御机制，但是我们这里为了测试方便，先将 CSRF 防御机制关闭，本书第 9 章将会详细介绍 CSRF 攻击与防御问题。

配置完成后，启动 Spring Boot 项目，浏览器地址栏中输入 http://localhost:8080/index，会自动跳转到 http://localhost:8080/login.html 页面，如图 2-5 所示。输入用户名和密码进行登录（用户名为 javaboy，密码为 123），登录成功之后，就可以访问到 index 页面了，如图 2-6 所示。

图 2-5　自定义的登录页面

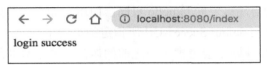

图 2-6　登录成功后的 index 页面

经过上面的配置，我们已经成功自定义了一个登录页面出来，用户在登录成功之后，就可以访问受保护的资源了。

2.2.2　配置细节

当然，前面的配置比较粗糙，这里还有一些配置的细节需要和读者分享一下。

在前面的配置中，我们用 defaultSuccessUrl 表示用户登录成功后的跳转地址，用 failureUrl 表示用户登录失败后的跳转地址。关于登录成功和登录失败，除了这两个方法可以配置之外，还有另外两个方法也可以配置。

2.2.2.1　登录成功

当用户登录成功之后，除了 defaultSuccessUrl 方法可以实现登录成功后的跳转之外，successForwardUrl 也可以实现登录成功后的跳转，代码如下：

```
@Configuration
public class SecurityConfig extends WebSecurityConfigurerAdapter {
    @Override
    protected void configure(HttpSecurity http) throws Exception {
        http.authorizeRequests()
                .anyRequest().authenticated()
                .and()
                .formLogin()
                .loginPage("/login.html")
                .loginProcessingUrl("/doLogin")
                .successForwardUrl("/index ")
                .failureUrl("/login.html")
                .usernameParameter("uname")
                .passwordParameter("passwd")
                .permitAll()
```

```
                .and()
                .csrf().disable();
    }
}
```

defaultSuccessUrl 和 successForwardUrl 的区别如下：

（1）defaultSuccessUrl 表示当用户登录成功之后，会自动重定向到登录之前的地址上，如果用户本身就是直接访问的登录页面，则登录成功后就会重定向到 defaultSuccessUrl 指定的页面中。例如，用户在未认证的情况下，访问了/hello 页面，此时会自动重定向到登录页面，当用户登录成功后，就会自动重定向到/hello 页面；而用户如果一开始就访问登录页面，则登录成功后就会自动重定向到 defaultSuccessUrl 所指定的页面中。

（2）successForwardUrl 则不会考虑用户之前的访问地址，只要用户登录成功，就会通过服务器端跳转到 successForwardUrl 所指定的页面。

（3）defaultSuccessUrl 有一个重载方法，如果重载方法的第二个参数传入 true，则 defaultSuccessUrl 的效果与 successForwardUrl 类似，即不考虑用户之前的访问地址，只要登录成功，就重定向到 defaultSuccessUrl 所指定的页面。不同之处在于，defaultSuccessUrl 是通过重定向实现的跳转（客户端跳转），而 successForwardUrl 则是通过服务器端跳转实现的。

无论是 defaultSuccessUrl 还是 successForwardUrl，最终所配置的都是 AuthenticationSuccessHandler 接口的实例。

Spring Security 中专门提供了 AuthenticationSuccessHandler 接口用来处理登录成功事项：

```
public interface AuthenticationSuccessHandler {
    default void onAuthenticationSuccess(HttpServletRequest request,
                                         HttpServletResponse response,
                                         FilterChain chain,
                                         Authentication authentication)
                                throws IOException, ServletException{
        onAuthenticationSuccess(request, response, authentication);
        chain.doFilter(request, response);
    }
    void onAuthenticationSuccess(HttpServletRequest request,
                                 HttpServletResponse response,
                                 Authentication authentication)
                            throws IOException, ServletException;
}
```

由上述代码可以看到，AuthenticationSuccessHandler 接口中一共定义了两个方法，其中一个是 default 方法，此方法是 Spring Security 5.2 开始加入进来的，在处理特定的认证请求 Authentication Filter 中会用到；另外一个非 default 方法，则用来处理登录成功的具体事项，其中 request 和 response 参数好理解，authentication 参数保存了登录成功的用户信息。我们将在后面的章节中详细介绍 authentication 参数。

AuthenticationSuccessHandler 接口共有三个实现类，如图 2-7 所示。

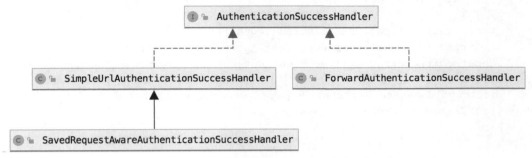

图 2-7　AuthenticationSuccessHandler 的三个实现类

（1）SimpleUrlAuthenticationSuccessHandler 继承自 AbstractAuthenticationTargetUrlRequestHandler，通过 AbstractAuthenticationTargetUrlRequestHandler 中的 handle 方法实现请求重定向。

（2）SavedRequestAwareAuthenticationSuccessHandler 在 SimpleUrlAuthenticationSuccessHandler 的基础上增加了请求缓存的功能，可以记录之前请求的地址，进而在登录成功后重定向到一开始访问的地址。

（3）ForwardAuthenticationSuccessHandler 的实现则比较容易，就是一个服务端跳转。

我们来重点分析 SavedRequestAwareAuthenticationSuccessHandler 和 ForwardAuthenticationSuccessHandler 的实现。

当通过 defaultSuccessUrl 来设置登录成功后重定向的地址时，实际上对应的实现类就是 SavedRequestAwareAuthenticationSuccessHandler，由于该类的源码比较长，这里列出来一部分核心代码：

```java
public class SavedRequestAwareAuthenticationSuccessHandler extends
        SimpleUrlAuthenticationSuccessHandler {
    private RequestCache requestCache = new HttpSessionRequestCache();
    @Override
    public void onAuthenticationSuccess(HttpServletRequest request,
                                        HttpServletResponse response,
                                        Authentication authentication)
                                throws ServletException, IOException {
        SavedRequest savedRequest = requestCache.getRequest(request, response);
        if (savedRequest == null) {
            super.onAuthenticationSuccess(request, response, authentication);
            return;
        }
        String targetUrlParameter = getTargetUrlParameter();
        if (isAlwaysUseDefaultTargetUrl()
                || (targetUrlParameter != null && StringUtils.hasText(request
                    .getParameter(targetUrlParameter)))) {
            requestCache.removeRequest(request, response);
            super.onAuthenticationSuccess(request, response, authentication);
```

```
        return;
    }
    clearAuthenticationAttributes(request);
    String targetUrl = savedRequest.getRedirectUrl();
    getRedirectStrategy().sendRedirect(request, response, targetUrl);
}
public void setRequestCache(RequestCache requestCache) {
    this.requestCache = requestCache;
}
}
```

这里的核心方法就是 onAuthenticationSuccess：

（1）首先从 requestCache 中获取缓存下来的请求，如果没有获取到缓存请求，就说明用户在访问登录页面之前并没有访问其他页面，此时直接调用父类的 onAuthenticationSuccess 方法来处理，最终会重定向到 defaultSuccessUrl 指定的地址。

（2）接下来会获取一个 targetUrlParameter，这个是用户显式指定的、希望登录成功后重定向的地址，例如用户发送的登录请求是 http://localhost:8080/doLogin?target=/hello，这就表示当用户登录成功之后，希望自动重定向到/hello 这个接口。getTargetUrlParameter 就是要获取重定向地址参数的 key，也就是上面的 target，拿到 target 之后，就可以获取到重定向地址了。

（3）如果 targetUrlParameter 存在，或者用户设置了 alwaysUseDefaultTargetUrl 为 true，这个时候缓存下来的请求就没有意义了。此时会直接调用父类的 onAuthenticationSuccess 方法完成重定向。targetUrlParameter 存在，则直接重定向到 targetUrlParameter 指定的地址；alwaysUseDefaultTargetUrl 为 true，则直接重定向到 defaultSuccessUrl 指定的地址；如果 targetUrlParameter 存在并且 alwaysUseDefaultTargetUrl 为 true，则重定向到 defaultSuccessUrl 指定的地址。

（4）如果前面的条件都不满足，那么最终会从缓存请求 savedRequest 中获取重定向地址，然后进行重定向操作。

这就是 SavedRequestAwareAuthenticationSuccessHandler 的实现逻辑，开发者也可以配置自己的 SavedRequestAwareAuthenticationSuccessHandler，代码如下：

```
@Configuration
public class SecurityConfig extends WebSecurityConfigurerAdapter {
    @Override
    protected void configure(HttpSecurity http) throws Exception {
        http.authorizeRequests()
                .anyRequest().authenticated()
                .and()
                .formLogin()
                .loginPage("/login.html")
                .loginProcessingUrl("/doLogin")
                .successHandler(successHandler())
                .failureUrl("/login.html")
                .usernameParameter("uname")
```

```
                .passwordParameter("passwd")
                .permitAll()
                .and()
                .csrf().disable();
    }
    SavedRequestAwareAuthenticationSuccessHandler successHandler() {
        SavedRequestAwareAuthenticationSuccessHandler handler =
                    new SavedRequestAwareAuthenticationSuccessHandler();
        handler.setDefaultTargetUrl("/index");
        handler.setTargetUrlParameter("target");
        return handler;
    }
}
```

注意在配置时指定了 targetUrlParameter 为 target，这样用户就可以在登录请求中，通过 target 来指定跳转地址了，然后我们修改一下前面 login.html 中的 form 表单：

```
<form id="login-form" class="form"
                       action="/doLogin?target=/hello" method="post">
    <h3 class="text-center text-info">登录</h3>
    <div class="form-group">
        <label for="username" class="text-info">用户名:</label><br>
        <input type="text" name="uname" id="username" class="form-control">
    </div>
    <div class="form-group">
        <label for="password" class="text-info">密码:</label><br>
        <input type="text" name="passwd" id="password" class="form-control">
    </div>
    <div class="form-group">
        <input type="submit" name="submit" class="btn btn-info btn-md"
            value="登录">
    </div>
</form>
```

在 form 表单中，action 修改为/doLogin?target=/hello，这样当用户登录成功之后，就始终跳转到/hello 接口了。

当我们通过 successForwardUrl 来设置登录成功后重定向的地址时，实际上对应的实现类就是 ForwardAuthenticationSuccessHandler，ForwardAuthenticationSuccessHandler 的源码特别简单，就是一个服务端转发，代码如下：

```
public class ForwardAuthenticationSuccessHandler
    implements AuthenticationSuccessHandler {
    private final String forwardUrl;
    public ForwardAuthenticationSuccessHandler(String forwardUrl) {
        this.forwardUrl = forwardUrl;
    }
    public void onAuthenticationSuccess(HttpServletRequest request,
                              HttpServletResponse response,
```

```
                                   Authentication authentication)
                throws IOException, ServletException {
        request.getRequestDispatcher(forwardUrl).forward(request, response);
    }
}
```

由上述代码可以看到，主要功能就是调用 getRequestDispatcher 方法进行服务端转发。

AuthenticationSuccessHandler 默认的三个实现类，无论是哪一个，都是用来处理页面跳转的。有时候页面跳转并不能满足我们的需求，特别是现在流行的前后端分离开发中，用户登录成功后，就不再需要页面跳转了，只需要给前端返回一个 JSON 数据即可，告诉前端登录成功还是登录失败，前端收到消息之后自行处理。像这样的需求，我们可以通过自定义 AuthenticationSuccessHandler 的实现类来完成：

```
public class MyAuthenticationSuccessHandler implements
    AuthenticationSuccessHandler{
    @Override
    public void onAuthenticationSuccess(HttpServletRequest request,
                                        HttpServletResponse response,
                                        Authentication authentication)
            throws IOException, ServletException {
        response.setContentType("application/json;charset=utf-8");
        Map<String, Object> resp = new HashMap<>();
        resp.put("status", 200);
        resp.put("msg", "登录成功!");
        ObjectMapper om = new ObjectMapper();
        String s = om.writeValueAsString(resp);
        response.getWriter().write(s);
    }
}
```

在自定义的 MyAuthenticationSuccessHandler 中，重写 onAuthenticationSuccess 方法，在该方法中，通过 HttpServletResponse 对象返回一段登录成功的 JSON 字符串给前端即可。最后，在 SecurityConfig 中配置自定义的 MyAuthenticationSuccessHandler，代码如下：

```
@Configuration
public class SecurityConfig extends WebSecurityConfigurerAdapter {
    @Override
    protected void configure(HttpSecurity http) throws Exception {
        http.authorizeRequests()
                .anyRequest().authenticated()
                .and()
                .formLogin()
                .loginPage("/login.html")
                .loginProcessingUrl("/doLogin")
                .successHandler(new MyAuthenticationSuccessHandler())
                .failureUrl("/login.html")
                .usernameParameter("uname")
```

```
                .passwordParameter("passwd")
                .permitAll()
                .and()
                .csrf().disable();
    }
}
```

配置完成后,重启项目。此时,当用户成功登录之后,就不会进行页面跳转了,而是返回一段 JSON 字符串。

2.2.2.2 登录失败

接下来看登录失败的处理逻辑。为了方便在前端页面展示登录失败的异常信息,我们首先在项目的 pom.xml 文件中引入 thymeleaf 依赖,代码如下:

```
<dependency>
    <groupId>org.springframework.boot</groupId>
    <artifactId>spring-boot-starter-thymeleaf</artifactId>
</dependency>
```

然后在 resources/templates 目录下新建 mylogin.html,代码如下:

```
<!DOCTYPE html>
<html lang="en" xmlns:th="http://www.thymeleaf.org">
<head>
    <meta charset="UTF-8">
    <title>登录</title>
    <link
        href="//maxcdn.bootstrapcdn.com/bootstrap/4.1.1/css/bootstrap.min.css"
        rel="stylesheet" id="bootstrap-css">
    <script src="//maxcdn.bootstrapcdn.com/bootstrap/4.1.1/js/bootstrap.min.js">
    </script>
    <script src="//cdnjs.cloudflare.com/ajax/libs/jquery/3.2.1/jquery.min.js">
    </script>
</head>
<style>
    #login .container #login-row #login-column #login-box {
        border: 1px solid #9C9C9C;
        background-color: #EAEAEA;
    }
</style>
<body>
<div id="login">
    <div class="container">
        <div id="login-row"
                    class="row justify-content-center align-items-center">
            <div id="login-column" class="col-md-6">
                <div id="login-box" class="col-md-12">
                    <form id="login-form" class="form"
                                        action="/doLogin" method="post">
```

```html
                <h3 class="text-center text-info">登录</h3>
                <div th:text="${SPRING_SECURITY_LAST_EXCEPTION}">
                                                                </div>
                <div class="form-group">
                   <label for="username"
                                class="text-info">用户名:</label><br>
                   <input type="text" name="uname"
                                id="username" class="form-control">
                </div>
                <div class="form-group">
                   <label for="password"
                                class="text-info">密码:</label><br>
                   <input type="text" name="passwd"
                                id="password" class="form-control">
                </div>
                <div class="form-group">
                   <input type="submit" name="submit"
                          class="btn btn-info btn-md" value="登录">
                </div>
            </form>
         </div>
       </div>
    </div>
</div>
</body>
```

　　mylogin.html 和前面的 login.html 基本类似，前面的 login.html 是静态页面，这里的 mylogin.html 是 thymeleaf 模板页面，mylogin.html 页面在 form 中多了一个 div，用来展示登录失败时候的异常信息，登录失败的异常信息会放在 request 中返回到前端，开发者可以将其直接提取出来展示。

　　既然 mylogin.html 是动态页面，就不能像静态页面那样直接访问了，需要我们给 mylogin.html 页面提供一个访问控制器：

```java
@Controller
public class MyLoginController {
    @RequestMapping("/mylogin.html")
    public String mylogin() {
        return "mylogin";
    }
}
```

　　最后再在 SecurityConfig 中配置登录页面，代码如下：

```java
@Configuration
public class SecurityConfig extends WebSecurityConfigurerAdapter {
    @Override
    protected void configure(HttpSecurity http) throws Exception {
```

```
        http.authorizeRequests()
                .anyRequest().authenticated()
                .and()
                .formLogin()
                .loginPage("/mylogin.html")
                .loginProcessingUrl("/doLogin")
                .defaultSuccessUrl("/index.html")
                .failureUrl("/mylogin.html")
                .usernameParameter("uname")
                .passwordParameter("passwd")
                .permitAll()
                .and()
                .csrf().disable();
    }
}
```

failureUrl 表示登录失败后重定向到 mylogin.html 页面。重定向是一种客户端跳转，重定向不方便携带请求失败的异常信息（只能放在 URL 中）。

如果希望能够在前端展示请求失败的异常信息，可以使用下面这种方式：

```
@Configuration
public class SecurityConfig extends WebSecurityConfigurerAdapter {
    @Override
    protected void configure(HttpSecurity http) throws Exception {
        http.authorizeRequests()
                .anyRequest().authenticated()
                .and()
                .formLogin()
                .loginPage("/mylogin.html")
                .loginProcessingUrl("/doLogin")
                .defaultSuccessUrl("/index.html")
                .failureForwardUrl("/mylogin.html")
                .usernameParameter("uname")
                .passwordParameter("passwd")
                .permitAll()
                .and()
                .csrf().disable();
    }
}
```

failureForwardUrl 方法从名字上就可以看出，这种跳转是一种服务器端跳转，服务器端跳转的好处是可以携带登录异常信息。如果登录失败，自动跳转回登录页面后，就可以将错误信息展示出来，如图 2-8 所示。

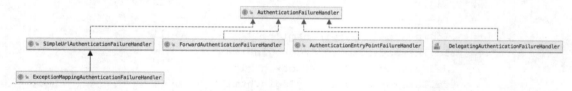

图 2-8　登录失败后展示异常信息

无论是 failureUrl 还是 failureForwardUrl，最终所配置的都是 AuthenticationFailureHandler 接口的实现。Spring Security 中提供了 AuthenticationFailureHandler 接口，用来规范登录失败的实现：

```
public interface AuthenticationFailureHandler {
    void onAuthenticationFailure(HttpServletRequest request,
        HttpServletResponse response, AuthenticationException exception)
        throws IOException, ServletException;
}
```

AuthenticationFailureHandler 接口中只有一个 onAuthenticationFailure 方法，用来处理登录失败请求，request 和 response 参数很好理解，最后的 exception 则表示登录失败的异常信息。Spring Security 中为 AuthenticationFailureHandler 一共提供了五个实现类，如图 2-9 所示。

图 2-9　AuthenticationFailureHandler 的实现类

（1）SimpleUrlAuthenticationFailureHandler 默认的处理逻辑就是通过重定向跳转到登录页面，当然也可以通过配置 forwardToDestination 属性将重定向改为服务器端跳转，failureUrl 方法的底层实现逻辑就是 SimpleUrlAuthenticationFailureHandler。

（2）ExceptionMappingAuthenticationFailureHandler 可以实现根据不同的异常类型，映射到不同的路径。

（3）ForwardAuthenticationFailureHandler 表示通过服务器端跳转来重新回到登录页面，failureForwardUrl 方法的底层实现逻辑就是 ForwardAuthenticationFailureHandler。

（4）AuthenticationEntryPointFailureHandler 是 Spring Security 5.2 新引进的处理类，可以通过 AuthenticationEntryPoint 来处理登录异常。

（5）DelegatingAuthenticationFailureHandler 可以实现为不同的异常类型配置不同的登录失败处理回调。

这里举一个简单的例子。假如不使用 failureForwardUrl 方法，同时又想在登录失败后通过服务器端跳转回到登录页面，那么可以自定义 SimpleUrlAuthenticationFailureHandler 配置，并将 forwardToDestination 属性设置为 true，代码如下：

```java
@Configuration
public class SecurityConfig extends WebSecurityConfigurerAdapter {
    @Override
    protected void configure(HttpSecurity http) throws Exception {
        http.authorizeRequests()
                .anyRequest().authenticated()
                .and()
                .formLogin()
                .loginPage("/mylogin.html")
                .loginProcessingUrl("/doLogin")
                .defaultSuccessUrl("/index.html")
                .failureHandler(failureHandler())
                .usernameParameter("uname")
                .passwordParameter("passwd")
                .permitAll()
                .and()
                .csrf().disable();
    }
    SimpleUrlAuthenticationFailureHandler failureHandler() {
        SimpleUrlAuthenticationFailureHandler handler =
                new SimpleUrlAuthenticationFailureHandler("/mylogin.html");
        handler.setUseForward(true);
        return handler;
    }
}
```

这样配置之后，如果用户再次登录失败，就会通过服务端跳转重新回到登录页面，登录页面也会展示相应的错误信息，效果和 failureForwardUrl 一致。

SimpleUrlAuthenticationFailureHandler 的源码也很简单，我们一起来看一下实现逻辑（源码比较长，这里列出来核心部分）：

```java
public class SimpleUrlAuthenticationFailureHandler implements
        AuthenticationFailureHandler {
    private String defaultFailureUrl;
    private boolean forwardToDestination = false;
    private boolean allowSessionCreation = true;
    private RedirectStrategy redirectStrategy = new DefaultRedirectStrategy();
    public SimpleUrlAuthenticationFailureHandler() {
    }
    public SimpleUrlAuthenticationFailureHandler(String defaultFailureUrl) {
        setDefaultFailureUrl(defaultFailureUrl);
    }
    public void onAuthenticationFailure(HttpServletRequest request,
            HttpServletResponse response, AuthenticationException exception)
```

```java
        throws IOException, ServletException {
    if (defaultFailureUrl == null) {
        response.sendError(HttpStatus.UNAUTHORIZED.value(),
            HttpStatus.UNAUTHORIZED.getReasonPhrase());
    }
    else {
        saveException(request, exception);
        if (forwardToDestination) {
            request.getRequestDispatcher(defaultFailureUrl)
                .forward(request, response);
        }
        else {
            redirectStrategy
                .sendRedirect(request, response, defaultFailureUrl);
        }
    }
}
protected final void saveException(HttpServletRequest request,
        AuthenticationException exception) {
    if (forwardToDestination) {
        request
            .setAttribute(WebAttributes.AUTHENTICATION_EXCEPTION, exception);
    }
    else {
        HttpSession session = request.getSession(false);
        if (session != null || allowSessionCreation) {
            request.getSession()
                .setAttribute(WebAttributes.AUTHENTICATION_EXCEPTION,
                                                exception);
        }
    }
}
public void setUseForward(boolean forwardToDestination) {
    this.forwardToDestination = forwardToDestination;
}
}
```

从这段源码中可以看到，当用户构造 SimpleUrlAuthenticationFailureHandler 对象的时候，就传入了 defaultFailureUrl，也就是登录失败时要跳转的地址。在 onAuthenticationFailure 方法中，如果发现 defaultFailureUrl 为 null，则直接通过 response 返回异常信息，否则调用 saveException 方法。在 saveException 方法中，如果 forwardToDestination 属性设置为 true，表示通过服务器端跳转回到登录页面，此时就把异常信息放到 request 中。再回到 onAuthenticationFailure 方法中，如果用户设置了 forwardToDestination 为 true，就通过服务器端跳转回到登录页面，否则通过重定向回到登录页面。

如果是前后端分离开发，登录失败时就不需要页面跳转了，只需要返回 JSON 字符串给前端即可，此时可以通过自定义 AuthenticationFailureHandler 的实现类来完成，代码如下：

```java
public class MyAuthenticationFailureHandler implements
    AuthenticationFailureHandler {
    @Override
    public void onAuthenticationFailure(HttpServletRequest request,
                                        HttpServletResponse response,
                                        AuthenticationException exception)
                                throws IOException, ServletException {
        response.setContentType("application/json;charset=utf-8");
        Map<String, Object> resp = new HashMap<>();
        resp.put("status", 500);
        resp.put("msg", "登录失败!" + exception.getMessage());
        ObjectMapper om = new ObjectMapper();
        String s = om.writeValueAsString(resp);
        response.getWriter().write(s);
    }
}
```

然后在 SecurityConfig 中进行配置即可：

```java
@Configuration
public class SecurityConfig extends WebSecurityConfigurerAdapter {
    @Override
    protected void configure(HttpSecurity http) throws Exception {
        http.authorizeRequests()
                .anyRequest().authenticated()
                .and()
                .formLogin()
                .loginPage("/mylogin.html")
                .loginProcessingUrl("/doLogin")
                .defaultSuccessUrl("/index.html")
                .failureHandler(new MyAuthenticationFailureHandler())
                .usernameParameter("uname")
                .passwordParameter("passwd")
                .permitAll()
                .and()
                .csrf().disable();
    }
}
```

配置完成后，当用户再次登录失败，就不会进行页面跳转了，而是直接返回 JSON 字符串，如图 2-10 所示。

图 2-10 用户登录失败后直接返回 JSON 字符串

2.2.2.3 注销登录

Spring Security 中提供了默认的注销页面,当然开发者也可以根据自己的需求对注销登录进行定制。

```java
@Configuration
public class SecurityConfig extends WebSecurityConfigurerAdapter {
    @Override
    protected void configure(HttpSecurity http) throws Exception {
        http.authorizeRequests()
                .anyRequest().authenticated()
                .and()
                .formLogin()
                //省略其他配置
                .and()
                .logout()
                .logoutUrl("/logout")
                .invalidateHttpSession(true)
                .clearAuthentication(true)
                .logoutSuccessUrl("/mylogin.html")
                .and()
                .csrf().disable();
    }
}
```

(1)通过.logout()方法开启注销登录配置。

(2)logoutUrl 指定了注销登录请求地址,默认是 GET 请求,路径为/logout。

(3)invalidateHttpSession 表示是否使 session 失效,默认为 true。

(4)clearAuthentication 表示是否清除认证信息,默认为 true。

(5)logoutSuccessUrl 表示注销登录后的跳转地址。

配置完成后,再次启动项目,登录成功后,在浏览器中输入 http://localhost:8080/logout 就可以发起注销登录请求了。注销成功后,会自动跳转到 mylogin.html 页面。

如果项目有需要,开发者也可以配置多个注销登录的请求,同时还可以指定请求的方法:

```java
@Configuration
public class SecurityConfig extends WebSecurityConfigurerAdapter {
    @Override
    protected void configure(HttpSecurity http) throws Exception {
        http.authorizeRequests()
                .anyRequest().authenticated()
                .and()
                .formLogin()
                //省略其他配置
                .and()
                .logout()
                .logoutRequestMatcher(new OrRequestMatcher(
                        new AntPathRequestMatcher("/logout1", "GET"),
```

```
                    new AntPathRequestMatcher("/logout2", "POST")))
            .invalidateHttpSession(true)
            .clearAuthentication(true)
            .logoutSuccessUrl("/mylogin.html")
            .and()
            .csrf().disable();
    }
}
```

上面这个配置表示注销请求路径有两个:

- 第一个是/logout1,请求方法是 GET。
- 第二个是/logout2,请求方法是 POST。

使用任意一个请求都可以完成登录注销。

如果项目是前后端分离的架构,注销成功后就不需要页面跳转了,只需将注销成功的信息返回给前端即可,此时我们可以自定义返回内容:

```
@Configuration
public class SecurityConfig extends WebSecurityConfigurerAdapter {
    @Override
    protected void configure(HttpSecurity http) throws Exception {
        http.authorizeRequests()
            .anyRequest().authenticated()
            .and()
            .formLogin()
                //省略其他配置
            .and()
            .logout()
            .logoutRequestMatcher(new OrRequestMatcher(
                    new AntPathRequestMatcher("/logout1", "GET"),
                    new AntPathRequestMatcher("/logout2", "POST")))
            .invalidateHttpSession(true)
            .clearAuthentication(true)
            .logoutSuccessHandler((req,resp,auth)->{
                resp.setContentType("application/json;charset=utf-8");
                Map<String, Object> result = new HashMap<>();
                result.put("status", 200);
                result.put("msg", "注销成功!");
                ObjectMapper om = new ObjectMapper();
                String s = om.writeValueAsString(result);
                resp.getWriter().write(s);
            })
            .and()
            .csrf().disable();
    }
}
```

配置 logoutSuccessHandler 和 logoutSuccessUrl 类似于前面所介绍的 successHandler 和

defaultSuccessUrl 之间的关系,只是类不同而已,因此这里不再赘述,读者可以按照我们前面的分析思路自行分析。

配置完成后,重启项目,登录成功后再去注销登录,无论是使用/logout1 还是/logout2 进行注销,只要注销成功后,就会返回一段 JSON 字符串。

如果开发者希望为不同的注销地址返回不同的结果,也是可以的,配置如下:

```java
@Configuration
public class SecurityConfig extends WebSecurityConfigurerAdapter {
    @Override
    protected void configure(HttpSecurity http) throws Exception {
        http.authorizeRequests()
                .anyRequest().authenticated()
                .and()
                .formLogin()
                //省略其他配置
                .and()
                .logout()
                .logoutRequestMatcher(new OrRequestMatcher(
                        new AntPathRequestMatcher("/logout1", "GET"),
                        new AntPathRequestMatcher("/logout2", "POST")))
                .invalidateHttpSession(true)
                .clearAuthentication(true)
                .defaultLogoutSuccessHandlerFor((req,resp,auth)->{
                    resp.setContentType("application/json;charset=utf-8");
                    Map<String, Object> result = new HashMap<>();
                    result.put("status", 200);
                    result.put("msg", "使用 logout1 注销成功!");
                    ObjectMapper om = new ObjectMapper();
                    String s = om.writeValueAsString(result);
                    resp.getWriter().write(s);
                },new AntPathRequestMatcher("/logout1","GET"))
                .defaultLogoutSuccessHandlerFor((req,resp,auth)->{
                    resp.setContentType("application/json;charset=utf-8");
                    Map<String, Object> result = new HashMap<>();
                    result.put("status", 200);
                    result.put("msg", "使用 logout2 注销成功!");
                    ObjectMapper om = new ObjectMapper();
                    String s = om.writeValueAsString(result);
                    resp.getWriter().write(s);
                },new AntPathRequestMatcher("/logout2","POST"))
                .and()
                .csrf().disable();
    }
}
```

通过 defaultLogoutSuccessHandlerFor 方法可以注册多个不同的注销成功回调函数,该方法第一个参数是注销成功回调,第二个参数则是具体的注销请求。当用户注销成功后,使用了哪个注销请求，就给出对应的响应信息。

2.3 登录用户数据获取

登录成功之后，在后续的业务逻辑中，开发者可能还需要获取登录成功的用户对象，如果不使用任何安全管理框架，那么可以将用户信息保存在 HttpSession 中，以后需要的时候直接从 HttpSession 中获取数据。在 Spring Security 中，用户登录信息本质上还是保存在 HttpSession 中，但是为了方便使用，Spring Security 对 HttpSession 中的用户信息进行了封装，封装之后，开发者若再想获取用户登录数据就会有两种不同的思路：

（1）从 SecurityContextHolder 中获取。
（2）从当前请求对象中获取。

这里列出来的两种方式是主流的做法，开发者也可以使用一些非主流的方式获取登录成功后的用户信息，例如直接从 HttpSession 中获取用户登录数据。

无论是哪种获取方式，都离不开一个重要的对象：Authentication。在 Spring Security 中，Authentication 对象主要有两方面的功能：

（1）作为 AuthenticationManager 的输入参数，提供用户身份认证的凭证，当它作为一个输入参数时，它的 isAuthenticated 方法返回 false，表示用户还未认证。
（2）代表已经经过身份认证的用户，此时的 Authentication 可以从 SecurityContext 中获取。

一个 Authentication 对象主要包含三个方面的信息：

（1）principal：定义认证的用户。如果用户使用用户名/密码的方式登录，principal 通常就是一个 UserDetails 对象。
（2）credentials：登录凭证，一般就是指密码。当用户登录成功之后，登录凭证会被自动擦除，以防止泄漏。
（3）authorities：用户被授予的权限信息。

Java 中本身提供了 Principal 接口用来描述认证主体，Principal 可以代表一个公司、个人或者登录 ID。Spring Security 中定义了 Authentication 接口用来规范登录用户信息，Authentication 继承自 Principal：

```
public interface Authentication extends Principal, Serializable {
    Collection<? extends GrantedAuthority> getAuthorities();
    Object getCredentials();
    Object getDetails();
    Object getPrincipal();
    boolean isAuthenticated();
    void setAuthenticated(boolean isAuthenticated);
```

}
```

这里接口中定义的方法都很好理解：

- getAuthorities 方法：用来获取用户权限。
- getCredentials 方法：用来获取用户凭证，一般来说就是密码。
- getDetails 方法：用来获取用户的详细信息，可能是当前的请求之类。
- getPrincipal 方法：用来获取当前用户信息，可能是一个用户名，也可能是一个用户对象。
- isAuthenticated 方法：当前用户是否认证成功。

可以看到，在 Spring Security 中，只要获取到 Authentication 对象，就可以获取到登录用户的详细信息。

不同的认证方式对应不同的 Authentication 实例，Spring Security 中的 Authentication 实现类如图 2-11 所示。

这些实现类现看起来可能会觉得陌生，不过没关系，在后续的章节中，这些实现类基本上都会涉及。现在我们先对每个类的功能做一个大概介绍：

（1）AbstractAuthenticationToken：该类实现了 Authentication 和 CredentialsContainer 两个接口，在 AbstractAuthenticationToken 中对 Authentication 接口定义的各个数据获取方法进行了实现，CredentialsContainer 则提供了登录凭证擦除方法。一般在登录成功后，为了防止用户信息泄漏，可以将登录凭证（例如密码）擦除。

（2）RememberMeAuthenticationToken：如果用户使用 RememberMe 的方式登录，登录信息将封装在 RememberMeAuthenticationToken 中。

（3）TestingAuthenticationToken：单元测试时封装的用户对象。

（4）AnonymousAuthenticationToken：匿名登录时封装的用户对象。

（5）RunAsUserToken：替换验证身份时封装的用户对象。

（6）UsernamePasswordAuthenticationToken：表单登录时封装的用户对象。

（7）JaasAuthenticationToken：JAAS 认证时封装的用户对象。

（8）PreAuthenticatedAuthenticationToken：Pre-Authentication 场景下封装的用户对象。

在这些 Authentication 的实例中，最常用的有两个：UsernamePasswordAuthenticationToken 和 RememberMeAuthenticationToken。在 2.1 节中的案例对应的用户认证对象就是 UsernamePasswordAuthenticationToken。

了解了 Authentication 对象之后，接下来我们来看一下如何在登录成功后获取用户登录信息，即 Authentication 对象。
```

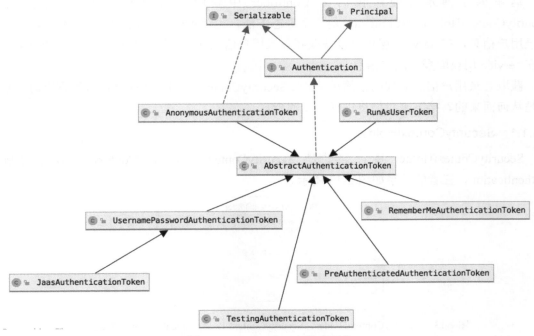

图 2-11　Authentication 的实现类

2.3.1　从 SecurityContextHolder 中获取

我们在 2.2 节案例的基础上，再添加一个 UserController，内容如下：

```
@RestController
public class UserController {
    @GetMapping("/user")
    public void userInfo() {
        Authentication authentication =
            SecurityContextHolder.getContext().getAuthentication();
        String name = authentication.getName();
        Collection<? extends GrantedAuthority> authorities =
                                    authentication.getAuthorities();
        System.out.println("name = " + name);
        System.out.println("authorities = " + authorities);
    }
}
```

配置完成后，启动项目，登录成功后，访问/user 接口，控制台就会打印出登录用户信息，当然，由于我们目前没有给用户配置角色，所以默认的用户角色为空数组，如图 2-12 所示。

```
name = javaboy
authorities = []
```

图 2-12　登录成功后打印出来的用户名和用户角色

这里为了演示方便，我们在 Controller 中获取登录用户信息，可以发现，SecurityContextHolder.getContext() 是一个静态方法，也就意味着我们随时随地都可以获取到登录用户信息，在 service 层也可以获取到登录用户信息（在实际项目中，大部分情况下也都是在 service 层获取登录用户信息）。

获取登录用户信息的代码很简单，那么 SecurityContextHolder 到底是什么？它里边的数据又是从何而来的？接下来我们将进行一一解析。

2.3.1.1　SecurityContextHolder

SecurityContextHolder 中存储的是 SecurityContext，SecurityContext 中存储的则是 Authentication，三者的关系如图 2-13 所示。

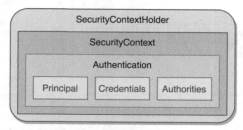

图 2-13　SecurityContextHolder、SecurityContext 以及 Authentication 之间的关系

这幅图清晰地描述了 SecurityContextHolder、SecurityContext 以及 Authentication 三者之间的关系。

首先在 SecurityContextHolder 中存放的是 SecurityContext，SecurityContextHolder 中定义了三种不同的数据存储策略，这实际上是一种典型的策略模式：

（1）MODE_THREADLOCAL：这种存放策略是将 SecurityContext 存放在 ThreadLocal 中，大家知道 ThreadLocal 的特点是在哪个线程中存储就要在哪个线程中读取，这其实非常适合 Web 应用，因为在默认情况下，一个请求无论经过多少 Filter 到达 Servlet，都是由一个线程来处理的。这也是 SecurityContextHolder 的默认存储策略，这种存储策略意味着如果在具体的业务处理代码中，开启了子线程，在子线程中去获取登录用户数据，就会获取不到。

（2）MODE_INHERITABLETHREADLOCAL：这种存储模式适用于多线程环境，如果希望在子线程中也能够获取到登录用户数据，那么可以使用这种存储模式。

（3）MODE_GLOBAL：这种存储模式实际上是将数据保存在一个静态变量中，在 Java Web 开发中，这种模式很少使用到。

Spring Security 中定义了 SecurityContextHolderStrategy 接口用来规范存储策略中的方法，我们来看一下：

```
public interface SecurityContextHolderStrategy {
    void clearContext();
    SecurityContext getContext();
    void setContext(SecurityContext context);
    SecurityContext createEmptyContext();
```

}
```

接口中一共定义了四个方法：

（1）clearContext：该方法用来清除存储的 SecurityContext 对象。
（2）getContext：该方法用来获取存储的 SecurityContext 对象。
（3）setContext：该方法用来设置存储的 SecurityContext 对象。
（4）createEmptyContext：该方法则用来创建一个空的 SecurityContext 对象。

在 Spring Security 中，SecurityContextHolderStrategy 接口一共有三个实现类，对应了三种不同的存储策略，如图 2-14 所示。

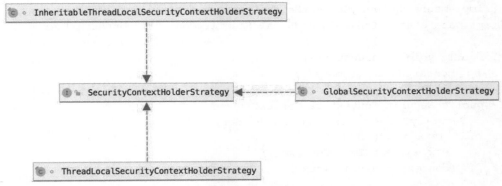

图 2-14　SecurityContextHolderStrategy 的三个实现类

每一个实现类都对应了不同的实现策略，我们先来看一下 ThreadLocalSecurityContextHolderStrategy：

```java
final class ThreadLocalSecurityContextHolderStrategy implements
    SecurityContextHolderStrategy {
    private static final ThreadLocal<SecurityContext> contextHolder =
                                                    new ThreadLocal<>();
    public void clearContext() {
        contextHolder.remove();
    }
    public SecurityContext getContext() {
        SecurityContext ctx = contextHolder.get();
        if (ctx == null) {
            ctx = createEmptyContext();
            contextHolder.set(ctx);
        }
        return ctx;
    }
    public void setContext(SecurityContext context) {
        contextHolder.set(context);
    }
    public SecurityContext createEmptyContext() {
```
```

```
        return new SecurityContextImpl();
    }
}
```

ThreadLocalSecurityContextHolderStrategy 实现了 SecurityContextHolderStrategy 接口，并实现了接口中的方法，存储数据的载体就是一个 ThreadLocal，所以针对 SecurityContext 的清空、获取以及存储，都是在 ThreadLocal 中进行操作，例如清空就是调用 ThreadLocal 的 remove 方法。SecurityContext 是一个接口，它只有一个实现类 SecurityContextImpl，所以创建就直接新建一个 SecurityContextImpl 对象即可。

再来看 InheritableThreadLocalSecurityContextHolderStrategy：

```
final class InheritableThreadLocalSecurityContextHolderStrategy
    implements SecurityContextHolderStrategy {
    private static final ThreadLocal<SecurityContext> contextHolder =
                                      new InheritableThreadLocal<>();
    public void clearContext() {
        contextHolder.remove();
    }
    public SecurityContext getContext() {
      SecurityContext ctx = contextHolder.get();
      if (ctx == null) {
        ctx = createEmptyContext();
        contextHolder.set(ctx);
      }
      return ctx;
    }
    public void setContext(SecurityContext context) {
        contextHolder.set(context);
    }
    public SecurityContext createEmptyContext() {
        return new SecurityContextImpl();
    }
}
```

InheritableThreadLocalSecurityContextHolderStrategy 和 ThreadLocalSecurityContextHolderStrategy 的实现策略基本一致，不同的是存储数据的载体变了，在 InheritableThreadLocalSecurityContextHolderStrategy 中存储数据的载体变成了 InheritableThreadLocal。InheritableThreadLocal 继承自 ThreadLocal，但是多了一个特性，就是在子线程创建的一瞬间，会自动将父线程中的数据复制到子线程中。该存储策略正是利用了这一特性，实现了在子线程中获取登录用户信息的功能。

最后再来看一下 GlobalSecurityContextHolderStrategy：

```
final class GlobalSecurityContextHolderStrategy implements
    SecurityContextHolderStrategy {
    private static SecurityContext contextHolder;
    public void clearContext() {
```

```
        contextHolder = null;
    }
    public SecurityContext getContext() {
        if (contextHolder == null) {
            contextHolder = new SecurityContextImpl();
        }
        return contextHolder;
    }
    public void setContext(SecurityContext context) {
        contextHolder = context;
    }
    public SecurityContext createEmptyContext() {
        return new SecurityContextImpl();
    }
}
```

GlobalSecurityContextHolderStrategy 的实现就更简单了，用一个静态变量来保存 SecurityContext，所以它也可以在多线程环境下使用。但是一般在 Web 开发中，这种存储策略使用得较少。

最后我们再来看一下 SecurityContextHolder 的源码：

```
public class SecurityContextHolder {
    public static final String MODE_THREADLOCAL = "MODE_THREADLOCAL";
    public static final String MODE_INHERITABLETHREADLOCAL =
                                        "MODE_INHERITABLETHREADLOCAL";
    public static final String MODE_GLOBAL = "MODE_GLOBAL";
    public static final String SYSTEM_PROPERTY = "spring.security.strategy";
    private static String strategyName = System.getProperty(SYSTEM_PROPERTY);
    private static SecurityContextHolderStrategy strategy;
    private static int initializeCount = 0;
    static {
        initialize();
    }
    public static void clearContext() {
        strategy.clearContext();
    }
    public static SecurityContext getContext() {
        return strategy.getContext();
    }
    public static int getInitializeCount() {
        return initializeCount;
    }
    private static void initialize() {
        if (!StringUtils.hasText(strategyName)) {
            strategyName = MODE_THREADLOCAL;
        }
        if (strategyName.equals(MODE_THREADLOCAL)) {
            strategy = new ThreadLocalSecurityContextHolderStrategy();
```

```java
        }
        else if (strategyName.equals(MODE_INHERITABLETHREADLOCAL)) {
            strategy = new InheritableThreadLocalSecurityContextHolderStrategy();
        }
        else if (strategyName.equals(MODE_GLOBAL)) {
            strategy = new GlobalSecurityContextHolderStrategy();
        }
        else {
            try {
                Class<?> clazz = Class.forName(strategyName);
                Constructor<?> customStrategy = clazz.getConstructor();
                strategy =
                  (SecurityContextHolderStrategy) customStrategy.newInstance();
            }
            catch (Exception ex) {
                ReflectionUtils.handleReflectionException(ex);
            }
        }
        initializeCount++;
    }
    public static void setContext(SecurityContext context) {
        strategy.setContext(context);
    }
    public static void setStrategyName(String strategyName) {
        SecurityContextHolder.strategyName = strategyName;
        initialize();
    }
    public static SecurityContextHolderStrategy getContextHolderStrategy() {
        return strategy;
    }
    public static SecurityContext createEmptyContext() {
        return strategy.createEmptyContext();
    }
}
```

从这段源码中可以看到，SecurityContextHolder 定义了三个静态常量用来描述三种不同的存储策略；存储策略 strategy 会在静态代码块中进行初始化，根据不同的 strategyName 初始化不同的存储策略；strategyName 变量表示目前正在使用的存储策略，开发者可以通过配置系统变量或者调用 setStrategyName 来修改 SecurityContextHolder 中的存储策略，调用 setStrategyName 后会重新初始化 strategy。

默认情况下，如果开发者试图从子线程中获取当前登录用户数据，就会获取失败，代码如下：

```java
@RestController
public class UserController {
    @GetMapping("/user")
    public void userInfo() {
```

```
        Authentication authentication =
                SecurityContextHolder.getContext().getAuthentication();
        String name = authentication.getName();
        Collection<? extends GrantedAuthority> authorities =
                                authentication.getAuthorities();
        System.out.println("name = " + name);
        System.out.println("authorities = " + authorities);
        new Thread(new Runnable() {
           @Override
           public void run() {
              Authentication authentication =
                   SecurityContextHolder.getContext().getAuthentication();
              if (authentication == null) {
                 System.out.println("获取用户信息失败");
                  return;
              }
              String name = authentication.getName();
              Collection<? extends GrantedAuthority> authorities =
                                    authentication.getAuthorities();
              String threadName = Thread.currentThread().getName();
              System.out.println(threadName + ":name = " + name);
              System.out.println(threadName + ":authorities = " + authorities);
           }
        }).start();
    }
}
```

在子线程中尝试获取登录用户数据时，获取到的数据为 null，如图 2-15 所示。

```
name = javaboy
authorities = []
获取用户信息失败
```

图 2-15　子线程获取登录用户信息为 null

子线程之所以获取不到登录用户信息，就是因为数据存储在 ThreadLocal 中，存储和读取不是同一个线程，所以获取不到。如果希望子线程中也能够获取到登录用户信息，可以将 SecurityContextHolder 中的存储策略改为 MODE_INHERITABLETHREADLOCAL，这样就支持多线程环境下获取登录用户信息了。

默认的存储策略是通过 System.getProperty 加载的，因此我们可以通过配置系统变量来修改默认的存储策略，以 IntelliJ IDEA 为例，首先单击启动按钮，选择 Edit Configurations 按钮，如图 2-16 所示，然后在打开的选项中，配置 VM options 参数，添加如下一行，配置界面如图 2-17 所示。

```
-Dspring.security.strategy=MODE_INHERITABLETHREADLOCAL
```

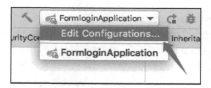

图 2-16　编辑启动配置

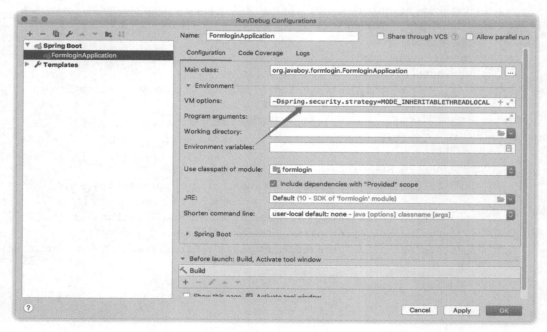

图 2-17　配置 SecurityContextHolder 中的存储策略

这样，在 SecurityContextHolder 中通过 System.getProperty 加载到的默认存储策略就支持多线程环境了。

配置完成之后，再次启动项目，此时访问/user 接口，即使在子线程中，也可以获取到登录用户信息了，如图 2-18 所示。

```
name = javaboy
authorities = []
Thread-132:name = javaboy
Thread-132:authorities = []
```

图 2-18　子线程中也可以获取到登录用户信息

看到这里读者不禁要问了，既然 SecurityContextHolder 默认是将用户信息存储在 ThreadLocal 中，在 Spring Boot 中不同的请求都是由不同的线程处理的，那为什么每一次请求都还能从 SecurityContextHolder 中获取到登录用户信息呢？这就不得不提到 Spring Security 过滤器链中重要的一环——SecurityContextPersistenceFilter。

2.3.1.2　SecurityContextPersistenceFilter

前面介绍了 Spring Security 中的常见过滤器，在这些过滤器中，存在一个非常重要的过滤器就是 SecurityContextPersistenceFilter。

默认情况下，在 Spring Security 过滤器链中，SecurityContextPersistenceFilter 是第二道防线，位于 WebAsyncManagerIntegrationFilter 之后。从 SecurityContextPersistenceFilter 这个过滤器的名字上就可以推断出来，它的作用是为了存储 SecurityContext 而设计的。

整体上来说，SecurityContextPersistenceFilter 主要做两件事情：

（1）当一个请求到来时，从 HttpSession 中获取 SecurityContext 并存入 SecurityContextHolder 中，这样在同一个请求的后续处理过程中，开发者始终可以通过 SecurityContextHolder 获取到当前登录用户信息。

（2）当一个请求处理完毕时，从 SecurityContextHolder 中获取 SecurityContext 并存入 HttpSession 中（主要针对异步 Servlet），方便下一个请求到来时，再从 HttpSession 中拿出来使用，同时擦除 SecurityContextHolder 中的登录用户信息。

> **注意**
>
> 在 SecurityContextPersistenceFilter 过滤器中，当一个请求处理完毕时，从 SecurityContextHolder 中获取 SecurityContext 存入 HttpSession 中，这一步的操作主要是针对异步 Servlet。如果不是异步 Servlet，在响应提交时，就会将 SecurityContext 保存到 HttpSession 中了，而不会等到在 SecurityContextPersistenceFilter 过滤器中再去存储。

这就是 SecurityContextPersistenceFilter 大致上做的事情，在正式开始介绍 SecurityContextPersistenceFilter 之前，需要先介绍另外一个接口，这就是 SecurityContextRepository 接口。

将 SecurityContext 存入 HttpSession，或者从 HttpSession 中加载数据并转为 SecurityContext 对象，这些事情都是由 SecurityContextRepository 接口的实现类完成的，因此这里我们就先从 SecurityContextRepository 接口开始看起。

首先我们来看一下 SecurityContextRepository 接口的定义：

```java
public interface SecurityContextRepository {
    SecurityContext loadContext(HttpRequestResponseHolder holder);
    void saveContext(SecurityContext context, HttpServletRequest request,
                     HttpServletResponse response);
    boolean containsContext(HttpServletRequest request);
}
```

SecurityContextRepository 接口中一共定义了三个方法：

（1）loadContext：这个方法用来加载 SecurityContext 对象出来，对于没有登录的用户，这里会返回一个空的 SecurityContext 对象，注意空的 SecurityContext 对象是指 SecurityContext 中不存在 Authentication 对象，而不是该方法返回 null。

（2）saveContext：该方法用来保存一个 SecurityContext 对象。

（3）containsContext：该方法可以判断 SecurityContext 对象是否存在。

在 Spring Security 框架中，为 SecurityContextRepository 接口一共提供了三个实现类，如

图 2-19 所示。

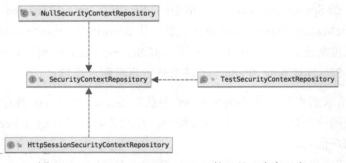

图 2-19　SecurityContextRepository 接口的三个实现类

在这三个实现类中，TestSecurityContextRepository 为单元测试提供支持；NullSecurityContextRepository 实现类中，loadContext 方法总是返回一个空的 SecurityContext 对象，saveContext 方法未做任何实现，containsContext 方法总是返回 false，所以 NullSecurityContextRepository 实现类实际上未做 SecurityContext 的存储工作。

在 Spring Security 中默认使用的实现类是 HttpSessionSecurityContextRepository，通过 HttpSessionSecurityContextRepository 实现了将 SecurityContext 存储到 HttpSession 以及从 HttpSession 中加载 SecurityContext 出来。这里我们来重点看一下 HttpSessionSecurityContextRepository 类。

在正式开始介绍 HttpSessionSecurityContextRepository 之前，首先来看一下 HttpSessionSecurityContextRepository 中定义的关于请求和封装的两个内部类。

首先是 HttpSessionSecurityContextRepository 中定义的对于响应的封装类 SaveToSessionResponseWrapper，我们先来看一下 SaveToSessionResponseWrapper 的继承关系图，如图 2-20 所示。

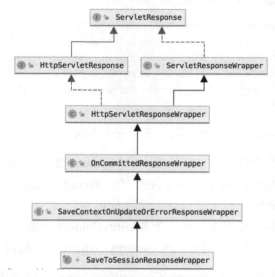

图 2-20　SaveToSessionResponseWrapper 继承关系图

从这幅继承关系图中可以看到，SaveToSessionResponseWrapper 实际上就是我们所熟知的 HttpServletResponse 功能的扩展。这里有三个关键的实现类：

（1）HttpServletResponseWrapper：HttpServletResponseWrapper 实现了 HttpServletResponse 接口，它是 HttpServletResponse 的装饰类，利用 HttpServletResponseWrapper 可以方便地操作参数和输出流等。

（2）OnCommittedResponseWrapper：OnCommittedResponseWrapper 继承自 HttpServletResponseWrapper，对其功能进行了增强，最重要的增强在于可以获取 HttpServletResponse 的提交行为。当 HttpServletResponse 的 sendError、sendRedirect、flushBuffer、flush 以及 close 等方法被调用时，onResponseCommitted 方法会被触发，开发者可以在 onResponseCommitted 方法中做一些数据保存操作，例如保存 SecurityContext。不过 OnCommittedResponseWrapper 中的 onResponseCommitted 方法只是一个抽象方法，并没有具体的实现，具体的实现则在它的实现类 SaveContextOnUpdateOrErrorResponseWrapper 中。

（3）SaveContextOnUpdateOrErrorResponseWrapper：该类继承自 OnCommittedResponseWrapper 并对 onResponseCommitted 方法做了实现。在 SaveContextOnUpdateOrErrorResponseWrapper 类中声明了一个 contextSaved 变量，表示 SecuirtyContext 是否已经存储成功。当 HttpServletResponse 提交时，会调用 onResponseCommitted 方法，在 onResponseCommitted 方法中调用 saveContext 方法，将 SecurityContext 保存到 HttpSession 中，同时将 contextSaved 变量标记为 true。saveContext 方法在这里也是一个抽象方法，具体的实现则在 SaveToSessionResponseWrapper 类中。

接下来看一下 HttpSessionSecurityContextRepository 中 SaveToSessionResponseWrapper 的定义：

```
final class SaveToSessionResponseWrapper extends
        SaveContextOnUpdateOrErrorResponseWrapper {
    private final HttpServletRequest request;
    private final boolean httpSessionExistedAtStartOfRequest;
    private final SecurityContext contextBeforeExecution;
    private final Authentication authBeforeExecution;
    SaveToSessionResponseWrapper(HttpServletResponse response,
            HttpServletRequest request,
        boolean httpSessionExistedAtStartOfRequest,
        SecurityContext context) {
        super(response, disableUrlRewriting);
        this.request = request;
        this.httpSessionExistedAtStartOfRequest =
                                httpSessionExistedAtStartOfRequest;
        this.contextBeforeExecution = context;
        this.authBeforeExecution = context.getAuthentication();
    }
    @Override
    protected void saveContext(SecurityContext context) {
```

```java
        final Authentication authentication = context.getAuthentication();
        HttpSession httpSession = request.getSession(false);
        if (authentication == null ||
                        trustResolver.isAnonymous(authentication)) {
            if (httpSession != null && authBeforeExecution != null) {
                httpSession.removeAttribute(springSecurityContextKey);
            }
            return;
        }
        if (httpSession == null) {
            httpSession = createNewSessionIfAllowed(context);
        }
        if (httpSession != null) {
            if (contextChanged(context)
                    ||
                httpSession.getAttribute(springSecurityContextKey) == null) {
                httpSession.setAttribute(springSecurityContextKey, context);
            }
        }
    }
    private boolean contextChanged(SecurityContext context) {
        return context != contextBeforeExecution
                || context.getAuthentication() != authBeforeExecution;
    }
    private HttpSession createNewSessionIfAllowed(SecurityContext context) {
        if (isTransientAuthentication(context.getAuthentication())) {
            return null;
        }
        if (httpSessionExistedAtStartOfRequest) {
            return null;
        }
        if (!allowSessionCreation) {
            return null;
        }
        if (contextObject.equals(context)) {
            return null;
        }
        try {
            return request.getSession(true);
        }
        catch (IllegalStateException e) {
        }
        return null;
    }
}
```

在 SaveToSessionResponseWrapper 中其实主要定义了三个方法：saveContext、contextChanged 以及 createNewSessionIfAllowed。

（1）saveContext：该方法主要是用来保存 SecurityContext，如果 authentication 对象为 null 或者它是一个匿名对象，则不需要保存 SecurityContext（参见 SEC-776：https://github.com/spring-projects/spring-security/issues/1036）；同时，如果 httpSession 不为 null 并且 authBeforeExecution 也不为 null，就从 httpSession 中将保存的登录用户数据移除，这个主要是为了防止开发者在注销成功的回调中继续调用 chain.doFilter 方法，进而导致原始的登录信息无法清除的问题（参见 SEC-1587：https://github.com/spring-projects/spring-security/issues/1826）；如果 httpSession 为 null，则去创建一个 HttpSession 对象；最后，如果 SecurityContext 发生了变化，或者 httpSession 中没有保存 SecurityContext，则调用 httpSession 中的 setAttribute 方法将 SecurityContext 保存起来。

（2）contextChanged：该方法主要用来判断 SecurityContext 是否发生变化，因为在程序运行过程中，开发者可能修改了 SecurityContext 中的 Authentication 对象。

（3）createNewSessionIfAllowed：该方法用来创建一个 HttpSession 对象。

这就是 HttpSessionSecurityContextRepository 中封装的 SaveToSessionResponseWrapper 对象，一个核心功能就是在 HttpServletResponse 提交的时候，将 SecurityContext 保存到 HttpSession 中。

接下来看一下 HttpSessionSecurityContextRepository 中关于 SaveToSessionRequestWrapper 的定义，SaveToSessionRequestWrapper 相对而言就要简单很多了：

```java
private static class SaveToSessionRequestWrapper extends
        HttpServletRequestWrapper {
    private final SaveContextOnUpdateOrErrorResponseWrapper response;
    SaveToSessionRequestWrapper(HttpServletRequest request,
            SaveContextOnUpdateOrErrorResponseWrapper response) {
        super(request);
        this.response = response;
    }
    @Override
    public AsyncContext startAsync() {
        response.disableSaveOnResponseCommitted();
        return super.startAsync();
    }
    @Override
    public AsyncContext startAsync(ServletRequest servletRequest,
            ServletResponse servletResponse) throws IllegalStateException {
        response.disableSaveOnResponseCommitted();
        return super.startAsync(servletRequest, servletResponse);
    }
}
```

SaveToSessionRequestWrapper 类实际上是在 Spring Security 3.2 之后出现的封装类，在 Spring Security 3.2 之前并不存在 SaveToSessionRequestWrapper 类。封装的 SaveToSessionRequestWrapper 类主要作用是禁止在异步 Servlet 提交时，自动保存 SecurityContext。

为什么要禁止呢？我们来看如下一段简单的代码：

```java
@GetMapping("/user2")
public void userInfo(HttpServletRequest req, HttpServletResponse resp) {
    AsyncContext asyncContext = req.startAsync();
    CompletableFuture.runAsync(() -> {
        try {
            PrintWriter out = asyncContext.getResponse().getWriter();
            out.write("hello javaboy!");
            asyncContext.complete();
        } catch (IOException e) {
            e.printStackTrace();
        }
    });
}
```

可以看到，在异步 Servlet 中，当任务执行完毕之后，HttpServletResponse 也会自动提交，在提交的过程中会自动保存 SecurityContext 到 HttpSession 中，但是由于是在子线程中，因此无法获取到 SecurityContext 对象（SecurityContextHolder 默认将数据存储在 ThreadLocal 中），所以会保存失败。如果开发者使用了异步 Servlet，则默认情况下会禁用 HttpServletResponse 提交时自动保存 SecurityContext 这一功能，改为在 SecurityContextPersistenceFilter 过滤器中完成 SecurityContext 保存操作。

看完了 HttpSessionSecurityContextRepository 中封装的两个请求/响应对象之后，接下来我们再来整体上看一下 HttpSessionSecurityContextRepository 类的功能：

```java
public class HttpSessionSecurityContextRepository implements
    SecurityContextRepository {
    public static final String SPRING_SECURITY_CONTEXT_KEY =
                                    "SPRING_SECURITY_CONTEXT";
    private final Object contextObject =
                        SecurityContextHolder.createEmptyContext();
    private boolean allowSessionCreation = true;
    private boolean disableUrlRewriting = false;
    private String springSecurityContextKey = SPRING_SECURITY_CONTEXT_KEY;
    private AuthenticationTrustResolver trustResolver =
                                new AuthenticationTrustResolverImpl();
    public SecurityContext loadContext(
                HttpRequestResponseHolder requestResponseHolder) {
        HttpServletRequest request = requestResponseHolder.getRequest();
        HttpServletResponse response = requestResponseHolder.getResponse();
        HttpSession httpSession = request.getSession(false);
        SecurityContext context = readSecurityContextFromSession(httpSession);
        if (context == null) {
            context = generateNewContext();
        }
        SaveToSessionResponseWrapper wrappedResponse =
            new SaveToSessionResponseWrapper(response,
```

```java
                                            request,
                                            httpSession != null,
                                            context);
    requestResponseHolder.setResponse(wrappedResponse);
    requestResponseHolder.setRequest(new SaveToSessionRequestWrapper(
            request, wrappedResponse));
    return context;
}
public void saveContext(SecurityContext context,
                        HttpServletRequest request,
                        HttpServletResponse response) {
    SaveContextOnUpdateOrErrorResponseWrapper responseWrapper = WebUtils
            .getNativeResponse(response,
                SaveContextOnUpdateOrErrorResponseWrapper.class);
    if (responseWrapper == null) {
        throw new IllegalStateException("");
    }
    if (!responseWrapper.isContextSaved()) {
        responseWrapper.saveContext(context);
    }
}
public boolean containsContext(HttpServletRequest request) {
    HttpSession session = request.getSession(false);
    if (session == null) {
        return false;
    }
    return session.getAttribute(springSecurityContextKey) != null;
}
private SecurityContext readSecurityContextFromSession(
                                        HttpSession httpSession) {
    if (httpSession == null) {
        return null;
    }
    Object contextFromSession =
                httpSession.getAttribute(springSecurityContextKey);
    if (contextFromSession == null) {
        return null;
    }
    if (!(contextFromSession instanceof SecurityContext)) {
        return null;
    }
    return (SecurityContext) contextFromSession;
}
protected SecurityContext generateNewContext() {
    return SecurityContextHolder.createEmptyContext();
}
public void setAllowSessionCreation(boolean allowSessionCreation) {
    this.allowSessionCreation = allowSessionCreation;
```

```java
    }
    public void setDisableUrlRewriting(boolean disableUrlRewriting) {
        this.disableUrlRewriting = disableUrlRewriting;
    }
    public void setSpringSecurityContextKey(String springSecurityContextKey) {
        this.springSecurityContextKey = springSecurityContextKey;
    }
    private static class SaveToSessionRequestWrapper extends
            HttpServletRequestWrapper {
        //省略
    }
    final class SaveToSessionResponseWrapper extends
            SaveContextOnUpdateOrErrorResponseWrapper {
        //省略
    }
    private boolean isTransientAuthentication(Authentication authentication) {
        return AnnotationUtils.getAnnotation(authentication.getClass(),
                            Transient.class) != null;
    }
    public void setTrustResolver(AuthenticationTrustResolver trustResolver) {
        this.trustResolver = trustResolver;
    }
}
```

（1）首先通过 SPRING_SECURITY_CONTEXT_KEY 变量定义了 SecurityContext 在 HttpSession 中存储的 key，如果开发者需要手动操作 HttpSession 中存储的 SecurityContext，可以通过该 key 来操作。

（2）trustResolver 是一个用户身份评估器，用来判断当前用户是匿名用户还是通过 RememberMe 登录的用户。

（3）在 loadContext 方法中，通过调用 readSecurityContextFromSession 方法来获取 SecurityContext 对象。如果获取到的对象为 null，则调用 generateNewContext 方法去生成一个空的 SecurityContext 对象，最后构造请求和响应的装饰类并存入 requestResponseHolder 对象中。

（4）saveContext 方法用来保存 SecurityContext，在保存之前，会先调用 isContextSaved 方法判断是否已经保存了，如果已经保存了，则不再保存。正常情况下，在 HttpServletResponse 提交时 SecurityContext 就已经保存到 HttpSession 中了；如果是异步 Servlet，则提交时不会自动将 SecurityContext 保存到 HttpSession，此时会在这里进行保存操作。

（5）containsContext 方法用来判断请求中是否存在 SecurityContext 对象。

（6）readSecurityContextFromSession 方法执行具体的 SecurityContext 读取逻辑，从 HttpSession 中获取 SecurityContext 并返回。

（7）generateNewContext 方法用来生成一个不包含 Authentication 的空的 SecurityContext 对象。

（8）setAllowSessionCreation 方法用来设置是否允许创建 HttpSession，默认是 true。

（9）setDisableUrlRewriting 方法表示是否禁用 URL 重写，默认是 false。

（10）setSpringSecurityContextKey 方法可以用来配置 HttpSession 中存储 SecurityContext 的 key。

（11）isTransientAuthentication 方法用来判断 Authentication 是否免于存储。

（12）setTrustResolver 方法用来配置身份评估器。

这就是 HttpSessionSecurityContextRepository 所提供的所有功能，这些功能都将在 SecurityContextPersistenceFilter 过滤器中进行调用，那么接下来我们就来看一下 SecurityContextPersistenceFilter 中的调用逻辑：

```java
public class SecurityContextPersistenceFilter extends GenericFilterBean {
    private SecurityContextRepository repo;
    private boolean forceEagerSessionCreation = false;
    public SecurityContextPersistenceFilter() {
        this(new HttpSessionSecurityContextRepository());
    }
    public SecurityContextPersistenceFilter(SecurityContextRepository repo) {
        this.repo = repo;
    }
    public void doFilter(ServletRequest req,
                    ServletResponse res, FilterChain chain)
                    throws IOException, ServletException {
        HttpServletRequest request = (HttpServletRequest) req;
        HttpServletResponse response = (HttpServletResponse) res;
        if (request.getAttribute(FILTER_APPLIED) != null) {
            chain.doFilter(request, response);
            return;
        }
        request.setAttribute(FILTER_APPLIED, Boolean.TRUE);
        if (forceEagerSessionCreation) {
            HttpSession session = request.getSession();
        }
        HttpRequestResponseHolder holder =
                    new HttpRequestResponseHolder(request, response);
        SecurityContext contextBeforeChainExecution =
                                            repo.loadContext(holder);
        try {
            SecurityContextHolder.setContext(contextBeforeChainExecution);
            chain.doFilter(holder.getRequest(), holder.getResponse());
        }
        finally {
            SecurityContext contextAfterChainExecution =
                                    SecurityContextHolder.getContext();
            SecurityContextHolder.clearContext();
            repo.saveContext(contextAfterChainExecution,
                        holder.getRequest(),
                        holder.getResponse());
            request.removeAttribute(FILTER_APPLIED);
```

```
        }
    }
    public void setForceEagerSessionCreation(
                            boolean forceEagerSessionCreation) {
        this.forceEagerSessionCreation = forceEagerSessionCreation;
    }
}
```

过滤器的核心方法当然是 doFilter，我们就从 doFilter 方法开始介绍：

（1）首先从 request 中获取 FILTER_APPLIED 属性，如果该属性值不为 null，则直接执行 chain.doFilter 方法，当前过滤器到此为止，这个判断主要是确保该请求只执行一次该过滤器。如果确实是该 request 第一次经过该过滤器，则给其设置上 FILTER_APPLIED 属性。

（2）forceEagerSessionCreation 变量表示是否要在过滤器链执行之前确保会话有效，由于这是一个比较耗费资源的操作，因此默认为 false。

（3）构造 HttpRequestResponseHolder 对象，将 HttpServletRequest 和 HttpServletResponse 都存储进去。

（4）调用 repo.loadContext 方法去加载 SecurityContext，repo 实际上就是我们前面所说 HttpSessionSecurityContextRepository 的实例，所以 loadContext 方法这里就不再赘述了。

（5）将读取到的 SecurityContext 存入 SecurityContextHolder 之中，这样，在接下来的处理逻辑中，开发者就可以直接通过 SecurityContextHolder 获取当前登录用户对象了。

（6）调用 chain.doFilter 方法使请求继续向下走，但是要注意，此时传递的 request 和 response 对象是在 HttpSessionSecurityContextRepository 中封装后的对象，即 SaveToSessionResponseWrapper 和 SaveToSessionRequestWrapper 的实例。

（7）当请求处理完毕后，在 finally 模块中，获取最新的 SecurityContext 对象（开发者可能在后续处理中修改了 SecurityContext 中的 Authentication 对象），然后清空 SecurityContextHolder 中的数据；再调用 repo.saveContext 方法保存 SecurityContext，具体的保存逻辑前面已经说过，这里就不再赘述了。

（8）最后，从 request 中移除 FILTER_APPLIED 属性。

这就是整个 SecurityContextPersistenceFilter 过滤器的工作逻辑。一言以蔽之，请求在到达 SecurityContextPersistenceFilter 过滤器之后，先从 HttpSession 中读取 SecurityContext 出来，并存入 SecurityContextHolder 之中以备后续使用；当请求离开 SecurityContextPersistenceFilter 过滤器的时候，获取最新的 SecurityContext 并存入 HttpSession 中，同时清空 SecurityContextHolder 中的登录用户信息。

这就是第一种登录数据的获取方式，即从 SecurityContextHolder 中获取。

2.3.2 从当前请求对象中获取

接下来我们来看一下第二种登录数据获取方式——从当前请求中获取。获取代码如下：

```
@RequestMapping("/authentication")
public void authentication(Authentication authentication) {
    System.out.println("authentication = " + authentication);
}
@RequestMapping("/principal")
public void principal(Principal principal) {
    System.out.println("principal = " + principal);
}
```

开发者可以直接在 Controller 的请求参数中放入 Authentication 对象来获取登录用户信息。通过前面的讲解，大家已经知道 Authentication 是 Principal 的子类，所以也可以直接在请求参数中放入 Principal 来接收当前登录用户信息。需要注意的是，即使参数是 Principal，真正的实例依然是 Authentication 的实例。

用过 Spring MVC 的读者都知道，Controller 中方法的参数都是当前请求 HttpServletRequest 带来的。毫无疑问，前面的 Authentication 和 Principal 参数也都是 HttpServletRequest 带来的，那么这些数据到底是何时放入 HttpServletRequest 的呢？又是以何种形式存在的呢？接下来我们一起分析一下。

在 Servlet 规范中，最早有三个和安全管理相关的方法：

```
public String getRemoteUser();
public boolean isUserInRole(String role);
public java.security.Principal getUserPrincipal();
```

（1）getRemoteUser 方法用来获取登录用户名。
（2）isUserInRole 方法用来判断当前登录用户是否具备某一个指定的角色。
（3）getUserPrincipal 方法用来获取当前认证主体。

从 Servlet 3.0 开始，在这三个方法的基础之上，又增加了三个和安全管理相关的方法：

```
public boolean authenticate(HttpServletResponse response)
        throws IOException, ServletException;
public void login(String username, String password) throws ServletException;
public void logout() throws ServletException;
```

（1）authenticate 方法可以判断当前请求是否认证成功。
（2）login 方法可以执行登录操作。
（3）logout 方法可以执行注销操作。

不过 HttpServletRequest 只是一个接口，这些安全认证相关的方法，在不同环境下会有不同的实现。

如果是一个普通的 Web 项目，不使用任何框架，HttpServletRequest 的默认实现类是 Tomcat 中的 RequestFacade，从这个类的名字上就可以看出来，这是一个使用了 Facade 模式（外观模式）的类，真正提供底层服务的是 Tomcat 中的 Request 对象，只不过这个 Request 对象在实现 Servlet 规范的同时，还定义了很多 Tomcat 内部的方法，为了避免开发者直接调用到这些内部方法，这里使用了外观模式。

在 Tomcat 的 Request 类中，对上面这些方法都做了实现，基本上都是基于 Tomcat 提供的 Realm 来实现的，这种认证方式非常冷门，项目中很少使用，因此这里不做过多介绍，感兴趣的读者可以查看 https://github.com/lenve/javaboy-code-samples 仓库中的 basiclogin 案例来了解其用法。

如果使用了 Spring Security 框架，那么我们在 Controller 参数中拿到的 HttpServletRequest 实例将是 Servlet3SecurityContextHolderAwareRequestWrapper，很明显，这是被 Spring Security 封装过的请求。

我们来看一下 Servlet3SecurityContextHolderAwareRequestWrapper 的继承关系，如图 2-21 所示。

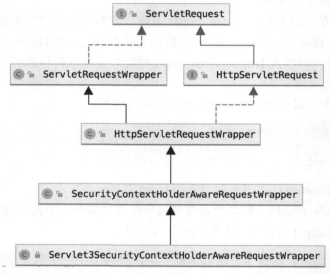

图 2-21　Servlet3SecurityContextHolderAwareRequestWrapper 继承关系图

HttpServletRequestWrapper 就不用过多介绍了，SecurityContextHolderAwareRequestWrapper 类主要实现了 Servlet 3.0 之前和安全管理相关的三个方法，也就是 getRemoteUser()、isUserInRole(String) 以及 getUserPrincipal()。Servlet 3.0 中新增的三个安全管理相关的方法，则在 Servlet3SecurityContextHolderAwareRequestWrapper 类中实现。获取用户登录信息主要和前面三个方法有关，因此这里我们主要来看一下 SecurityContextHolderAwareRequestWrapper 类中相关方法的实现。

```
public class SecurityContextHolderAwareRequestWrapper
                            extends HttpServletRequestWrapper {
```

```java
    private final AuthenticationTrustResolver trustResolver;
    private final String rolePrefix;
    public SecurityContextHolderAwareRequestWrapper(
                                    HttpServletRequest request,
                                    String rolePrefix) {
        this(request, new AuthenticationTrustResolverImpl(), rolePrefix);
    }
    public SecurityContextHolderAwareRequestWrapper(
                                    HttpServletRequest request,
            AuthenticationTrustResolver trustResolver, String rolePrefix) {
        super(request);
        this.rolePrefix = rolePrefix;
        this.trustResolver = trustResolver;
    }
    private Authentication getAuthentication() {
        Authentication auth =
                SecurityContextHolder.getContext().getAuthentication();
        if (!trustResolver.isAnonymous(auth)) {
            return auth;
        }
        return null;
    }
    @Override
    public String getRemoteUser() {
        Authentication auth = getAuthentication();
        if ((auth == null) || (auth.getPrincipal() == null)) {
            return null;
        }
        if (auth.getPrincipal() instanceof UserDetails) {
            return ((UserDetails) auth.getPrincipal()).getUsername();
        }
        return auth.getPrincipal().toString();
    }
    @Override
    public Principal getUserPrincipal() {
        Authentication auth = getAuthentication();
        if ((auth == null) || (auth.getPrincipal() == null)) {
            return null;
        }
        return auth;
    }
    private boolean isGranted(String role) {
        Authentication auth = getAuthentication();
        if (rolePrefix != null && role != null
                                && !role.startsWith(rolePrefix)) {
            role = rolePrefix + role;
        }
        if ((auth == null) || (auth.getPrincipal() == null)) {
```

```
            return false;
        }
        Collection<? extends GrantedAuthority> authorities =
                                                auth.getAuthorities();
        if (authorities == null) {
            return false;
        }
        for (GrantedAuthority grantedAuthority : authorities) {
            if (role.equals(grantedAuthority.getAuthority())) {
                return true;
            }
        }
        return false;
    }
    @Override
    public boolean isUserInRole(String role) {
        return isGranted(role);
    }
}
```

SecurityContextHolderAwareRequestWrapper 类其实非常好理解:

（1）getAuthentication：该方法用来获取当前登录对象 Authentication，获取方式就是我们前面所讲的从 SecurityContextHolder 中获取。如果不是匿名对象就返回，否则就返回 null。

（2）getRemoteUser：该方法返回了当前登录用户的用户名，如果 Authentication 对象中存储的 Principal 是当前登录用户对象，则返回用户名；如果 Authentication 对象中存储的 Principal 是当前登录用户名（字符串），则直接返回即可。

（3）getUserPrincipal：该方法返回当前登录用户对象，其实就是 Authentication 的实例。

（4）isGranted：该方法是一个私有方法，作用是判断当前登录用户是否具备某一个指定的角色。判断逻辑也很简单，先对传入进来的角色进行预处理，有的情况下可能需要添加 ROLE_ 前缀，角色前缀的问题在本书后面的章节中会做详细介绍，这里先不做过多的展开。然后调用 Authentication#getAuthorities 方法，获取当前登录用户所具备的所有角色，最后再和传入进来的参数进行比较。

（5）isUserInRole：该方法调用 isGranted 方法，进而实现判断当前用户是否具备某一个指定角色的功能。

看到这里，相信读者已经明白了，在使用了 Spring Security 之后，我们通过 HttpServletRequest 就可以获取到很多当前登录用户信息了，代码如下：

```
@RequestMapping("/info")
public void info(HttpServletRequest req) {
    String remoteUser = req.getRemoteUser();
    Authentication auth = ((Authentication) req.getUserPrincipal());
    boolean admin = req.isUserInRole("admin");
    System.out.println("remoteUser = " + remoteUser);
```

```java
    System.out.println("auth.getName() = " + auth.getName());
    System.out.println("admin = " + admin);
}
```

执行该方法，打印结果如下：

```
remoteUser = javaboy
auth.getName() = javaboy
admin = false
```

前面我们直接将 Authentication 或者 Principal 写到 Controller 参数中，实际上就是 Spring MVC 框架从 Servlet3SecurityContextHolderAwareRequestWrapper 中提取的用户信息。

那么 Spring Security 是如何将默认的请求对象转化为 Servlet3SecurityContextHolderAwareRequestWrapper 的呢？这就涉及 Spring Security 过滤器链中另外一个重要的过滤器——SecurityContextHolderAwareRequestFilter。

前面我们提到 Spring Security 过滤器中，有一个 SecurityContextHolderAwareRequest Filter 过滤器，该过滤器的主要作用就是对 HttpServletRequest 请求进行再包装，重写 HttpServletRequest 中和安全管理相关的方法。HttpServletRequest 在整个请求过程中会被包装多次，每一次的包装都会给它增添新的功能，例如在经过 SecurityContextPersistenceFilter 请求时就会对它进行包装。

我们来看一下 SecurityContextHolderAwareRequestFilter 过滤器的源码（部分）：

```java
public class SecurityContextHolderAwareRequestFilter
                                        extends GenericFilterBean {
    public void doFilter(ServletRequest req,
                    ServletResponse res, FilterChain chain)
                    throws IOException, ServletException {
        chain.doFilter(this.requestFactory.create((HttpServletRequest) req,
            (HttpServletResponse) res), res);
    }
private HttpServletRequestFactory createServlet3Factory(String rolePrefix) {
    HttpServlet3RequestFactory factory =
                        new HttpServlet3RequestFactory(rolePrefix);
    factory.setTrustResolver(this.trustResolver);
    factory.setAuthenticationEntryPoint(this.authenticationEntryPoint);
    factory.setAuthenticationManager(this.authenticationManager);
    factory.setLogoutHandlers(this.logoutHandlers);
    return factory;
    }
}
final class HttpServlet3RequestFactory implements HttpServletRequestFactory {
    @Override
    public HttpServletRequest create(HttpServletRequest request,
        HttpServletResponse response) {
        return new Servlet3SecurityContextHolderAwareRequestWrapper(request,
                                        this.rolePrefix, response);
    }
```

```
private class Servlet3SecurityContextHolderAwareRequestWrapper
        extends SecurityContextHolderAwareRequestWrapper {
    //......
}
}
```

从这段源码中可以看到,在 SecurityContextHolderAwareRequestFilter#doFilter 方法中,会调用 requestFactory.create 方法对请求重新进行包装。requestFactory 就是 HttpServletRequestFactory 类的实例,它的 create 方法里边就直接创建了一个 Servlet3SecurityContextHolderAwareRequestWrapper 实例。

对请求的 HttpServletRequest 包装之后,接下来在过滤器链中传递的 HttpServletRequest 对象,它的 getRemoteUser()、isUserInRole(String)以及 getUserPrincipal()方法就可以直接使用了。

HttpServletRequest 中 getUserPrincipal()方法有了返回值之后,最终在 Spring MVC 的 ServletRequestMethodArgumentResolver#resolveArgument(Class<?>, HttpServletRequest)方法中进行默认参数解析,自动解析出 Principal 对象。开发者在 Controller 中既可以通过 Principal 来接收参数,也可以通过 Authentication 对象来接收。

经过前面的介绍,相信读者对于 Spring Security 中两种获取登录用户信息的方式,以及这两种获取方式的原理,都有一定的了解了。

2.4 用户定义

在前面的案例中,我们的登录用户是基于配置文件来配置的(本质是基于内存),但是在实际开发中,这种方式肯定是不可取的,在实际项目中,用户信息肯定要存入数据库之中。

Spring Security 支持多种用户定义方式,接下来我们就逐个来看一下这些定义方式。通过前面的介绍(参见 2.1.3 小节),大家对于 UserDetailsService 以及它的子类都有了一定的了解,自定义用户其实就是使用 UserDetailsService 的不同实现类来提供用户数据,同时将配置好的 UserDetailsService 配置给 AuthenticationManagerBuilder,系统再将 UserDetailsService 提供给 AuthenticationProvider 使用。

2.4.1 基于内存

前面案例中用户的定义本质上还是基于内存,只是我们没有将 InMemoryUserDetailsManager 类明确抽出来自定义,现在我们通过自定义 InMemoryUserDetailsManager 来看一下基于内存的用户是如何自定义的。

重写 WebSecurityConfigurerAdapter 类的 configure(AuthenticationManagerBuilder)方法,内容如下:

```
@Override
```

```java
protected void configure(AuthenticationManagerBuilder auth) throws Exception {
    InMemoryUserDetailsManager manager = new InMemoryUserDetailsManager();
    manager.createUser(User.withUsername("javaboy")
                        .password("{noop}123").roles("admin").build());
    manager.createUser(User.withUsername("sang")
                        .password("{noop}123").roles("user").build());
    auth.userDetailsService(manager);
}
```

首先构造了一个 InMemoryUserDetailsManager 实例，调用该实例的 createUser 方法来创建用户对象，我们在这里分别设置了用户名、密码以及用户角色。需要注意的是，用户密码加了一个 {noop} 前缀，表示密码不加密，明文存储（关于密码加密问题，会在后面的章节中专门介绍）。

配置完成后，启动项目，此时就可以使用这里配置的两个用户登录了。

InMemoryUserDetailsManager 的实现原理很简单，它间接实现了 UserDetailsService 接口并重写了它里边的 loadUserByUsername 方法，同时它里边维护了一个 HashMap 变量，Map 的 key 就是用户名，value 则是用户对象，createUser 就是往这个 Map 中存储数据，loadUserByUsername 方法则是从该 Map 中读取数据，这里的源码比较简单，就不贴出来了，读者可以自行查看。

2.4.2 基于 JdbcUserDetailsManager

JdbcUserDetailsManager 支持将用户数据持久化到数据库，同时它封装了一系列操作用户的方法，例如用户的添加、更新、查找等。

Spring Security 中为 JdbcUserDetailsManager 提供了数据库脚本，位置在 org/springframework/security/core/userdetails/jdbc/users.ddl，内容如下：

```sql
create table users(username varchar_ignorecase(50) not null
primary key,
password varchar_ignorecase(500) not null,
enabled boolean not null);

create table authorities (username varchar_ignorecase(50) not null,
authority varchar_ignorecase(50) not null,
constraint fk_authorities_users
foreign key(username) references users(username));

create unique index ix_auth_username on authorities (username,authority);
```

可以看到这里一共创建了两张表，users 表就是存放用户信息的表，authorities 则是存放用户角色的表。但是大家注意 SQL 的数据类型中有一个 varchar_ignorecase，这个其实是针对 HSQLDB 的数据类型，我们这里使用的是 MySQL 数据库，所以这里手动将 varchar_ignorecase 类型修改为 varchar 类型，然后去数据库中执行修改后的脚本。

另一方面，由于要将数据存入数据库中，所以我们的项目也要提供数据库支持，JdbcUserDetailsManager 底层实际上是使用 JdbcTemplate 来完成的，所以这里主要添加两个依赖：

```xml
<dependency>
    <groupId>org.springframework.boot</groupId>
    <artifactId>spring-boot-starter-jdbc</artifactId>
</dependency>
<dependency>
    <groupId>mysql</groupId>
    <artifactId>mysql-connector-java</artifactId>
</dependency>
```

然后在 resources/application.properties 中配置数据库连接信息：

```
spring.datasource.username=root
spring.datasource.password=123
spring.datasource.url=jdbc:mysql:///security?useUnicode=true&characterEncoding=UTF-8&serverTimezone=Asia/Shanghai
```

配置完成后，我们重写 WebSecurityConfigurerAdapter 类的 configure(AuthenticationManagerBuilder) 方法，内容如下：

```java
@Configuration
public class SecurityConfig extends WebSecurityConfigurerAdapter {
    @Autowired
    DataSource dataSource;
    @Override
    protected void configure(AuthenticationManagerBuilder auth)
                                                    throws Exception {
        JdbcUserDetailsManager manager =
                        new JdbcUserDetailsManager(dataSource);
        if (!manager.userExists("javaboy")) {
            manager.createUser(User.withUsername("javaboy")
                    .password("{noop}123").roles("admin").build());
        }
        if (!manager.userExists("sang")) {
            manager.createUser(User.withUsername("sang")
                    .password("{noop}123").roles("user").build());
        }
        auth.userDetailsService(manager);
    }
    @Override
    protected void configure(HttpSecurity http) throws Exception {
        //省略
    }
}
```

（1）当引入 spring-boot-starter-jdbc 并配置了数据库连接信息后，一个 DataSource 实例就

有了，这里首先引入 DataSource 实例。

（2）在 configure 方法中，创建一个 JdbcUserDetailsManager 实例，在创建时传入 DataSource 实例。通过 userExists 方法可以判断一个用户是否存在，该方法本质上就是去数据库中查询对应的用户；如果用户不存在，则通过 createUser 方法可以创建一个用户，该方法本质上就是向数据库中添加一个用户。

（3）最后将 manager 实例设置到 auth 对象中。

配置完成后，重启项目，如果项目启动成功，数据库中就会自动添加进来两条数据，如图 2-22、图 2-23 所示。

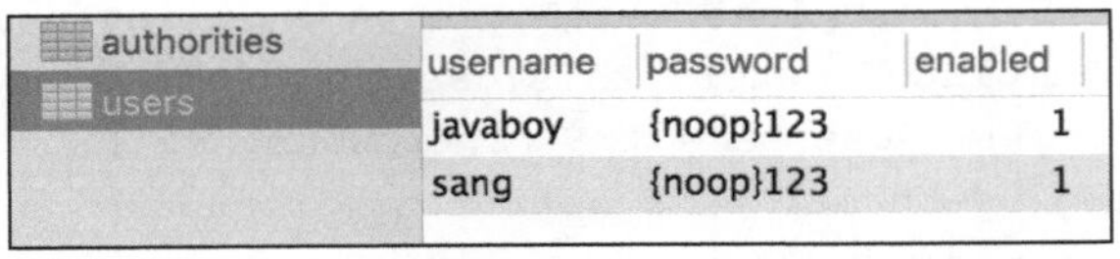

图 2-22　数据库中自动存入两条 users 数据

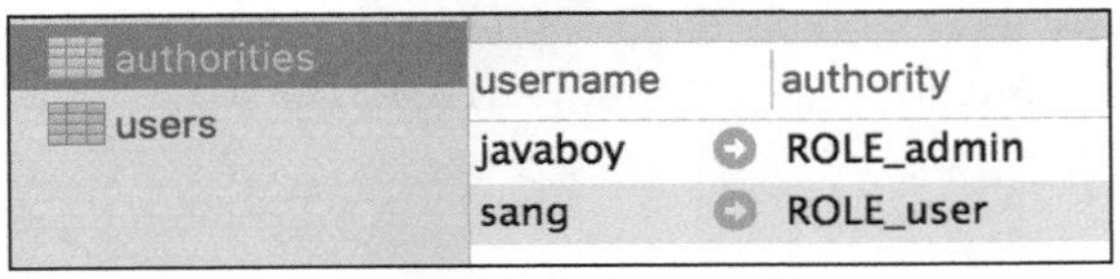

图 2-23　数据库中自动存入两条角色数据

此时，我们就可以使用 javaboy/123、sang/123 进行登录测试了。

在 JdbcUserDetailsManager 的继承体系中，首先是 JdbcDaoImpl 实现了 UserDetailsService 接口，并实现了基本的 loadUserByUsername 方法。JdbcUserDetailsManager 则继承自 JdbcDaoImpl，同时完善了数据库操作，又封装了用户的增删改查方法。这里，我们以 loadUserByUsername 为例，看一下源码，其余的增删改操作相对来说都比较容易，这里就不再赘述了。

JdbcDaoImpl#loadUserByUsername：

```java
@Override
public UserDetails loadUserByUsername(String username)
                                    throws UsernameNotFoundException {
    List<UserDetails> users = loadUsersByUsername(username);
    if (users.size() == 0) {
        throw new UsernameNotFoundException(
                this.messages.getMessage("JdbcDaoImpl.notFound",
                        new Object[] { username }, "Username {0} not found"));
    }
    UserDetails user = users.get(0);
    Set<GrantedAuthority> dbAuthsSet = new HashSet<>();
    if (this.enableAuthorities) {
        dbAuthsSet.addAll(loadUserAuthorities(user.getUsername()));
    }
    if (this.enableGroups) {
```

```
        dbAuthsSet.addAll(loadGroupAuthorities(user.getUsername()));
    }
    List<GrantedAuthority> dbAuths = new ArrayList<>(dbAuthsSet);
    addCustomAuthorities(user.getUsername(), dbAuths);
    if (dbAuths.size() == 0) {
        throw new UsernameNotFoundException(this.messages.getMessage(
                "JdbcDaoImpl.noAuthority", new Object[] { username },
                "User {0} has no GrantedAuthority"));
    }
    return createUserDetails(username, user, dbAuths);
}
protected List<UserDetails> loadUsersByUsername(String username) {
    return getJdbcTemplate().query(this.usersByUsernameQuery,
            new String[] { username }, (rs, rowNum) -> {
                String username1 = rs.getString(1);
                String password = rs.getString(2);
                boolean enabled = rs.getBoolean(3);
                return new User(username1, password, enabled, true, true, true,
                        AuthorityUtils.NO_AUTHORITIES);
            });
}
```

（1）首先根据用户名，调用 loadUsersByUsername 方法去数据库中查询用户，查询出来的是一个 List 集合，集合中如果没有数据，说明用户不存在，则直接抛出异常。

（2）如果集合中存在数据，则将集合中的第一条数据拿出来，然后再去查询用户角色，最后根据这些信息创建一个新的 UserDetails 出来。

（3）需要注意的是，这里还引入了分组的概念，不过考虑到 JdbcUserDetailsManager 并非我们实际项目中的主流方案，因此这里不做过多介绍。

这就是使用 JdbcUserDetailsManager 做数据持久化。这种方式看起来简单，都不用开发者自己写 SQL，但是局限性比较大，无法灵活地定义用户表、角色表等，而在实际开发中，我们还是希望能够灵活地掌控数据表结构，因此 JdbcUserDetailsManager 使用场景非常有限。

2.4.3 基于 MyBatis

使用 MyBatis 做数据持久化是目前大多数企业应用采取的方案，Spring Security 中结合 MyBatis 可以灵活地定制用户表以及角色表，我们对此进行详细介绍。

首先需要设计三张表，分别是用户表、角色表以及用户角色关联表，三张表的关系如图 2-24 所示。

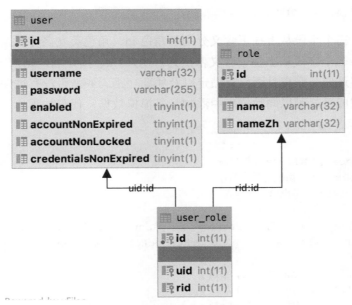

图 2-24　用户表、角色表以及用户角色关联表

用户和角色是多对多的关系，我们使用 user_role 来将两者关联起来。

数据库脚本如下：

```sql
CREATE TABLE `role` (
  `id` int(11) NOT NULL AUTO_INCREMENT,
  `name` varchar(32) DEFAULT NULL,
  `nameZh` varchar(32) DEFAULT NULL,
  PRIMARY KEY (`id`)
) ENGINE=InnoDB DEFAULT CHARSET=utf8;

CREATE TABLE `user` (
  `id` int(11) NOT NULL AUTO_INCREMENT,
  `username` varchar(32) DEFAULT NULL,
  `password` varchar(255) DEFAULT NULL,
  `enabled` tinyint(1) DEFAULT NULL,
  `accountNonExpired` tinyint(1) DEFAULT NULL,
  `accountNonLocked` tinyint(1) DEFAULT NULL,
  `credentialsNonExpired` tinyint(1) DEFAULT NULL,
  PRIMARY KEY (`id`)
) ENGINE=InnoDB DEFAULT CHARSET=utf8;

CREATE TABLE `user_role` (
  `id` int(11) NOT NULL AUTO_INCREMENT,
  `uid` int(11) DEFAULT NULL,
  `rid` int(11) DEFAULT NULL,
  PRIMARY KEY (`id`),
  KEY `uid` (`uid`),
  KEY `rid` (`rid`)
```

```
) ENGINE=InnoDB DEFAULT CHARSET=utf8;
```

对于角色表,三个字段从上往下含义分别为角色 id、角色英文名称以及角色中文名称。

对于用户表,七个字段从上往下含义依次为:用户 id、用户名、用户密码、账户是否可用、账户是否没有过期、账户是否没有锁定以及凭证(密码)是否没有过期。

数据库创建完成后,可以向数据库中添加几条模拟数据,代码如下:

```sql
INSERT INTO `role` (`id`, `name`, `nameZh`)
VALUES
    (1,'ROLE_dba','数据库管理员'),
    (2,'ROLE_admin','系统管理员'),
    (3,'ROLE_user','用户');

INSERT INTO `user` (`id`, `username`, `password`, `enabled`,
        `accountNonExpired`, `accountNonLocked`, `credentialsNonExpired`)
VALUES
    (1,'root','{noop}123',1,1,1,1),
    (2,'admin','{noop}123',1,1,1,1),
    (3,'sang','{noop}123',1,1,1,1);

INSERT INTO `user_role` (`id`, `uid`, `rid`)
VALUES
    (1,1,1),
    (2,1,2),
    (3,2,2),
    (4,3,3);
```

这样,数据库的准备工作就算完成了。

在 Spring Security 项目中,我们需要引入 MyBatis 和 MySQL 依赖,代码如下:

```xml
<dependency>
    <groupId>org.mybatis.spring.boot</groupId>
    <artifactId>mybatis-spring-boot-starter</artifactId>
    <version>2.1.3</version>
</dependency>
<dependency>
    <groupId>mysql</groupId>
    <artifactId>mysql-connector-java</artifactId>
</dependency>
```

同时在 resources/application.properties 中配置数据库基本连接信息:

```
spring.datasource.username=root
spring.datasource.password=123
spring.datasource.url=jdbc:mysql:///security02?useUnicode=true&characterEncoding=UTF-8&serverTimezone=Asia/Shanghai
```

接下来创建用户类和角色类:

```java
public class User implements UserDetails {
```

```java
    private Integer id;
    private String username;
    private String password;
    private Boolean enabled;
    private Boolean accountNonExpired;
    private Boolean accountNonLocked;
    private Boolean credentialsNonExpired;
    private List<Role> roles = new ArrayList<>();
    @Override
    public Collection<? extends GrantedAuthority> getAuthorities() {
        List<SimpleGrantedAuthority> authorities = new ArrayList<>();
        for (Role role : roles) {
            authorities.add(new SimpleGrantedAuthority(role.getName()));
        }
        return authorities;
    }
    @Override
    public String getPassword() {
        return password;
    }
    @Override
    public String getUsername() {
        return username;
    }
    @Override
    public boolean isAccountNonExpired() {
        return accountNonExpired;
    }
    @Override
    public boolean isAccountNonLocked() {
        return accountNonLocked;
    }
    @Override
    public boolean isCredentialsNonExpired() {
        return credentialsNonExpired;
    }
    @Override
    public boolean isEnabled() {
        return enabled;
    }
    //省略其他 getter/setter
}
public class Role {
    private Integer id;
    private String name;
    private String nameZh;
    //省略 getter/setter
}
```

自定义用户类需要实现 UserDetails 接口,并实现接口中的方法,这些方法的含义我们在 2.1.3 小节中已经介绍过了,这里不再赘述。其中 roles 属性用来保存用户所具备的角色信息,由于系统获取用户角色调用的方法是 getAuthorities,所以我们在 getAuthorities 方法中,将 roles 中的角色转为系统可识别的对象并返回。

> **注 意**
>
> User 类中的 isXXX 方法可以当成 get 方法对待,不需要再给这些属性生成 get 方法。

接下来我们自定义 UserDetailsService 以及对应的数据库查询方法:

```java
@Service
public class MyUserDetailsService implements UserDetailsService {
    @Autowired
    UserMapper userMapper;
    @Override
    public UserDetails loadUserByUsername(String username)
                                    throws UsernameNotFoundException {
        User user = userMapper.loadUserByUsername(username);
        if (user == null) {
            throw new UsernameNotFoundException("用户不存在");
        }
        user.setRoles(userMapper.getRolesByUid(user.getId()));
        return user;
    }
}
@Mapper
public interface UserMapper {
    List<Role> getRolesByUid(Integer id);
    User loadUserByUsername(String username);
}
```

自定义 MyUserDetailsService 实现 UserDetailsService 接口,并实现该接口中的方法。loadUserByUsername 方法经过前面章节的讲解,相信大家已经很熟悉了,该方法就是根据用户名去数据库中加载用户,如果从数据库中没有查到用户,则抛出 UsernameNotFoundException 异常;如果查询到用户了,则给用户设置 roles 属性。

UserMapper 中定义两个方法用于支持 MyUserDetailsService 中的查询操作。

最后,在 UserMapper.xml 中定义查询 SQL,代码如下:

```xml
<!DOCTYPE mapper
        PUBLIC "-//mybatis.org//DTD Mapper 3.0//EN"
        "http://mybatis.org/dtd/mybatis-3-mapper.dtd">
<mapper namespace="org.javaboy.formlogin.mapper.UserMapper">
    <select id="loadUserByUsername"
                        resultType="org.javaboy.formlogin.model.User">
        select * from user where username=#{username};
    </select>
```

```xml
    <select id="getRolesByUid" resultType="org.javaboy.formlogin.model.Role">
        select r.* from role r,user_role ur where r.`id`=ur.`rid`
    </select>
</mapper>
```

为了方便，我们将 UserMapper.xml 文件和 UserMapper 接口放在了相同的包下。为了防止 Maven 打包时自动忽略了 XML 文件，还需要在 pom.xml 中添加如下配置：

```xml
<build>
    <resources>
        <resource>
            <directory>src/main/java</directory>
            <includes>
                <include>**/*.xml</include>
            </includes>
        </resource>
        <resource>
            <directory>src/main/resources</directory>
        </resource>
    </resources>
</build>
```

最后一步，就是在 SecurityConfig 中注入 UserDetailsService：

```java
@Configuration
public class SecurityConfig extends WebSecurityConfigurerAdapter {
    @Autowired
    MyUserDetailsService myUserDetailsService;
    @Override
    protected void configure(AuthenticationManagerBuilder auth)
                                                    throws Exception {
        auth.userDetailsService(myUserDetailsService);
    }
    @Override
    protected void configure(HttpSecurity http) throws Exception {
        http.authorizeRequests()
                //省略
    }
}
```

配置 UserDetailsService 的方式和前面配置 JdbcUserDetailsManager 的方式基本一致，只不过配置对象变成了 myUserDetailsService 而已。

至此，整个配置工作就完成了。

接下来启动项目，利用数据库中添加的模拟用户进行登录测试，就可以成功登录了，测试方式和前面章节一致，这里不再赘述。

2.4.4 基于 Spring Data JPA

考虑到在 Spring Boot 技术栈中也有不少人使用 Spring Data JPA，因此这里针对 Spring Security+Spring Data JPA 也做一个简单介绍，具体思路和基于 MyBatis 的整合类似。

首先引入 Spring Data JPA 的依赖和 MySQL 依赖：

```xml
<dependency>
    <groupId>org.springframework.boot</groupId>
    <artifactId>spring-boot-starter-data-jpa</artifactId>
</dependency>
<dependency>
    <groupId>mysql</groupId>
    <artifactId>mysql-connector-java</artifactId>
</dependency>
```

然后在 resources/application.properties 中配置数据库和 JPA，代码如下：

```
spring.datasource.username=root
spring.datasource.password=123
spring.datasource.url=jdbc:mysql:///security03?useUnicode=true&characterEncoding=UTF-8&serverTimezone=Asia/Shanghai

spring.jpa.database=mysql
spring.jpa.database-platform=mysql
spring.jpa.hibernate.ddl-auto=update
spring.jpa.show-sql=true
spring.jpa.properties.hibernate.dialect=org.hibernate.dialect.MySQL8Dialect
```

数据库的配置还是和以前一样，JPA 的配置则主要配置了数据库平台，数据表更新方式、是否打印 SQL 以及对应的数据库方言。

使用 Spring Data JPA 的好处是我们不用提前准备 SQL 脚本，所以接下来配置两个数据库实体类即可：

```java
@Entity(name = "t_user")
public class User implements UserDetails {
    @Id
    @GeneratedValue(strategy = GenerationType.IDENTITY)
    private Long id;
    private String username;
    private String password;
    private boolean accountNonExpired;
    private boolean accountNonLocked;
    private boolean credentialsNonExpired;
    private boolean enabled;
    @ManyToMany(fetch = FetchType.EAGER,cascade = CascadeType.PERSIST)
    private List<Role> roles;
```

```java
    @Override
    public Collection<? extends GrantedAuthority> getAuthorities() {
        List<SimpleGrantedAuthority> authorities = new ArrayList<>();
        for (Role role : getRoles()) {
            authorities.add(new SimpleGrantedAuthority(role.getName()));
        }
        return authorities;
    }
    @Override
    public String getPassword() {
        return password;
    }
    @Override
    public String getUsername() {
        return username;
    }
    @Override
    public boolean isAccountNonExpired() {
        return accountNonExpired;
    }
    @Override
    public boolean isAccountNonLocked() {
        return accountNonLocked;
    }
    @Override
    public boolean isCredentialsNonExpired() {
        return credentialsNonExpired;
    }
    @Override
    public boolean isEnabled() {
        return enabled;
    }
    //省略 getter/setter
}
@Entity(name = "t_role")
public class Role {
    @Id
    @GeneratedValue(strategy = GenerationType.IDENTITY)
    private Long id;
    private String name;
    private String nameZh;
    //省略 getter/setter
}
```

这两个实体类和前面 MyBatis 中实体类的配置类似,需要注意的是 roles 属性上多了一个多对多配置。

接下来配置 UserDetailsService,并提供数据查询方法:

```java
@Service
public class MyUserDetailsService implements UserDetailsService {
    @Autowired
    UserDao userDao;
    @Override
    public UserDetails loadUserByUsername(String username)
                                    throws UsernameNotFoundException {
        User user = userDao.findUserByUsername(username);
        if (user == null) {
            throw new UsernameNotFoundException("用户不存在");
        }
        return user;
    }
}
public interface UserDao extends JpaRepository<User,Integer> {
    User findUserByUsername(String username);
}
```

MyUserDetailsService 的定义也和前面的类似，不同之处在于数据查询方法的变化。定义 UserDao 继承自 JpaRepository，并定义一个 findUserByUsername 方法，剩下的事情 Spring Data JPA 框架会帮我们完成。

最后，再在 SecurityConfig 中配置 MyUserDetailsService，配置方式和 MyBatis 一模一样，这里就不再把代码贴出来了。

使用了 Spring Data JPA 之后，当项目启动时，会自动在数据库中创建相关的表，而不用我们自己去写脚本，这也是使用 Spring Data JPA 的方便之处。

为了测试方便，我们可以在单元测试中执行如下代码，向数据库中添加测试数据：

```java
@Autowired
UserDao userDao;
@Test
void contextLoads() {
    User u1 = new User();
    u1.setUsername("javaboy");
    u1.setPassword("{noop}123");
    u1.setAccountNonExpired(true);
    u1.setAccountNonLocked(true);
    u1.setCredentialsNonExpired(true);
    u1.setEnabled(true);
    List<Role> rs1 = new ArrayList<>();
    Role r1 = new Role();
    r1.setName("ROLE_admin");
    r1.setNameZh("管理员");
    rs1.add(r1);
    u1.setRoles(rs1);
    userDao.save(u1);
}
```

测试数据添加成功之后，接下来启动项目，使用测试数据进行登录测试，具体测试过程就不再赘述了。

至此，四种不同的用户定义方式就介绍完了。这四种方式，异曲同工，只是数据存储的方式不一样而已，其他的执行流程都是一样的。

2.5 小 结

本章主要介绍了 Spring Security 认证的一些基本操作和原理，对于认证流程做了简单的梳理，登录表单进行了详细配置；同时还研究了获取当前登录用户信息的两种方式以及四种不同的用户定义方式。本章算是一个引子，让大家先来感受一下 Spring Security 的基本用法，接下来的章节中，我们将对这里的诸多登录细节做进一步的深入讲解。

第 3 章

认证流程分析

Spring Security 中默认的一套登录流程是非常完善并且严谨的。但是项目需求非常多样化，很多时候，我们可能还需要对 Spring Security 登录流程进行定制，定制的前提是开发者先深刻理解 Spring Security 登录流程，然后在此基础之上，完成对登录流程的定制。本章将从头梳理 Spring Security 登录流程，并通过几个常见的登录定制案例，让读者更加深刻地理解 Spring Security 登录流程。

本章涉及的主要知识点有：

- 登录流程分析。
- 配置多个数据源。
- 添加登录验证码。

3.1 登录流程分析

要搞清楚 Spring Security 认证流程，我们得先认识与之相关的三个基本组件（Authentication 对象在第 2 章已经做过介绍，这里不再赘述）：AuthenticationManager、ProviderManager 以及 AuthenticationProvider，同时还要去了解接入认证功能的过滤器 AbstractAuthenticationProcessingFilter，这四个类搞明白了，基本上认证流程也就清楚了，下面我们逐个分析一下。

3.1.1 AuthenticationManager

从名称上可以看出，AuthenticationManager 是一个认证管理器，它定义了 Spring Security

过滤器要如何执行认证操作。AuthenticationManager 在认证成功后，会返回一个 Authentication 对象，这个 Authentication 对象会被设置到 SecurityContextHolder 中。如果开发者不想用 Spring Security 提供的一套认证机制，那么也可以自定义认证流程，认证成功后，手动将 Authentication 存入 SecurityContextHolder 中。

```
public interface AuthenticationManager {
    Authentication authenticate(Authentication authentication)
            throws AuthenticationException;
}
```

从 AuthenticationManager 的源码中可以看到，AuthenticationManager 对传入的 Authentication 对象进行身份认证，此时传入的 Authentication 参数只有用户名/密码等简单的属性，如果认证成功，返回的 Authentication 的属性会得到完全填充，包括用户所具备的角色信息。

AuthenticationManager 是一个接口，它有着诸多的实现类，开发者也可以自定义 AuthenticationManager 的实现类，不过在实际应用中，我们使用最多的是 ProviderManager。在 Spring Security 框架中，默认也是使用 ProviderManager。

3.1.2 AuthenticationProvider

2.3 节介绍了 Spring Security 支持多种不同的认证方式，不同的认证方式对应不同的身份类型，AuthenticationProvider 就是针对不同的身份类型执行具体的身份认证。例如，常见的 DaoAuthenticationProvider 用来支持用户名/密码登录认证，RememberMeAuthenticationProvider 用来支持"记住我"的认证。

AuthenticationProvider 的源码如下：

```
public interface AuthenticationProvider {
    Authentication authenticate(Authentication authentication)
            throws AuthenticationException;
    boolean supports(Class<?> authentication);
}
```

（1）authenticate 方法用来执行具体的身份认证。
（2）supports 方法用来判断当前 AuthenticationProvider 是否支持对应的身份类型。

当使用用户名/密码的方式登录时，对应的 AuthenticationProvider 实现类是 DaoAuthenticationProvider，而 DaoAuthenticationProvider 继承自 AbstractUserDetailsAuthenticationProvider 并且没有重写 authenticate 方法，所以具体的认证逻辑在 AbstractUserDetailsAuthenticationProvider 的 authenticate 方法中。我们就从 AbstractUserDetailsAuthenticationProvider 开始看起：

```
public abstract class AbstractUserDetailsAuthenticationProvider implements
        AuthenticationProvider, InitializingBean, MessageSourceAware {
```

```java
private UserCache userCache = new NullUserCache();
private boolean forcePrincipalAsString = false;
protected boolean hideUserNotFoundExceptions = true;
private UserDetailsChecker preAuthenticationChecks =
                        new DefaultPreAuthenticationChecks();
private UserDetailsChecker postAuthenticationChecks =
                        new DefaultPostAuthenticationChecks();
protected abstract void additionalAuthenticationChecks(
                                        UserDetails userDetails,
            UsernamePasswordAuthenticationToken authentication)
                                    throws AuthenticationException;
public Authentication authenticate(Authentication authentication)
        throws AuthenticationException {
    String username =
        (authentication.getPrincipal() == null) ? "NONE_PROVIDED"
                                    : authentication.getName();
    boolean cacheWasUsed = true;
    UserDetails user = this.userCache.getUserFromCache(username);
    if (user == null) {
        cacheWasUsed = false;
        try {
            user = retrieveUser(username,
                    (UsernamePasswordAuthenticationToken)authentication);
        }
        catch (UsernameNotFoundException notFound) {
            if (hideUserNotFoundExceptions) {
                throw new BadCredentialsException(messages.getMessage(
                "AbstractUserDetailsAuthenticationProvider.badCredentials",
                    "Bad credentials"));
            }
            else {
                throw notFound;
            }
        }
    }
    try {
        preAuthenticationChecks.check(user);
        additionalAuthenticationChecks(user,
            (UsernamePasswordAuthenticationToken) authentication);
    }
    catch (AuthenticationException exception) {
        if (cacheWasUsed) {
            cacheWasUsed = false;
            user = retrieveUser(username,
                (UsernamePasswordAuthenticationToken)
                                                authentication);
            preAuthenticationChecks.check(user);
            additionalAuthenticationChecks(user,
```

```java
                    (UsernamePasswordAuthenticationToken) authentication);
        }
        else {
            throw exception;
        }
    }
    postAuthenticationChecks.check(user);
    if (!cacheWasUsed) {
        this.userCache.putUserInCache(user);
    }
    Object principalToReturn = user;
    if (forcePrincipalAsString) {
        principalToReturn = user.getUsername();
    }
    return createSuccessAuthentication(principalToReturn,
                                        authentication, user);
}
protected Authentication createSuccessAuthentication(Object principal,
        Authentication authentication, UserDetails user) {
    UsernamePasswordAuthenticationToken result =
                        new UsernamePasswordAuthenticationToken(
        principal, authentication.getCredentials(),
        authoritiesMapper.mapAuthorities(user.getAuthorities()));
    result.setDetails(authentication.getDetails());
    return result;
}
protected abstract UserDetails retrieveUser(String username,
                    UsernamePasswordAuthenticationToken authentication)
                    throws AuthenticationException;
public boolean supports(Class<?> authentication) {
    return (UsernamePasswordAuthenticationToken.class
        .isAssignableFrom(authentication));
}
private class DefaultPreAuthenticationChecks implements UserDetailsChecker {
    public void check(UserDetails user) {
        if (!user.isAccountNonLocked()) {
            throw new LockedException(messages.getMessage(
                "AbstractUserDetailsAuthenticationProvider.locked",
                "User account is locked"));
        }
        if (!user.isEnabled()) {
            throw new DisabledException(messages.getMessage(
                "AbstractUserDetailsAuthenticationProvider.disabled",
                "User is disabled"));
        }
        if (!user.isAccountNonExpired()) {
            throw new AccountExpiredException(messages.getMessage(
                "AbstractUserDetailsAuthenticationProvider.expired",
```

```
                "User account has expired"));
        }
    }
}
private class DefaultPostAuthenticationChecks implements
    UserDetailsChecker {
    public void check(UserDetails user) {
        if (!user.isCredentialsNonExpired()) {
            throw new CredentialsExpiredException(messages.getMessage(
            "AbstractUserDetailsAuthenticationProvider.credentialsExpired",
                "User credentials have expired"));
        }
    }
}
```

AbstractUserDetailsAuthenticationProvider 是一个抽象类，抽象方法在它的实现类 DaoAuthenticationProvider 中完成。AbstractUserDetailsAuthenticationProvider 本身逻辑很简单，我们一起来看一下：

（1）一开始先声明一个用户缓存对象 userCache，默认情况下没有启用缓存对象。

（2）hideUserNotFoundExceptions 表示是否隐藏用户名查找失败的异常，默认为 true。为了确保系统安全，用户在登录失败时只会给出一个模糊提示，例如"用户名或密码输入错误"。在 Spring Security 内部，如果用户名查找失败，则会抛出 UsernameNotFoundException 异常，但是该异常会被自动隐藏，转而通过一个 BadCredentialsException 异常来代替它，这样，开发者在处理登录失败异常时，无论是用户名输入错误还是密码输入错误，收到的总是 BadCredentialsException，这样做的一个好处是可以避免新手程序员将用户名输入错误和密码输入错误两个异常分开提示。

（3）forcePrincipalAsString 表示是否强制将 Principal 对象当成字符串来处理，默认是 false。Authentication 中的 principal 属性类型是一个 Object，正常来说，通过 principal 属性可以获取到当前登录用户对象（即 UserDetails），但是如果 forcePrincipalAsString 设置为 true，则 Authentication 中的 principal 属性返回就是当前登录用户名，而不是用户对象。

（4）preAuthenticationChecks 对象则是用于做用户状态检查，在用户认证过程中，需要检验用户状态是否正常，例如账户是否被锁定、账户是否可用、账户是否过期等。

（5）postAuthenticationChecks 对象主要负责在密码校验成功后，检查密码是否过期。

（6）additionalAuthenticationChecks 是一个抽象方法，主要就是校验密码，具体的实现在 DaoAuthenticationProvider 中。

（7）authenticate 方法就是核心的校验方法了。在方法中，首先从登录数据中获取用户名，然后根据用户名去缓存中查询用户对象，如果查询不到，则根据用户名调用 retrieveUser 方法从数据库中加载用户；如果没有加载到用户，则抛出异常（用户不存在异常会被隐藏）。拿到用户对象之后，首先调用 preAuthenticationChecks.check 方法进行用户状态检查，然后调用

additionalAuthenticationChecks 方法进行密码的校验操作，最后调用 postAuthenticationChecks.check 方法检查密码是否过期，当所有步骤都顺利完成后，调用 createSuccessAuthentication 方法创建一个认证后的 UsernamePasswordAuthenticationToken 对象并返回，认证后的对象中包含了认证主体、凭证以及角色等信息。

这就是 AbstractUserDetailsAuthenticationProvider 类的工作流程，有几个抽象方法是在 DaoAuthenticationProvider 中实现的，我们再来看一下 DaoAuthenticationProvider 中的定义：

```java
public class DaoAuthenticationProvider extends
                        AbstractUserDetailsAuthenticationProvider {
    private static final String USER_NOT_FOUND_PASSWORD =
                                            "userNotFoundPassword";
    private PasswordEncoder passwordEncoder;
    private volatile String userNotFoundEncodedPassword;
    private UserDetailsService userDetailsService;
    private UserDetailsPasswordService userDetailsPasswordService;
    public DaoAuthenticationProvider() {
    setPasswordEncoder(PasswordEncoderFactories.
                                    createDelegatingPasswordEncoder());
    }
    protected void additionalAuthenticationChecks(UserDetails userDetails,
                    UsernamePasswordAuthenticationToken authentication)
                            throws AuthenticationException {
        if (authentication.getCredentials() == null) {
            throw new BadCredentialsException(messages.getMessage(
                "AbstractUserDetailsAuthenticationProvider.badCredentials",
                "Bad credentials"));
        }
        String presentedPassword = authentication.getCredentials().toString();
        if (!passwordEncoder
                .matches(presentedPassword, userDetails.getPassword())) {
            throw new BadCredentialsException(messages.getMessage(
                "AbstractUserDetailsAuthenticationProvider.badCredentials",
                "Bad credentials"));
        }
    }
    protected final UserDetails retrieveUser(String username,
                    UsernamePasswordAuthenticationToken authentication)
                            throws AuthenticationException {
        prepareTimingAttackProtection();
        try {
            UserDetails loadedUser =
                this.getUserDetailsService().loadUserByUsername(username);
            if (loadedUser == null) {
                throw new InternalAuthenticationServiceException(
                    "UserDetailsService returned null, which is an interface
                    contract violation");
```

```java
            }
            return loadedUser;
        }
        catch (UsernameNotFoundException ex) {
            mitigateAgainstTimingAttack(authentication);
            throw ex;
        }
        catch (InternalAuthenticationServiceException ex) {
            throw ex;
        }
        catch (Exception ex) {
            throw new InternalAuthenticationServiceException(ex.getMessage(),
                                                             ex);
        }
    }
    @Override
    protected Authentication createSuccessAuthentication(Object principal,
                       Authentication authentication, UserDetails user) {
        boolean upgradeEncoding = this.userDetailsPasswordService != null
             && this.passwordEncoder.upgradeEncoding(user.getPassword());
        if (upgradeEncoding) {
            String presentedPassword =
                            authentication.getCredentials().toString();
            String newPassword =
                        this.passwordEncoder.encode(presentedPassword);
            user = this.userDetailsPasswordService
                                .updatePassword(user, newPassword);
        }
        return super
            .createSuccessAuthentication(principal, authentication, user);
    }
    private void prepareTimingAttackProtection() {
        if (this.userNotFoundEncodedPassword == null) {
            this.userNotFoundEncodedPassword =
                    this.passwordEncoder.encode(USER_NOT_FOUND_PASSWORD);
        }
    }
    private void mitigateAgainstTimingAttack(
                    UsernamePasswordAuthenticationToken authentication) {
        if (authentication.getCredentials() != null) {
            String presentedPassword =
                            authentication.getCredentials().toString();
            this.passwordEncoder.matches(presentedPassword,
                                this.userNotFoundEncodedPassword);
        }
    }
}
```

在 DaoAuthenticationProvider 中：

（1）首先定义了 USER_NOT_FOUND_PASSWORD 常量，这个是当用户查找失败时的默认密码；passwordEncoder 是一个密码加密和比对工具，这个在第 5 章会有专门的介绍，这里先不做过多解释；userNotFoundEncodedPassword 变量则用来保存默认密码加密后的值；userDetailsService 是一个用户查找工具，userDetailsService 在第 2 章已经讲过，这里不再赘述；userDetailsPasswordService 则用来提供密码修改服务。

（2）在 DaoAuthenticationProvider 的构造方法中，默认就会指定 PasswordEncoder，当然开发者也可以通过 set 方法自定义 PasswordEncoder。

（3）additionalAuthenticationChecks 方法主要进行密码校验，该方法第一个参数 userDetails 是从数据库中查询出来的用户对象，第二个参数 authentication 则是登录用户输入的参数。从这两个参数中分别提取出来用户密码，然后调用 passwordEncoder.matches 方法进行密码比对。

（4）retrieveUser 方法则是获取用户对象的方法，具体做法就是调用 UserDetailsService#loadUserByUsername 方法去数据库中查询。

（5）在 retrieveUser 方法中，有一个值得关注的地方。在该方法一开始，首先会调用 prepareTimingAttackProtection 方法，该方法的作用是使用 PasswordEncoder 对常量 USER_NOT_FOUND_PASSWORD 进行加密，将加密结果保存在 userNotFoundEncodedPassword 变量中。当根据用户名查找用户时，如果抛出了 UsernameNotFoundException 异常，则调用 mitigateAgainstTimingAttack 方法进行密码比对。有读者会说，用户都没查找到，怎么比对密码？需要注意，在调用 mitigateAgainstTimingAttack 方法进行密码比对时，使用了 userNotFoundEncodedPassword 变量作为默认密码和登录请求传来的用户密码进行比对。这是一个一开始就注定要失败的密码比对，那么为什么还要进行比对呢？这主要是为了避免旁道攻击（Side-channel attack）。如果根据用户名查找用户失败，就直接抛出异常而不进行密码比对，那么黑客经过大量的测试，就会发现有的请求耗费时间明显小于其他请求，那么进而可以得出该请求的用户名是一个不存在的用户名（因为用户名不存在，所以不需要密码比对，进而节省时间），这样就可以获取到系统信息。为了避免这一问题，所以当用户查找失败时，也会调用 mitigateAgainstTimingAttack 方法进行密码比对，这样就可以迷惑黑客。

（6）createSuccessAuthentication 方法则是在登录成功后，创建一个全新的 UsernamePasswordAuthenticationToken 对象，同时会判断是否需要进行密码升级，如果需要进行密码升级，就会在该方法中进行加密方案升级。

通过对 AbstractUserDetailsAuthenticationProvider 和 DaoAuthenticationProvider 的讲解，相信读者已经很明白 AuthenticationProvider 中的认证逻辑了。

提 示

在密码学中，旁道攻击（Side-channel attack）又称侧信道攻击、边信道攻击。这种攻击方式不是暴力破解或者是研究加密算法的弱点。它是基于从密码系统的物理实现中获取信息，比如时间、功率消耗、电磁泄漏等，这些信息可被利用于对系统的进一步破解。

3.1.3 ProviderManager

ProviderManager 是 AuthenticationManager 的一个重要实现类。在开始学习之前，我们先通过一幅图来了解一下 ProviderManager 和 AuthenticationProvider 之间的关系，如图 3-1 所示。

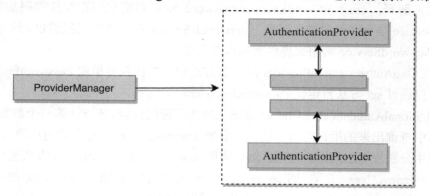

图 3-1　ProviderManager 和 AuthenticationProvider 的关系

在 Spring Security 中，由于系统可能同时支持多种不同的认证方式，例如同时支持用户名/密码认证、RememberMe 认证、手机号码动态认证等，而不同的认证方式对应了不同的 AuthenticationProvider，所以一个完整的认证流程可能由多个 AuthenticationProvider 来提供。

多个 AuthenticationProvider 将组成一个列表，这个列表将由 ProviderManager 代理。换句话说，在 ProviderManager 中存在一个 AuthenticationProvider 列表，在 ProviderManager 中遍历列表中的每一个 AuthenticationProvider 去执行身份认证，最终得到认证结果。

ProviderManager 本身也可以再配置一个 AuthenticationManager 作为 parent，这样当 ProviderManager 认证失败之后，就可以进入到 parent 中再次进行认证。理论上来说，ProviderManager 的 parent 可以是任意类型的 AuthenticationManager，但是通常都是由 ProviderManager 来扮演 parent 的角色，也就是 ProviderManager 是 ProviderManager 的 parent。

ProviderManager 本身也可以有多个，多个 ProviderManager 共用同一个 parent，当存在多个过滤器链的时候非常有用。当存在多个过滤器链时，不同的路径可能对应不同的认证方式，但是不同路径可能又会同时存在一些共有的认证方式,这些共有的认证方式可以在 parent 中统一处理。

根据上面的介绍，我们绘出新的 ProviderManager 和 AuthenticationProvider 关系图，如图 3-2 所示。

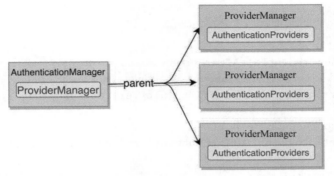

图 3-2　ProviderManager 中包含多个 AuthenticationProvider，并共享同一个 parent

我们重点看一下 ProviderManager 中的 authenticate 方法：

```java
public Authentication authenticate(Authentication authentication)
    throws AuthenticationException {
  Class<? extends Authentication> toTest = authentication.getClass();
  AuthenticationException lastException = null;
  AuthenticationException parentException = null;
  Authentication result = null;
  Authentication parentResult = null;
  for (AuthenticationProvider provider : getProviders()) {
     if (!provider.supports(toTest)) {
        continue;
     }
     try {
         result = provider.authenticate(authentication);
         if (result != null) {
            copyDetails(authentication, result);
            break;
         }
     }
     catch (AccountStatusException |
                       InternalAuthenticationServiceException e) {
        prepareException(e, authentication);
         throw e;
     } catch (AuthenticationException e) {
         lastException = e;
     }
  }
  if (result == null && parent != null) {
     try {
         result = parentResult = parent.authenticate(authentication);
     }
     catch (ProviderNotFoundException e) {
     }
     catch (AuthenticationException e) {
         lastException = parentException = e;
```

```java
            }
        }
        if (result != null) {
            if (eraseCredentialsAfterAuthentication
                    && (result instanceof CredentialsContainer)) {
                ((CredentialsContainer) result).eraseCredentials();
            }
            if (parentResult == null) {
                eventPublisher.publishAuthenticationSuccess(result);
            }
            return result;
        }
        if (lastException == null) {
            lastException = new ProviderNotFoundException(messages.getMessage(
                    "ProviderManager.providerNotFound",
                    new Object[] { toTest.getName() },
                    "No AuthenticationProvider found for {0}"));
        }
        if (parentException == null) {
            prepareException(lastException, authentication);
        }
        throw lastException;
    }
```

这段源码的逻辑还是非常清晰的，我们分析一下：

（1）首先获取 authentication 对象的类型。

（2）分别定义当前认证过程抛出的异常、parent 中认证时抛出的异常、当前认证结果以及 parent 中认证结果对应的变量。

（3）getProviders 方法用来获取当前 ProviderManager 所代理的所有 AuthenticationProvider 对象，遍历这些 AuthenticationProvider 对象进行身份认证。

（4）判断当前 AuthenticationProvider 是否支持当前 Authentication 对象，要是不支持，则继续处理列表中的下一个 AuthenticationProvider 对象。

（5）调用 provider.authenticate 方法进行身份认证，如果认证成功，返回认证后的 Authentication 对象，同时调用 copyDetails 方法给 Authentication 对象的 details 属性赋值。由于可能是多个 AuthenticationProvider 执行认证操作，所以如果抛出异常，则通过 lastException 变量来记录。

（6）在 for 循环执行完成后，如果 result 还是没有值，说明所有的 AuthenticationProvider 都认证失败，此时如果 parent 不为空，则调用 parent 的 authenticate 方法进行认证。

（7）接下来，如果 result 不为空，就将 result 中的凭证擦除，防止泄漏。如果使用了用户名/密码的方式登录，那么所谓的擦除实际上就是将密码字段设置为 null，同时将登录成功的事件发布出去（发布登录成功事件需要 parentResult 为 null。如果 parentResult 不为 null，表示在 parent 中已经认证成功了，认证成功的事件也已经在 parent 中发布出去了，这样会导致认证成功的事件重复发布）。如果用户认证成功，此时就将 result 返回，后面的代码也就不再执行了。

（8）如果前面没能返回 result，说明认证失败。如果 lastException 为 null，说明 parent 为 null 或者没有认证亦或者认证失败了但是没有抛出异常，此时构造 ProviderNotFoundException 异常赋值给 lastException。

（9）如果 parentException 为 null，发布认证失败事件（如果 parentException 不为 null，则说明认证失败事件已经发布过了）。

（10）最后抛出 lastException 异常。

这就是 ProviderManager 中 authenticate 方法的身份认证逻辑，其他方法的源码要相对简单很多，在这里就不一一解释了。

现在，大家已经熟悉了 Authentication、AuthenticationManager、AuthenticationProvider 以及 ProviderManager 的工作原理了，接下来的问题就是这些组件如何跟登录关联起来？这就涉及一个重要的过滤器——AbstractAuthenticationProcessingFilter。

3.1.4　AbstractAuthenticationProcessingFilter

作为 Spring Security 过滤器链中的一环，AbstractAuthenticationProcessingFilter 可以用来处理任何提交给它的身份认证，图 3-3 描述了 AbstractAuthenticationProcessingFilter 的工作流程。

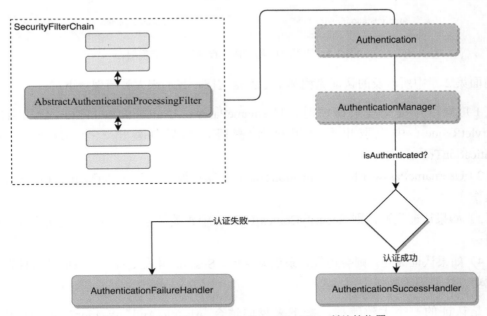

图 3-3　AbstractAuthenticationProcessingFilter 所处的位置

图中显示的流程是一个通用的架构。

AbstractAuthenticationProcessingFilter 作为一个抽象类，如果使用用户名/密码的方式登录，那么它对应的实现类是 UsernamePasswordAuthenticationFilter，构造出来的 Authentication 对象

则是 UsernamePasswordAuthenticationToken。至于 AuthenticationManager，前面已经说过，一般情况下它的实现类就是 ProviderManager，这里在 ProviderManager 中进行认证，认证成功就会进入认证成功的回调，否则进入认证失败的回调。因此，我们可以对上面的流程图再做进一步细化，如图 3-4 所示。

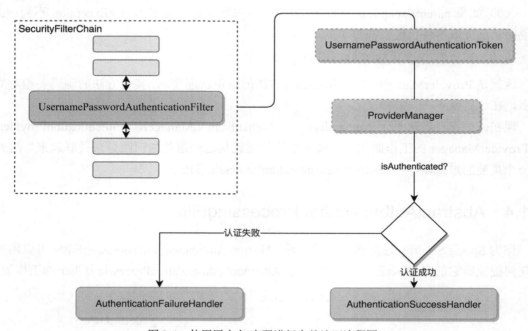

图 3-4　使用用户名/密码进行身份认证流程图

前面第 2 章中所涉及的认证流程基本上就是这样，我们来大致梳理一下：

（1）当用户提交登录请求时，UsernamePasswordAuthenticationFilter 会从当前请求 HttpServletRequest 中提取出登录用户名/密码，然后创建出一个 UsernamePasswordAuthenticationToken 对象。

（2）UsernamePasswordAuthenticationToken 对象将被传入 ProviderManager 中进行具体的认证操作。

（3）如果认证失败，则 SecurityContextHolder 中相关信息将被清除，登录失败回调也会被调用。

（4）如果认证成功，则会进行登录信息存储、Session 并发处理、登录成功事件发布以及登录成功方法回调等操作。

这是认证的一个大致流程。接下来我们结合 AbstractAuthenticationProcessingFilter 和 UsernamePasswordAuthenticationFilter 的源码来看一下。

先来看 AbstractAuthenticationProcessingFilter 源码（部分核心代码）：

```
public abstract class AbstractAuthenticationProcessingFilter
    extends GenericFilterBean
```

```java
        implements ApplicationEventPublisherAware, MessageSourceAware {
public void doFilter(ServletRequest req,
                ServletResponse res, FilterChain chain)
                throws IOException, ServletException {
    HttpServletRequest request = (HttpServletRequest) req;
    HttpServletResponse response = (HttpServletResponse) res;
    if (!requiresAuthentication(request, response)) {
        chain.doFilter(request, response);
        return;
    }
    Authentication authResult;
    try {
        authResult = attemptAuthentication(request, response);
        if (authResult == null) {
            return;
        }
        sessionStrategy.onAuthentication(authResult, request, response);
    }
    catch (InternalAuthenticationServiceException failed) {
        unsuccessfulAuthentication(request, response, failed);
        return;
    }
    catch (AuthenticationException failed) {
        unsuccessfulAuthentication(request, response, failed);
        return;
    }
    if (continueChainBeforeSuccessfulAuthentication) {
        chain.doFilter(request, response);
    }
    successfulAuthentication(request, response, chain, authResult);
}
protected boolean requiresAuthentication(HttpServletRequest request,
        HttpServletResponse response) {
    return requiresAuthenticationRequestMatcher.matches(request);
}
public abstract Authentication attemptAuthentication(
                            HttpServletRequest request,
                            HttpServletResponse response)
        throws AuthenticationException, IOException, ServletException;
protected void successfulAuthentication(HttpServletRequest request,
                    HttpServletResponse response, FilterChain chain,
                    Authentication authResult)
                    throws IOException, ServletException {
    SecurityContextHolder.getContext().setAuthentication(authResult);
    rememberMeServices.loginSuccess(request, response, authResult);
    if (this.eventPublisher != null) {
        eventPublisher
            .publishEvent(new InteractiveAuthenticationSuccessEvent(
```

```
                                    authResult, this.getClass()));
        }
        successHandler.onAuthenticationSuccess(request, response, authResult);
    }
    protected void unsuccessfulAuthentication(HttpServletRequest request,
            HttpServletResponse response, AuthenticationException failed)
            throws IOException, ServletException {
        SecurityContextHolder.clearContext();
        rememberMeServices.loginFail(request, response);
        failureHandler.onAuthenticationFailure(request, response, failed);
    }
}
```

（1）首先通过 requiresAuthentication 方法来判断当前请求是不是登录认证请求，如果是认证请求，就执行接下来的认证代码；如果不是认证请求，则直接继续走剩余的过滤器即可。

（2）调用 attemptAuthentication 方法来获取一个经过认证后的 Authentication 对象，attemptAuthentication 方法是一个抽象方法，具体实现在它的子类 UsernamePasswordAuthenticationFilter 中。

（3）认证成功后，通过 sessionStrategy.onAuthentication 方法来处理 session 并发问题。

（4）continueChainBeforeSuccessfulAuthentication 变量用来判断请求是否还需要继续向下走。默认情况下该参数的值为 false，即认证成功后，后续的过滤器将不再执行了。

（5）unsuccessfulAuthentication 方法用来处理认证失败事宜，主要做了三件事：①从 SecurityContextHolder 中清除数据；②清除 Cookie 等信息；③调用认证失败的回调方法。

（6）successfulAuthentication 方法主要用来处理认证成功事宜，主要做了四件事：①向 SecurityContextHolder 中存入用户信息；②处理 Cookie；③发布认证成功事件，这个事件类型是 InteractiveAuthenticationSuccessEvent，表示通过一些自动交互的方式认证成功，例如通过 RememberMe 的方式登录；④调用认证成功的回调方法。

这就是 AbstractAuthenticationProcessingFilter 大致上所做的事情，还有一个抽象方法 attemptAuthentication 是在它的继承类 UsernamePasswordAuthenticationFilter 中实现的，接下来我们来看一下 UsernamePasswordAuthenticationFilter 类：

```
public class UsernamePasswordAuthenticationFilter extends
        AbstractAuthenticationProcessingFilter {
    public static final String SPRING_SECURITY_FORM_USERNAME_KEY = "username";
    public static final String SPRING_SECURITY_FORM_PASSWORD_KEY = "password";
    private String usernameParameter = SPRING_SECURITY_FORM_USERNAME_KEY;
    private String passwordParameter = SPRING_SECURITY_FORM_PASSWORD_KEY;
    private boolean postOnly = true;
    public UsernamePasswordAuthenticationFilter() {
        super(new AntPathRequestMatcher("/login", "POST"));
    }
    public Authentication attemptAuthentication(HttpServletRequest request,
            HttpServletResponse response) throws AuthenticationException {
```

```java
        if (postOnly && !request.getMethod().equals("POST")) {
            throw new AuthenticationServiceException(
                "Authentication method not supported: "
                                        + request.getMethod());
        }
        String username = obtainUsername(request);
        String password = obtainPassword(request);
        if (username == null) {
            username = "";
        }
        if (password == null) {
            password = "";
        }
        username = username.trim();
        UsernamePasswordAuthenticationToken authRequest =
            new UsernamePasswordAuthenticationToken(username, password);
        setDetails(request, authRequest);
        return this.getAuthenticationManager().authenticate(authRequest);
    }
    @Nullable
    protected String obtainPassword(HttpServletRequest request) {
        return request.getParameter(passwordParameter);
    }
    @Nullable
    protected String obtainUsername(HttpServletRequest request) {
        return request.getParameter(usernameParameter);
    }
    protected void setDetails(HttpServletRequest request,
            UsernamePasswordAuthenticationToken authRequest) {
    authRequest.setDetails(authenticationDetailsSource.buildDetails(request));
    }
    public void setUsernameParameter(String usernameParameter) {
        this.usernameParameter = usernameParameter;
    }
    public void setPasswordParameter(String passwordParameter) {
        this.passwordParameter = passwordParameter;
    }
    public void setPostOnly(boolean postOnly) {
        this.postOnly = postOnly;
    }
    public final String getUsernameParameter() {
        return usernameParameter;
    }
    public final String getPasswordParameter() {
        return passwordParameter;
    }
}
```

（1）首先声明了默认情况下登录表单的用户名字段和密码字段，用户名字段的 key 默认是 username，密码字段的 key 默认是 password。当然，这两个字段也可以自定义，自定义的方式就是我们在 SecurityConfig 中配置的 .usernameParameter("uname") 和 .passwordParameter("passwd")（参考 2.2 节）。

（2）在 UsernamePasswordAuthenticationFilter 过滤器构建的时候，指定了当前过滤器只用来处理登录请求，默认的登录请求是/login，当然开发者也可以自行配置。

（3）接下来就是最重要的 attemptAuthentication 方法了，在该方法中，首先确认请求是 post 类型；然后通过 obtainUsername 和 obtainPassword 方法分别从请求中提取出用户名和密码，具体的提取过程就是调用 request.getParameter 方法；拿到登录请求传来的用户名/密码之后，构造出一个 authRequest，然后调用 getAuthenticationManager().authenticate 方法进行认证，这就进入到我们前面所说的 ProviderManager 的流程中了，具体认证过程就不再赘述了。

以上就是整个认证流程。

搞懂了认证流程，那么接下来如果想要自定义一些认证方式，就会非常容易了，比如定义多个数据源、添加登录校验码等。下面，我们将通过两个案例，来活学活用上面所讲的认证流程。

3.2 配置多个数据源

多个数据源是指在同一个系统中，用户数据来自不同的表，在认证时，如果第一张表没有查找到用户，那就去第二张表中查询，依次类推。

看了前面的分析，要实现这个需求就很容易了。认证要经过 AuthenticationProvider，每一个 AuthenticationProvider 中都配置了一个 UserDetailsService，而不同的 UserDetailsService 则可以代表不同的数据源。所以我们只需要手动配置多个 AuthenticationProvider，并为不同的 AuthenticationProvider 提供不同的 UserDetailsService 即可。

为了方便起见，这里通过 InMemoryUserDetailsManager 来提供 UserDetailsService 实例，在实际开发中，只需要将 UserDetailsService 换成自定义的即可，具体配置如下：

```java
@Configuration
public class SecurityConfig extends WebSecurityConfigurerAdapter {
    @Bean
    @Primary
    UserDetailsService us1() {
        return new InMemoryUserDetailsManager(User.builder()
        .username("javaboy").password("{noop}123").roles("admin").build());
    }
    @Bean
    UserDetailsService us2() {
        return new InMemoryUserDetailsManager(User.builder()
```

```
            .username("sang").password("{noop}123").roles("user").build());
}
@Override
@Bean
public AuthenticationManager authenticationManagerBean()
                                                throws Exception {
    DaoAuthenticationProvider dao1 = new DaoAuthenticationProvider();
    dao1.setUserDetailsService(us1());
    DaoAuthenticationProvider dao2 = new DaoAuthenticationProvider();
    dao2.setUserDetailsService(us2());
    ProviderManager manager = new ProviderManager(dao1, dao2);
    return manager;
}
@Override
protected void configure(HttpSecurity http) throws Exception {
    http.authorizeRequests()
            .anyRequest().authenticated()
            //省略
    }
}
```

首先定义了两个 UserDetailsService 实例，不同实例中存储了不同的用户；然后重写 authenticationManagerBean 方法，在该方法中，定义了两个 DaoAuthenticationProvider 实例并分别设置了不同的 UserDetailsService；最后构建 ProviderManager 实例并传入两个 DaoAuthenticationProvider。当系统进行身份认证操作时，就会遍历 ProviderManager 中不同的 DaoAuthenticationProvider，进而调用到不同的数据源。

> **提示**
>
> 在本书的配套案例中，笔者提供了一个基于 MyBatis 配置多数据源的案例，读者可以参考。

3.3 添加登录验证码

登录验证码也是项目中一个常见的需求，但是 Spring Security 对此并未提供自动化配置方案，需要开发者自行定义。一般来说，有两种实现登录验证码的思路：

（1）自定义过滤器。
（2）自定义认证逻辑。

通过自定义过滤器来实现登录验证码，这种方案我们会在本书后面的过滤器章节中介绍，这里先来看如何通过自定义认证逻辑实现添加登录验证码功能。

生成验证码，可以自定义一个生成工具类，也可以使用一些现成的开源库来实现。这里采用开源库 kaptcha，首先引入 kaptcha 依赖，代码如下：

```xml
<dependency>
    <groupId>com.github.penggle</groupId>
    <artifactId>kaptcha</artifactId>
    <version>2.3.2</version>
</dependency>
```

然后对 kaptcha 进行配置：

```java
@Configuration
public class KaptchaConfig {
    @Bean
    Producer kaptcha() {
        Properties properties = new Properties();
        properties.setProperty("kaptcha.image.width", "150");
        properties.setProperty("kaptcha.image.height", "50");
        properties.setProperty("kaptcha.textproducer.char.string",
                                                    "0123456789");
        properties.setProperty("kaptcha.textproducer.char.length", "4");
        Config config = new Config(properties);
        DefaultKaptcha defaultKaptcha = new DefaultKaptcha();
        defaultKaptcha.setConfig(config);
        return defaultKaptcha;
    }
}
```

配置一个 Producer 实例，主要配置一下生成的图片验证码的宽度、长度、生成字符、验证码的长度等信息。配置完成后，我们就可以在 Controller 中定义一个验证码接口了：

```java
@Autowired
Producer producer;
@GetMapping("/vc.jpg")
public void getVerifyCode(HttpServletResponse resp, HttpSession session)
                                                    throws IOException {
    resp.setContentType("image/jpeg");
    String text = producer.createText();
    session.setAttribute("kaptcha", text);
    BufferedImage image = producer.createImage(text);
    try(ServletOutputStream out = resp.getOutputStream()) {
        ImageIO.write(image, "jpg", out);
    }
}
```

在这个验证码接口中，我们主要做了两件事：

（1）生成验证码文本，并将文本存入 HttpSession 中。
（2）根据验证码文本生成验证码图片，并通过 IO 流写出到前端。

接下来修改登录表单，加入验证码，代码如下：

```html
<form id="login-form" class="form" action="/doLogin" method="post">
```

```html
    <h3 class="text-center text-info">登录</h3>
    <div th:text="${SPRING_SECURITY_LAST_EXCEPTION}"></div>
    <div class="form-group">
        <label for="username" class="text-info">用户名:</label><br>
        <input type="text" name="uname" id="username" class="form-control">
    </div>
    <div class="form-group">
        <label for="password" class="text-info">密码:</label><br>
        <input type="text" name="passwd" id="password" class="form-control">
    </div>
    <div class="form-group">
        <label for="kaptcha" class="text-info">验证码:</label><br>
        <input type="text" name="kaptcha" id="kaptcha" class="form-control">
        <img src="/vc.jpg" alt="">
    </div>
    <div class="form-group">
        <input type="submit" name="submit" class="btn btn-info btn-md"
                                                            value="登录">
    </div>
</form>
```

在登录表单中增加一个验证码输入框,验证码的图片地址就是我们在 Controller 中定义的验证码接口地址。

接下来就是验证码的校验了。经过前面的介绍,读者已经了解到,身份认证实际上就是在 AuthenticationProvider#authenticate 方法中完成的。所以,验证码的校验,我们可以在该方法执行之前进行,需要配置如下类:

```java
public class KaptchaAuthenticationProvider extends DaoAuthenticationProvider {
    @Override
    public Authentication authenticate(Authentication authentication)
                                    throws AuthenticationException {
        HttpServletRequest req = ((ServletRequestAttributes) RequestContextHolder
                            .getRequestAttributes()).getRequest();
        String kaptcha = req.getParameter("kaptcha");
        String sessionKaptcha =
                    (String) req.getSession().getAttribute("kaptcha");
        if (kaptcha != null && sessionKaptcha != null
                        && kaptcha.equalsIgnoreCase(sessionKaptcha)) {
            return super.authenticate(authentication);
        }
        throw new AuthenticationServiceException("验证码输入错误");
    }
}
```

这里重写 authenticate 方法,在 authenticate 方法中,从 RequestContextHolder 中获取当前请求,进而获取到验证码参数和存储在 HttpSession 中的验证码文本进行比较,比较通过则继续执行父类的 authenticate 方法,比较不通过,就抛出异常。

> **注　意**
>
> 有的读者可能会想到通过重写 DaoAuthenticationProvider 类的 additionalAuthenticationChecks 方法来完成验证码的校验，这个从技术上来说是没有问题的，但是这会让验证码失去存在的意义，因为当 additionalAuthenticationChecks 方法被调用时，数据库查询已经做了，仅仅剩下密码没有校验，此时，通过验证码来拦截恶意登录的功能就已经失效了。

最后，在 SecurityConfig 中配置 AuthenticationManager，代码如下：

```java
@Configuration
public class SecurityConfig extends WebSecurityConfigurerAdapter {
    @Bean
    UserDetailsService us1() {
        return new InMemoryUserDetailsManager(User.builder()
            .username("javaboy").password("{noop}123").roles("admin").build());
    }
    @Bean
    AuthenticationProvider kaptchaAuthenticationProvider() {
        KaptchaAuthenticationProvider provider =
                                    new KaptchaAuthenticationProvider();
        provider.setUserDetailsService(us1());
        return provider;
    }
    @Override
    @Bean
    public AuthenticationManager authenticationManagerBean()throws Exception {
        ProviderManager manager =
                new ProviderManager(kaptchaAuthenticationProvider());
        return manager;
    }
    @Override
    protected void configure(HttpSecurity http) throws Exception {
        http.authorizeRequests()
                .antMatchers("/vc.jpg").permitAll()
                .anyRequest().authenticated()
                //省略
    }
}
```

这里配置分三步：首先配置 UserDetailsService 提供数据源；然后提供一个 AuthenticationProvider 实例并配置 UserDetailsService；最后重写 authenticationManagerBean 方法，提供一个自己的 ProviderManager 并使用自定义的 AuthenticationProvider 实例。

另外需要注意，在 configure(HttpSecurity)方法中给验证码接口配置放行，permitAll 表示这个接口不需要登录就可以访问。

配置完成后，启动项目，浏览器中输入任意地址都会跳转到登录页面，如图 3-5 所示。

图 3-5　包含有验证码的登录页面

此时，输入用户名、密码以及验证码就可以成功登录。如果验证码输入错误，则登录页面会自动展示错误信息。

3.4　小　结

本章主要分析了 Spring Security 认证流程中涉及的几个重要的类，AuthenticationManager、AuthenticationProvider、ProviderManager 以及 AbstractAuthenticationProcessingFilter，通过这几个类让大家理解 Spring Security 的认证流程到底是什么样子的，同时我们通过两个案例来深化读者对这几个类的理解，并提示目前流行的短信验证码登录的实现思路，也可以通过自定义 AuthenticationProvider 来实现，感兴趣的读者可以自行尝试。

第 4 章

过滤器链分析

提起 Spring Security 的实现原理,很多读者都会想到过滤器链。因为 Spring Security 中的所有功能都是通过过滤器来实现的,这些过滤器组成一个完整的过滤器链。那么,这些过滤器链是如何初始化的?我们前面反复提到的 AuthenticationManager 又是如何初始化的?通过前面章节的学习,相信读者已经有了一些认识。本章我们将从头开始,分析 Spring Security 的初始化流程,同时再通过六个案例来让读者深入理解并且学会如何制作过滤器链。由于初始化流程相对复杂,因此我们没有选择在第 1 章一开始就讲解 Spring Security 初始化流程,而是放到本章。当读者对于 Spring Security 有一个基本的认知之后再来讲解,此时相对来说就会比较容易理解。

本章涉及的主要知识点有:

- 初始化流程分析。
- ObjectPostProcessor 的使用。
- 多种用户定义方式。
- 定义多个过滤器链。
- 静态资源过滤。
- 使用 JSON 格式登录。
- 添加登录验证码。

4.1 初始化流程分析

Spring Security 初始化流程整体上来说理解起来并不难,但是这里涉及许多零碎的知识点,把这些零碎的知识点搞懂了，再来梳理初始化流程就会容易很多。因此，这里先介绍一下 Spring Security 中一些常见的关键组件，在理解这些组件的基础上，再来分析初始化流程，就能加深对其的理解。

4.1.1 ObjectPostProcessor

ObjectPostProcessor 是 Spring Security 中使用频率最高的组件之一，它是一个对象后置处理器，也就是当一个对象创建成功后，如果还有一些额外的事情需要补充，那么可以通过 ObjectPostProcessor 来进行处理。这个接口中默认只有一个方法 postProcess，该方法用来完成对对象的二次处理，代码如下：

```
public interface ObjectPostProcessor<T> {
    <O extends T> O postProcess(O object);
}
```

ObjectPostProcessor 默认有两个继承类，如图 4-1 所示。

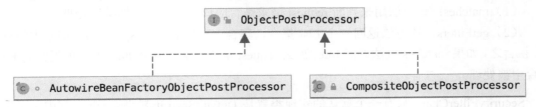

图 4-1　ObjectPostProcessor 的继承类

- AutowireBeanFactoryObjectPostProcessor：由于 Spring Security 中大量采用了 Java 配置，许多过滤器都是直接 new 出来的，这些直接 new 出来的对象并不会自动注入到 Spring 容器中。Spring Security 这样做的本意是为了简化配置，但是却带来了另外一个问题就是，大量 new 出来的对象需要我们手动注册到 Spring 容器中去。AutowireBeanFactoryObjectPostProcessor 对象所承担的就是这件事，一个对象 new 出来之后，只要调用 AutowireBeanFactoryObjectPostProcessor#postProcess 方法，就可以成功注入到 Spring 容器中，它的实现原理就是通过调用 Spring 容器中的 AutowireCapableBeanFactory 对象将一个 new 出来的对象注入到 Spring 容器中去。
- CompositeObjectPostProcessor：这是 ObjectPostProcessor 的另一个实现，一个对象可以有一个后置处理器，开发者也可以自定义多个对象后置处理器。CompositeObjectPostProcessor 是一个组合的对象后置处理器，它里边维护了一个 List 集合，集合中存放了某一个对象的所有后置处理器，当需要执行对象的后置处理器时，会遍历集合中的所有 ObjectPostProcessor 实例，分别调用实例的 postProcess 方法进行对象后置处理。在 Spring Security 框架中，最终使用的对象后置处理器其实就是 CompositeObjectPostProcessor，它里边的集合默认只有一个对象，就是 AutowireBeanFactoryObjectPostProcessor。

在 Spring Security 中，开发者可以灵活地配置项目中需要哪些 Spring Security 过滤器，一且选定过滤器之后，每一个过滤器都会有一个对应的配置器，叫作 xxxConfigurer（例如 CorsConfigurer、CsrfConfigurer 等），过滤器都是在 xxxConfigurer 中 new 出来的，然后在 postProcess 方法中处理一遍，就将这些过滤器注入到 Spring 容器中了。

这是对象后置处理器 ObjectPostProcessor 的主要作用。

4.1.2 SecurityFilterChain

从名称上可以看出，SecurityFilterChain 就是 Spring Security 中的过滤器链对象。下面来看一下 SecurityFilterChain 的源码：

```java
public interface SecurityFilterChain {
    boolean matches(HttpServletRequest request);
    List<Filter> getFilters();
}
```

可以看到，SecurityFilterChain 中有两个方法：

（1）matches：该方法用来判断 request 请求是否应该被当前过滤器链所处理。

（2）getFilters：该方法返回一个 List 集合，集合中存放的就是 Spring Security 中的过滤器。换言之，如果 matches 方法返回 true，那么 request 请求就会在 getFilters 方法所返回的 Filter 集合中被处理。

SecurityFilterChain 只有一个默认的实现类就是 DefaultSecurityFilterChain，其中定义了两个属性，并具体实现了 SecurityFilterChain 中的两个方法：

```java
public final class DefaultSecurityFilterChain implements SecurityFilterChain {
    private final RequestMatcher requestMatcher;
    private final List<Filter> filters;
    public DefaultSecurityFilterChain(RequestMatcher requestMatcher,
                                      Filter... filters) {
        this(requestMatcher, Arrays.asList(filters));
    }
    public DefaultSecurityFilterChain(RequestMatcher requestMatcher,
                                      List<Filter> filters) {
        this.requestMatcher = requestMatcher;
        this.filters = new ArrayList<>(filters);
    }
    public RequestMatcher getRequestMatcher() {
        return requestMatcher;
    }
    public List<Filter> getFilters() {
        return filters;
    }
    public boolean matches(HttpServletRequest request) {
        return requestMatcher.matches(request);
    }
}
```

```
    @Override
    public String toString() {
        return "[ " + requestMatcher + ", " + filters + "]";
    }
}
```

可以看到，在 DefaultSecurityFilterChain 的构造方法中，需要传入两个对象，一个是请求匹配器 requestMatcher，另一个则是过滤器集合或者过滤器数组 filters。这个实现类比较简单，这里就不再赘述了。

> **注　意**
>
> 需要注意的是，在一个 Spring Security 项目中，SecurityFilterChain 的实例可能会有多个，我们在本章后面的小节中会详细分析，并演示多个 SecurityFilterChain 实例的情况。

4.1.3　SecurityBuilder

Spring Security 中所有需要构建的对象都可以通过 SecurityBuilder 来实现，默认的过滤器链、代理过滤器、AuthenticationManager 等，都可以通过 SecurityBuilder 来构建。SecurityBuilder 的实现类如图 4-2 所示。

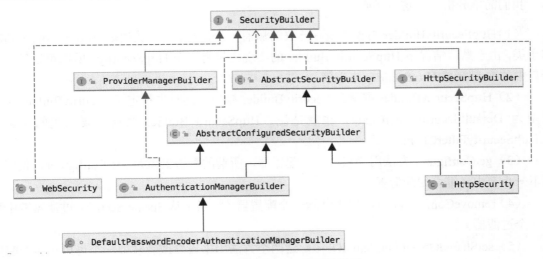

图 4-2　SecurityBuilder 继承关系图

SecurityBuilder

我们先来看 SecurityBuilder 的源码：

```
public interface SecurityBuilder<O> {
    O build() throws Exception;
}
```

由上述代码可以看到，SecurityBuilder 中只有一个 build 方法，就是对象构建方法。build

方法的返回值，就是具体构建的对象泛型 O，也就是说不同的 SecurityBuilder 将来会构建出不同的对象。

HttpSecurityBuilder

HttpSecurityBuilder 是用来构建 HttpSecurity 对象的，HttpSecurityBuilder 的定义如下：

```java
public interface HttpSecurityBuilder<H extends HttpSecurityBuilder<H>>
            extends SecurityBuilder<DefaultSecurityFilterChain> {
    <C extends SecurityConfigurer<DefaultSecurityFilterChain, H>> C
                            getConfigurer(Class<C> clazz);
    <C extends SecurityConfigurer<DefaultSecurityFilterChain, H>> C
                            removeConfigurer(<C> clazz);
    <C> void setSharedObject(Class<C> sharedType, C object);
    <C> C getSharedObject(Class<C> sharedType);
    H authenticationProvider(AuthenticationProvider authenticationProvider);
    H userDetailsService(UserDetailsService userDetailsService)
                                                throws Exception;
    H addFilterAfter(Filter filter, Class<? extends Filter> afterFilter);
    H addFilterBefore(Filter filter, Class<? extends Filter> beforeFilter);
    H addFilter(Filter filter);
}
```

我们简单分析一下这段源码：

（1）HttpSecurityBuilder 对象本身在定义时就有一个泛型，这个泛型是 HttpSecurityBuilder 的子类，由于默认情况下 HttpSecurityBuilder 的实现类只有一个 HttpSecurity，所以可以暂且把接口中的 H 都当成 HttpSecurity 来理解。

（2）HttpSecurityBuilder 继承自 SecurityBuilder 接口，同时也指定了 SecurityBuilder 中的泛型为 DefaultSecurityFilterChain，也就是说，HttpSecurityBuilder 最终想要构建的对象是 DefaultSecurityFilterChain。

（3）getConfigurer 方法用来获取一个配置器，所谓的配置器就是 xxxConfigurer，我们将在下一小节中详细介绍配置器。

（4）removeConfigurer 方法用来移除一个配置器（相当于从 Spring Security 过滤器链中移除一个过滤器）。

（5）setSharedObject/getSharedObject 这两个方法用来设置或者获取一个可以在多个配置器之间共享的对象。

（6）authenticationProvider 方法可以用来配置一个认证器 AuthenticationProvider。

（7）userDetailsService 方法可以用来配置一个数据源 UserDetailsService。

（8）addFilterAfter/addFilterBefore 方法表示在某一个过滤器之后或者之前添加一个自定义的过滤器。

（9）addFilter 方法可以添加一个过滤器，这个过滤器必须是 Spring Security 框架提供的过滤器的一个实例或者其扩展，添加完成后，会自动进行过滤器的排序。

AbstractSecurityBuilder

AbstractSecurityBuilder 实现了 SecurityBuilder 接口，并对 build 做了完善，确保只 build 一次。我们来看一下 AbstractSecurityBuilder 源码：

```java
public abstract class AbstractSecurityBuilder<O>
                                           implements SecurityBuilder<O> {
    private AtomicBoolean building = new AtomicBoolean();
    private O object;
    public final O build() throws Exception {
        if (this.building.compareAndSet(false, true)) {
            this.object = doBuild();
            return this.object;
        }
        throw new AlreadyBuiltException("This object has already been built");
    }
    public final O getObject() {
        if (!this.building.get()) {
            throw new IllegalStateException("This object has not been built");
        }
        return this.object;
    }
    protected abstract O doBuild() throws Exception;
}
```

由上述代码可以看到，在 AbstractSecurityBuilder 类中：

（1）首先声明了 building 变量，可以确保即使在多线程环境下，配置类也只构建一次。

（2）对 build 方法进行重写，并且设置为 final，这样在 AbstractSecurityBuilder 的子类中将不能再次重写 build 方法。在 build 方法内部，通过 building 变量来控制配置类只构建一次，具体的构建工作则交给 doBuild 方法去完成。

（3）getObject 方法用来返回构建的对象。

（4）doBuild 方法则是具体的构建方法，该方法在 AbstractSecurityBuilder 中是一个抽象方法，具体的实现在其子类中。

一言以蔽之，AbstractSecurityBuilder 的作用是确保目标对象只被构建一次。

AbstractConfiguredSecurityBuilder

AbstractConfiguredSecurityBuilder 类的源码就稍微长一点，我们分别来看。

首先在 AbstractConfiguredSecurityBuilder 中声明了一个枚举类，用来描述构建过程的不同状态：

```java
private enum BuildState {
    UNBUILT(0),
    INITIALIZING(1),
    CONFIGURING(2),
    BUILDING(3),
```

```
    BUILT(4);
    private final int order;
    BuildState(int order) {
        this.order = order;
    }
    public boolean isInitializing() {
        return INITIALIZING.order == order;
    }
    public boolean isConfigured() {
        return order >= CONFIGURING.order;
    }
}
```

可以看到，整个构建过程一共有五种不同的状态：

- UNBUILT：配置类构建前。
- INITIALIZING：初始化中（初始化完成之前是这个状态）。
- CONFIGURING：配置中（开始构建之前是这个状态）。
- BUILDING：构建中。
- BUILT：构建完成。

这个枚举类里边还提供了两个判断方法，isInitializing 表示是否正在初始化中，isConfigured 方法表示是否已完成配置。

AbstractConfiguredSecurityBuilder 中还声明了 configurers 变量，用来保存所有的配置类。针对 configurers 变量，我们可以进行添加配置、移除配置等操作，相关方法如下：

```
public abstract class AbstractConfiguredSecurityBuilder<O,
B extends SecurityBuilder<O>> extends AbstractSecurityBuilder<O> {
    private final LinkedHashMap<Class<? extends SecurityConfigurer<O, B>>,
List<SecurityConfigurer<O, B>>> configurers = new LinkedHashMap<>();
    public <C extends SecurityConfigurerAdapter<O, B>> C apply(C configurer)
            throws Exception {
        configurer.addObjectPostProcessor(objectPostProcessor);
        configurer.setBuilder((B) this);
        add(configurer);
        return configurer;
    }
    public <C extends SecurityConfigurer<O, B>> C apply(C configurer)
                                                        throws Exception {
        add(configurer);
        return configurer;
    }
    private <C extends SecurityConfigurer<O, B>> void add(C configurer) {
        Class<? extends SecurityConfigurer<O, B>> clazz =
(Class<? extends SecurityConfigurer<O, B>>) configurer.getClass();
        synchronized (configurers) {
            if (buildState.isConfigured()) {
```

```java
            throw new IllegalStateException("Cannot apply " + configurer
                    + " to already built object");
        }
        List<SecurityConfigurer<O, B>> configs =
allowConfigurersOfSameType ? this.configurers.get(clazz) : null;
        if (configs == null) {
            configs = new ArrayList<>(1);
        }
        configs.add(configurer);
        this.configurers.put(clazz, configs);
        if (buildState.isInitializing()) {
            this.configurersAddedInInitializing.add(configurer);
        }
    }
}
public <C extends SecurityConfigurer<O, B>> List<C>
                                    getConfigurers(Class<C> clazz) {
    List<C> configs = (List<C>) this.configurers.get(clazz);
    if (configs == null) {
        return new ArrayList<>();
    }
    return new ArrayList<>(configs);
}
public <C extends SecurityConfigurer<O, B>> List<C>
                                    removeConfigurers(Class<C> clazz) {
    List<C> configs = (List<C>) this.configurers.remove(clazz);
    if (configs == null) {
        return new ArrayList<>();
    }
    return new ArrayList<>(configs);
}
public <C extends SecurityConfigurer<O, B>> C
                                    getConfigurer(Class<C> clazz) {
    List<SecurityConfigurer<O, B>> configs = this.configurers.get(clazz);
    if (configs == null) {
        return null;
    }
    if (configs.size() != 1) {
        throw new IllegalStateException("Only one configurer expected for
                        type " + clazz + ", but got " + configs);
    }
    return (C) configs.get(0);
}
public <C extends SecurityConfigurer<O, B>> C
                                    removeConfigurer(Class<C> clazz) {
    List<SecurityConfigurer<O, B>> configs =
                                    this.configurers.remove(clazz);
    if (configs == null) {
```

```java
            return null;
        }
        if (configs.size() != 1) {
            throw new IllegalStateException("Only one configurer expected for
                        type " + clazz + ", but got " + configs);
        }
        return (C) configs.get(0);
    }
    private Collection<SecurityConfigurer<O, B>> getConfigurers() {
        List<SecurityConfigurer<O, B>> result = new ArrayList<>();
        for (List<SecurityConfigurer<O, B>> configs :
                                    this.configurers.values()) {
            result.addAll(configs);
        }
        return result;
    }
}
```

我们解析一下这段源码：

（1）首先声明了一个 configurers 变量，用来保存所有的配置类，key 是配置类 Class 对象，值是一个 List 集合中放着配置类。

（2）apply 方法有两个，参数类型略有差异，主要功能基本一致，都是向 configurers 变量中添加配置类，具体的添加过程则是调用 add 方法。

（3）add 方法用来将所有的配置类保存到 configurers 中，在添加的过程中，如果 allowConfigurersOfSameType 变量为 true，则表示允许相同类型的配置类存在，也就是 List 集合中可以存在多个相同类型的配置类。默认情况下，如果是普通配置类，allowConfigurersOfSameType 是 false，所以 List 集合中的配置类始终只有一个配置类；如果在 AuthenticationManagerBuilder 中设置 allowConfigurersOfSameType 为 true，此时相同类型的配置类可以有多个（下文会详细分析 AuthenticationManagerBuilder）。

（4）getConfigurers(Class<C>)方法可以从 configurers 中返回某一个配置类对应的所有实例。

（5）removeConfigurers 方法可以从 configurers 中移除某一个配置类对应的所有实例，并返回被移除掉的配置类实例集合。

（6）getConfigurer 方法也是获取配置类实例，但是只获取集合中第一项。

（7）removeConfigurer 方法可以从 configurers 中移除某一个配置类对应的所有配置类实例，并返回被移除掉的配置类实例中的第一项。

（8）getConfigurers 方法是一个私有方法，主要是把所有的配置类实例放到一个集合中返回。在配置类初始化和配置的时候，会调用到该方法。

这些就是 AbstractConfiguredSecurityBuilder 中关于 configurers 的所有操作。

接下来就是 AbstractConfiguredSecurityBuilder 中的 doBuild 方法了，这是核心的构建方法，

我们一起来看一下与之相关的方法：

```
@Override
protected final O doBuild() throws Exception {
    synchronized (configurers) {
        buildState = BuildState.INITIALIZING;
        beforeInit();
        init();
        buildState = BuildState.CONFIGURING;
        beforeConfigure();
        configure();
        buildState = BuildState.BUILDING;
        O result = performBuild();
        buildState = BuildState.BUILT;
        return result;
    }
}
protected void beforeInit() throws Exception {
}
protected void beforeConfigure() throws Exception {
}
protected abstract O performBuild() throws Exception;
private void init() throws Exception {
    Collection<SecurityConfigurer<O, B>> configurers = getConfigurers();
    for (SecurityConfigurer<O, B> configurer : configurers) {
        configurer.init((B) this);
    }
    for (SecurityConfigurer<O, B> configurer :
                                    configurersAddedInInitializing) {
        configurer.init((B) this);
    }
}
private void configure() throws Exception {
    Collection<SecurityConfigurer<O, B>> configurers = getConfigurers();
    for (SecurityConfigurer<O, B> configurer : configurers) {
        configurer.configure((B) this);
    }
}
```

（1）在 doBuild 方法中，一边更新构建状态，一边执行构建方法。构建方法中，beforeInit 是一个空的初始化方法，如果需要在初始化之前做一些准备工作，可以通过重写该方法实现。

（2）init 方法是所有配置类的初始化方法，在该方法中，遍历所有的配置类，并调用其 init 方法完成初始化操作。

（3）beforeConfigure 方法可以在 configure 方法执行之前做一些准备操作。该方法默认也是一个空方法。

（4）configure 方法用来完成所有配置类的配置，在 configure 方法中，遍历所有的配置类，分别调用其 configure 方法完成配置。

（5）performBuild 方法用来做最终的构建操作，前面的准备工作完成后，最后在 performBuild 方法中完成构建，这是一个抽象方法，具体的实现则在不同的配置类中。

这些就是 AbstractConfiguredSecurityBuilder 中最主要的几个方法，其他一些方法比较简单，这里就不一一赘述了。

ProviderManagerBuilder

ProviderManagerBuilder 继承自 SecurityBuilder 接口，并制定了构建的对象是 AuthenticationManager，代码如下：

```java
public interface ProviderManagerBuilder<B extends ProviderManagerBuilder<B>>
    extends SecurityBuilder<AuthenticationManager> {
    B authenticationProvider(AuthenticationProvider authenticationProvider);
}
```

可以看到，ProviderManagerBuilder 中增加了一个 authenticationProvider 方法，同时通过泛型指定了构建的对象为 AuthenticationManager。

AuthenticationManagerBuilder

AuthenticationManagerBuilder 用来构建 AuthenticationManager 对象，它继承自 AbstractConfiguredSecurityBuilder，并且实现了 ProviderManagerBuilder 接口，源码比较长，我们截取部分常用代码，代码如下：

```java
public class AuthenticationManagerBuilder extends
        AbstractConfiguredSecurityBuilder<AuthenticationManager,
                                        AuthenticationManagerBuilder>
        implements ProviderManagerBuilder<AuthenticationManagerBuilder> {
    public AuthenticationManagerBuilder(ObjectPostProcessor<Object>
                                                objectPostProcessor) {
        super(objectPostProcessor, true);
    }
    public AuthenticationManagerBuilder parentAuthenticationManager(
            AuthenticationManager authenticationManager) {
        if (authenticationManager instanceof ProviderManager) {
            eraseCredentials(((ProviderManager) authenticationManager)
                .isEraseCredentialsAfterAuthentication());
        }
        this.parentAuthenticationManager = authenticationManager;
        return this;
    }
    public InMemoryUserDetailsManagerConfigurer<AuthenticationManagerBuilder>
            inMemoryAuthentication() throws Exception {
        return apply(new InMemoryUserDetailsManagerConfigurer<>());
    }
    public JdbcUserDetailsManagerConfigurer<AuthenticationManagerBuilder>
                                                jdbcAuthentication()
```

```java
        throws Exception {
    return apply(new JdbcUserDetailsManagerConfigurer<>());
}
public <T extends UserDetailsService>
DaoAuthenticationConfigurer<AuthenticationManagerBuilder, T>
        userDetailsService(T userDetailsService) throws Exception {
    this.defaultUserDetailsService = userDetailsService;
    return apply(new DaoAuthenticationConfigurer<>(
            userDetailsService));
}
public AuthenticationManagerBuilder authenticationProvider(
        AuthenticationProvider authenticationProvider) {
    this.authenticationProviders.add(authenticationProvider);
    return this;
}
@Override
protected ProviderManager performBuild() throws Exception {
    if (!isConfigured()) {
        return null;
    }
    ProviderManager providerManager =
                    new ProviderManager(authenticationProviders,
                                    parentAuthenticationManager);
    if (eraseCredentials != null) {
      providerManager
          .setEraseCredentialsAfterAuthentication(eraseCredentials);
    }
    if (eventPublisher != null) {
        providerManager.setAuthenticationEventPublisher(eventPublisher);
    }
    providerManager = postProcess(providerManager);
    return providerManager;
}
```

（1）首先在 AuthenticationManagerBuilder 的构造方法中，调用了父类的构造方法，注意第二个参数传递了 true，表示允许相同类型的配置类同时存在（结合 AbstractConfiguredSecurityBuilder 的源码来理解）。

（2）parentAuthenticationManager 方法用来给一个 AuthenticationManager 设置 parent。

（3）inMemoryAuthentication 方法用来配置基于内存的数据源，该方法会自动创建 InMemoryUserDetailsManagerConfigurer 配置类，并最终将该配置类添加到父类的 configurers 变量中。由于设置了允许相同类型的配置类同时存在，因此 inMemoryAuthentication 方法可以反复调用多次。

（4）jdbcAuthentication 以及 userDetailsService 方法与 inMemoryAuthentication 方法类似，也是用来配置数据源的，这里不再赘述。

（5）authenticationProvider 方法用来向 authenticationProviders 集合中添加 AuthenticationProvider 对象，根据前面第 3 章的介绍，我们已经知道一个 AuthenticationManager 实例中包含多个 AuthenticationProvider 实例，那么多个 AuthenticationProvider 实例可以通过 authenticationProvider 方法进行添加。

（6）performBuild 方法则执行具体的构建工作，常用的 AuthenticationManager 实例就是 ProviderManager，所以这里创建 ProviderManager 对象，并且配置 authenticationProviders 和 parentAuthenticationManager 对象，ProviderManager 对象创建成功之后，再去对象后置处理器中处理一遍再返回。

这就是 AuthenticationManagerBuilder 中的一个大致逻辑。

HttpSecurity

HttpSecurity 的主要作用是用来构建一条过滤器链，并反映到代码上，也就是构建一个 DefaultSecurityFilterChain 对象。一个 DefaultSecurityFilterChain 对象包含一个路径匹配器和多个 Spring Security 过滤器，HttpSecurity 中通过收集各种各样的 xxxConfigurer，将 Spring Security 过滤器对应的配置类收集起来，并保存到父类 AbstractConfiguredSecurityBuilder 的 configurers 变量中，在后续的构建过程中，再将这些 xxxConfigurer 构建为具体的 Spring Security 过滤器，同时添加到 HttpSecurity 的 filters 对象中。

由于 HttpSecurity 中存在大量功能类似的方法，因此这里挑选一个作为例子用来说明 HttpSecurity 的配置原理，代码如下：

```java
public final class HttpSecurity extends
AbstractConfiguredSecurityBuilder<DefaultSecurityFilterChain, HttpSecurity>
        implements SecurityBuilder<DefaultSecurityFilterChain>,
        HttpSecurityBuilder<HttpSecurity> {
    public FormLoginConfigurer<HttpSecurity> formLogin() throws Exception {
        return getOrApply(new FormLoginConfigurer<>());
    }
    public HttpSecurity formLogin(
        Customizer<FormLoginConfigurer<HttpSecurity>> formLoginCustomizer)
                                                throws Exception {
        formLoginCustomizer
                .customize(getOrApply(new FormLoginConfigurer<>()));
        return HttpSecurity.this;
    }
    public HttpSecurity authenticationProvider(
            AuthenticationProvider authenticationProvider) {
        getAuthenticationRegistry()
                    .authenticationProvider(authenticationProvider);
        return this;
    }
    public HttpSecurity userDetailsService(UserDetailsService
                        userDetailsService) throws Exception {
        getAuthenticationRegistry().userDetailsService(userDetailsService);
```

```java
        return this;
    }
    private AuthenticationManagerBuilder getAuthenticationRegistry() {
        return getSharedObject(AuthenticationManagerBuilder.class);
    }
    @Override
    protected void beforeConfigure() throws Exception {
        setSharedObject(AuthenticationManager.class,
                            getAuthenticationRegistry().build());
    }
    @Override
    protected DefaultSecurityFilterChain performBuild() {
        filters.sort(comparator);
        return new DefaultSecurityFilterChain(requestMatcher, filters);
    }
    public HttpSecurity addFilterAfter(Filter filter,
                                Class<? extends Filter> afterFilter) {
        comparator.registerAfter(filter.getClass(), afterFilter);
        return addFilter(filter);
    }
    public HttpSecurity addFilterBefore(Filter filter,
            Class<? extends Filter> beforeFilter) {
        comparator.registerBefore(filter.getClass(), beforeFilter);
        return addFilter(filter);
    }
    public HttpSecurity addFilter(Filter filter) {
        Class<? extends Filter> filterClass = filter.getClass();
        if (!comparator.isRegistered(filterClass)) {
            throw new IllegalArgumentException(
                "The Filter class "
                    + filterClass.getName()
                    + " does not have a registered order and cannot be added without a specified order. Consider using addFilterBefore or addFilterAfter instead.");
        }
        this.filters.add(filter);
        return this;
    }
    public HttpSecurity addFilterAt(Filter filter,
                                Class<? extends Filter> atFilter) {
        this.comparator.registerAt(filter.getClass(), atFilter);
        return addFilter(filter);
    }
    private <C extends SecurityConfigurerAdapter<DefaultSecurityFilterChain, HttpSecurity>> C getOrApply(configurer) throws Exception {
        C existingConfig = (C) getConfigurer(configurer.getClass());
        if (existingConfig != null) {
            return existingConfig;
```

```
        }
        return apply(configurer);
    }
}
```

（1）以 form 表单登录配置为例，在 HttpSecurity 中有两个重载方法可以进行配置：第一个是一个无参的 formLogin 方法，该方法的返回值是一个 FormLoginConfigurer<HttpSecurity> 对象，开发者可以在该对象的基础上继续完善对 form 表单的配置，我们在前面章节中配置的表单登录都是通过这种方式来进行配置的。第二个是一个有参的 formLogin 方法，该方法的参数是一个 FormLoginConfigurer 对象，返回值则是一个 HttpSecurity 对象，也就是说开发者可以提前在外面配置好 FormLoginConfigurer 对象，然后直接传进来进行配置即可，返回值 HttpSecurity 对象则可以在方法返回后直接进行其他过滤器的配置。无论是有参还是无参，最终都会调用到 getOrApply 方法，该方法会调用父类的 getConfigurer 方法去查看是否已经有对应的配置类了，如果有，则直接返回；如果没有，则调用 apply 方法添加到父类的 configurers 变量中。HttpSecurity 中其他过滤器的配置都和 form 表单登录配置类似，这里就不再赘述了。

（2）每一套过滤器链都会有一个 AuthenticationManager 对象来进行认证操作（如果认证失败，则会调用 AuthenticationManager 的 parent 再次进行认证），主要是通过 authenticationProvider 方法配置执行认证的 authenticationProvider 对象，通过 userDetailsService 方法配置 UserDetailsService，最后在 beforeConfigure 方法中触发 AuthenticationManager 对象的构建。

（3）performBuild 方法则是进行 DefaultSecurityFilterChain 对象的构建，传入请求匹配器和过滤器集合 filters，在构建之前，会先按照既定的顺序对 filters 进行排序。

（4）通过 addFilterAfter、addFilterBefore 两个方法，我们可以在某一个过滤器之后或者之前添加一个自定义的过滤器（该方法已在 HttpSecurityBuilder 中声明，此处是具体实现）。

（5）addFilter 方法可以向过滤器链中添加一个过滤器，这个过滤器必须是 Spring Security 框架提供的过滤器的一个实例或者其扩展。实际上，在每一个 xxxConfigurer 的 configure 方法中，都会调用 addFilter 方法将构建好的过滤器添加到 HttpSecurity 中的 filters 集合中（addFilter 方法已在 HttpSecurityBuilder 中声明，此处是具体实现）。

（6）addFilterAt 方法可以在指定位置添加一个过滤器。需要注意的是，在同一个位置添加多个过滤器并不会覆盖现有的过滤器。

这便是 HttpSecurity 的基本功能。

WebSecurity

相比于 HttpSecurity，WebSecurity 是在一个更大的层面上去构建过滤器。一个 HttpSecurity 对象可以构建一个过滤器链，也就是一个 DefaultSecurityFilterChain 对象，而一个项目中可以存在多个 HttpSecurity 对象，也就可以构建多个 DefaultSecurityFilterChain 过滤器链。

WebSecurity 负责将 HttpSecurity 所构建的 DefaultSecurityFilterChain 对象（可能有多个），以及其他一些需要忽略的请求，再次重新构建为一个 FilterChainProxy 对象，同时添加上 HTTP 防火墙。

我们来看一下 WebSecurity 中的几个关键方法：

```java
public final class WebSecurity extends
        AbstractConfiguredSecurityBuilder<Filter, WebSecurity> implements
        SecurityBuilder<Filter>, ApplicationContextAware {
    private final List<RequestMatcher> ignoredRequests = new ArrayList<>();
    private final List<SecurityBuilder<? extends SecurityFilterChain>>
                        securityFilterChainBuilders = new ArrayList<>();
    public WebSecurity httpFirewall(HttpFirewall httpFirewall) {
        this.httpFirewall = httpFirewall;
        return this;
    }
    public WebSecurity addSecurityFilterChainBuilder(
                        SecurityBuilder<? extends SecurityFilterChain>
                                        securityFilterChainBuilder) {
        this.securityFilterChainBuilders.add(securityFilterChainBuilder);
        return this;
    }
    @Override
    protected Filter performBuild() throws Exception {
        int chainSize =
            ignoredRequests.size() + securityFilterChainBuilders.size();
        List<SecurityFilterChain> securityFilterChains =
                                        new ArrayList<>(chainSize);
        for (RequestMatcher ignoredRequest : ignoredRequests) {
            securityFilterChains
                    .add(new DefaultSecurityFilterChain(ignoredRequest));
        }
        for (SecurityBuilder<? extends SecurityFilterChain>
                securityFilterChainBuilder : securityFilterChainBuilders) {
            securityFilterChains.add(securityFilterChainBuilder.build());
        }
        FilterChainProxy filterChainProxy =
                        new FilterChainProxy(securityFilterChains);
        if (httpFirewall != null) {
            filterChainProxy.setFirewall(httpFirewall);
        }
        filterChainProxy.afterPropertiesSet();
        Filter result = filterChainProxy;
        postBuildAction.run();
        return result;
    }
}
```

（1）首先在 WebSecurity 中声明了 ignoredRequests 集合，这个集合中保存了所有被忽略的请求，因为在实际项目中，并非所有的请求都需要经过 Spring Security 过滤器链，有一些静态资源可能不需要权限认证，直接返回给客户端即可，那么这些需要忽略的请求可以直接保存在 ignoredRequests 变量中。

（2）接下来声明了一个 securityFilterChainBuilders 集合，该集合用来保存所有的 HttpSecurity 对象，每一个 HttpSecurity 对象创建成功之后，通过 addSecurityFilterChainBuilder 方法将 HttpSecurity 对象添加到 securityFilterChainBuilders 集合中。

（3）httpFirewall 方法可以用来配置请求防火墙，关于请求防火墙，我们会在后面的章节中专门讲解。

（4）performBuild 方法则是具体的构建方法，在该方法中，首先统计出过滤器链的总个数（被忽略的请求个数+通过 HttpSecurity 创建出来的过滤器链个数），然后创建一个集合 securityFilterChains，遍历被忽略的请求并分别构建成 DefaultSecurityFilterChain 对象保存到 securityFilterChains 集合中。需要注意的是，对于被忽略的请求，在构建 DefaultSecurityFilterChain 对象时，只是传入了请求匹配器，而没有传入对应的过滤器链，这就意味着这些被忽略掉的请求，将来不必经过 Spring Security 过滤器链；接下来再遍历 securityFilterChainBuilders 集合，调用每个对象的 build 方法构建 DefaultSecurityFilterChain 并存入 securityFilterChains 集合中，然后传入 securityFilterChains 集合构建 FilterChainProxy 对象，最后再设置 HTTP 防火墙。所有设置完成之后，最后返回 filterChainProxy 对象。

FilterChainProxy 就是我们最终构建出来的代理过滤器链，通过 Spring 提供的 DelegatingFilterProxy 将 FilterChainProxy 对象嵌入到 Web Filter 中（原生过滤器链中）。

读者可以回忆一下前面我们绘制的 FilterChainProxy 架构图，对照着来理解上面的源码应该就很容易了，如图 4-3 所示。

至此，关于 SecurityBuilder 体系中的几个关键类就介绍完了，至于 HttpSecurity 和 WebSecurity 是怎么配置到一起的，我们将在后面的章节中进行分析。

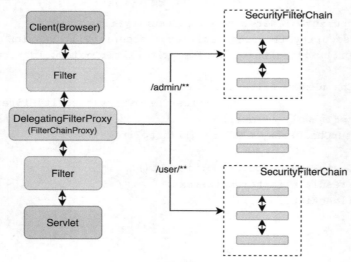

图 4-3　FilterChainProxy 架构图

4.1.4　FilterChainProxy

FilterChainProxy 通过 DelegatingFilterProxy 代理过滤器被集成到 Web Filter 中，DelegatingFilterProxy 作为一个代理对象，相信很多读者可能都用过（例如在 Spring 中整合 Shiro 就会用到），它不承载具体的业务。

所以，Spring Security 中的过滤器链的最终执行，就是在 FilterChainProxy 中，因此这里也来分析一下 FilterChainProxy 的源码。

FilterChainProxy 的源码比较长，我们一段一段来看：

```java
private List<SecurityFilterChain> filterChains;
private FilterChainValidator filterChainValidator =
                                    new NullFilterChainValidator();
private HttpFirewall firewall = new StrictHttpFirewall();
public FilterChainProxy() {
}
public FilterChainProxy(SecurityFilterChain chain) {
    this(Arrays.asList(chain));
}
public FilterChainProxy(List<SecurityFilterChain> filterChains) {
    this.filterChains = filterChains;
}
```

首先声明了三个变量：

（1）由于在 Spring Security 中可以同时存在多个过滤器链，filterChains 就是用来保存过滤器链的，注意保存的是过滤器链，而不是一个个具体的过滤器。

（2）filterChainValidator 是一个过滤器链配置完成后的验证器，默认使用 NullFilterChainValidator 其实没有做任何验证。

（3）创建了一个默认的防火墙对象 firewall。

在构造方法中传入过滤器链的集合，并赋值给 filterChains 变量。

由于 FilterChainProxy 本质上就是一个过滤器，因此它的核心方法就是 doFilter 方法，接下来我们来看一下 doFilter 方法：

```java
@Override
public void doFilter(ServletRequest request, ServletResponse response,
            FilterChain chain) throws IOException, ServletException {
    boolean clearContext = request.getAttribute(FILTER_APPLIED) == null;
    if (clearContext) {
        try {
            request.setAttribute(FILTER_APPLIED, Boolean.TRUE);
            doFilterInternal(request, response, chain);
        }
        finally {
            SecurityContextHolder.clearContext();
            request.removeAttribute(FILTER_APPLIED);
```

```
        }
    }
    else {
        doFilterInternal(request, response, chain);
    }
}
```

doFilter 方法相当于是整个 Spring Security 过滤器链的入口，我们在前面章节中所涉及的一些具体的过滤器如 SecurityContextPersistenceFilter，都是在该 doFilter 方法之后执行的。作为整个过滤器链的入口，这里多了一个 clearContext 变量，如果是第一次执行该 doFilter 方法，执行完成后，在 finally 代码块中需要从 SecurityContextHolder 里清除用户信息，这个主要是为了防止用户没有正确配置 SecurityContextPersistenceFilter，从而导致登录用户信息没有被正确清除，进而发生内存泄漏。

在 doFilter 方法中，过滤器的具体执行则交给了 doFilterInternal 方法：

```
private void doFilterInternal(ServletRequest request,
                              ServletResponse response,
                              FilterChain chain)
                    throws IOException, ServletException {
    FirewalledRequest fwRequest = firewall
            .getFirewalledRequest((HttpServletRequest) request);
    HttpServletResponse fwResponse = firewall
            .getFirewalledResponse((HttpServletResponse) response);
    List<Filter> filters = getFilters(fwRequest);
    if (filters == null || filters.size() == 0) {
        fwRequest.reset();
        chain.doFilter(fwRequest, fwResponse);
        return;
    }
    VirtualFilterChain vfc = new VirtualFilterChain(fwRequest, chain, filters);
    vfc.doFilter(fwRequest, fwResponse);
}
private List<Filter> getFilters(HttpServletRequest request) {
    for (SecurityFilterChain chain : filterChains) {
        if (chain.matches(request)) {
            return chain.getFilters();
        }
    }
    return null;
}
```

在 doFilterInternal 方法中，首先会将 request 对象转换为一个 FirewalledRequest 对象，这个转换过程会进行 Http 防火墙处理（Http 防火墙将在第 8 章详细介绍），同时将 response 对象也转为 HttpServletResponse。接下来调用 getFilters 方法获取当前请求对应的过滤器链，getFilters 方法会遍历 filterChains 集合，进而判断出当前请求和哪一个过滤器链是对应的，如果找到的过滤器链 filters 为 null，或者 filters 中没有元素，说明当前请求并不需要经过 Spring

Security 过滤器链，此时执行 fwRequest.reset 方法对 Http 防火墙中的属性进行重置，再执行 chain.doFilter 方法，回到 Web Filter 中，Spring Security 过滤器链将被跳过（回忆上一小结 WebSecurity 中配置的忽略请求）。如果 filters 集合中是有元素的，也就是说当前请求需要经过 filters 集合中元素所构成的过滤器链，那么构建一个虚拟的过滤器链对象 VirtualFilterChain，并执行其 doFilter 方法。

```java
private static class VirtualFilterChain implements FilterChain {
    private final FilterChain originalChain;
    private final List<Filter> additionalFilters;
    private final FirewalledRequest firewalledRequest;
    private final int size;
    private int currentPosition = 0;
    private VirtualFilterChain(FirewalledRequest firewalledRequest,
            FilterChain chain, List<Filter> additionalFilters) {
        this.originalChain = chain;
        this.additionalFilters = additionalFilters;
        this.size = additionalFilters.size();
        this.firewalledRequest = firewalledRequest;
    }
    @Override
    public void doFilter(ServletRequest request, ServletResponse response)
            throws IOException, ServletException {
        if (currentPosition == size) {
            this.firewalledRequest.reset();
            originalChain.doFilter(request, response);
        }
        else {
            currentPosition++;
            Filter nextFilter = additionalFilters.get(currentPosition - 1);
            nextFilter.doFilter(request, response, this);
        }
    }
}
```

VirtualFilterChain 中首先声明了五个变量：

（1）originalChain：表示原生的过滤器链，执行它的 doFilter 方法会回到 Web Filter 中。
（2）additionalFilters：这个 List 集合中存储的 Filter 就是本次请求的 Filter。
（3）firewalledRequest：当前请求对象。
（4）size：过滤器链的大小。
（5）currentPosition：过滤器链执行的下标。

在 VirtualFilterChain 的构造方法中，会给相应的变量赋值。

在 doFilter 方法中，会首先判断当前执行的下标是否等于过滤器链的大小，如果相等，则说明整个过滤器链中的所有过滤器都已经挨个走一遍了，此时先对 Http 防火墙中的属性进行重置，然后调用 originalChain.doFilter 方法跳出 Spring Security Filter，回到 Web Filter；如果不相等，则 currentPosition 自增，然后从过滤器链集合中取出一个过滤器去执行，注意执行的时候第三个参数 this 表示当前对象（即 VirtualFilterChain），这样在每一个过滤器执行完之后，最后的 chain.doFilter 方法又会回到当前 doFilter 方法中，继续下一个过滤器的调用。

这就是 FilterChainProxy 的一个大致工作原理。

4.1.5 SecurityConfigurer

SecurityConfigurer 中有两个核心方法，一个是 init 方法，用来完成配置类的初始化操作，另外一个是 configure 方法，进行配置类的配置。上一小结介绍的 AbstractConfiguredSecurityBuilder 类，里边的 init 方法和 configure 其实就是在遍历执行不同配置类的 init 和 configure 方法。

SecurityConfigurer 的实现类比较多，这里主要梳理一下常见的 SecurityConfigurer 实现类，我们分别来看一下。

SecurityConfigurer

先来看 SecurityConfigurer 源码，代码如下：

```java
public interface SecurityConfigurer<O, B extends SecurityBuilder<O>> {
    void init(B builder) throws Exception;
    void configure(B builder) throws Exception;
}
```

可以看到，SecurityConfigurer 只有两个方法：init 和 configure，两个方法的参数都是 SecurityBuilder 对象，也就是说在这两个方法中对 SecurityBuilder 进行初始化和配置。

SecurityConfigurer 的子类非常多，因为每一个过滤器都有自己对应的 xxxConfigurer，这里着重介绍几个关键的实现类，如图 4-4 所示。

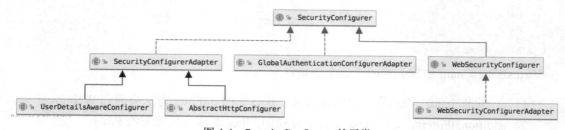

图 4-4　SecurityConfigurer 的子类

我们分别来看这几个实现类。

SecurityConfigurerAdapter

SecurityConfigurerAdapter 实现了 SecurityConfigurer 接口，它的源码如下：

```java
public abstract class
        SecurityConfigurerAdapter<O, B extends SecurityBuilder<O>>
                    implements SecurityConfigurer<O, B> {
```

```java
private B securityBuilder;
private CompositeObjectPostProcessor objectPostProcessor =
                                new CompositeObjectPostProcessor();
public void init(B builder) throws Exception {
}
public void configure(B builder) throws Exception {
}
public B and() {
    return getBuilder();
}
protected final B getBuilder() {
    if (securityBuilder == null) {
        throw new IllegalStateException("securityBuilder cannot be null");
    }
    return securityBuilder;
}
protected <T> T postProcess(T object) {
    return (T) this.objectPostProcessor.postProcess(object);
}
public void addObjectPostProcessor(ObjectPostProcessor<?>
                                            objectPostProcessor) {
this.objectPostProcessor.addObjectPostProcessor(objectPostProcessor);
}
public void setBuilder(B builder) {
    this.securityBuilder = builder;
}
private static final class CompositeObjectPostProcessor implements
        ObjectPostProcessor<Object> {
    private List<ObjectPostProcessor<?>> postProcessors =
                                                new ArrayList<>();
    public Object postProcess(Object object) {
        for (ObjectPostProcessor opp : postProcessors) {
            Class<?> oppClass = opp.getClass();
            Class<?> oppType = GenericTypeResolver
                .resolveTypeArgument(oppClass, ObjectPostProcessor.class);
            if (oppType == null ||
                    oppType.isAssignableFrom(object.getClass())) {
                object = opp.postProcess(object);
            }
        }
        return object;
    }
    private boolean addObjectPostProcessor(
            ObjectPostProcessor<?> objectPostProcessor) {
        boolean result = this.postProcessors.add(objectPostProcessor);
        postProcessors.sort(AnnotationAwareOrderComparator.INSTANCE);
        return result;
    }
```

```
        }
    }
```

从这段源码中,我们可以分析出 SecurityConfigurerAdapter 主要做了如下几件事:

(1) 提供了一个 SecurityBuilder 对象,为每一个配置类都提供一个 SecurityBuilder 对象,将来通过 SecurityBuilder 构建出具体的配置对象;通过 and 方法返回 SecurityBuilder 对象,这样方便不同的配置类在配置时,可以进行链式配置(第 2 章中我们在定义 SecurityConfig 时所使用的 and 方法)。

(2) 定义了内部类 CompositeObjectPostProcessor,这是一个复合的对象后置处理器。

(3) 提供了一个 addObjectPostProcessor 方法,通过该方法可以向复合的对象后置处理器中添加新的 ObjectPostProcessor 实例。

这是 SecurityConfigurerAdapter 提供的主要功能。

UserDetailsAwareConfigurer

UserDetailsAwareConfigurer 的子类主要负责配置用户认证相关的组件,如 UserDetailsService 等,UserDetailsAwareConfigurer 中提供了获取 UserDetailsService 的抽象方法,具体实现则在它的子类中,UserDetailsAwareConfigurer 的子类如图 4-5 所示。

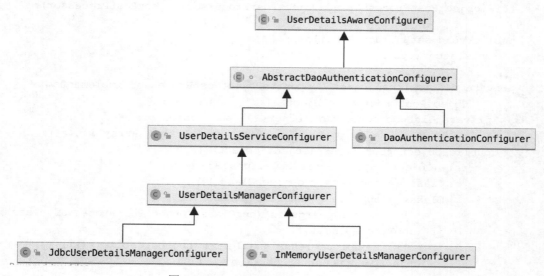

图 4-5 UserDetailsAwareConfigurer 的子类

- AbstractDaoAuthenticationConfigurer:完成对 DaoAuthenticationProvider 的配置。
- UserDetailsServiceConfigurer:完成对 UserDetailsService 的配置。
- UserDetailsManagerConfigurer:使用 UserDetailsManager 构建用户对象,完成对 AuthenticationManagerBuilder 的填充。
- JdbcUserDetailsManagerConfigurer:配置 JdbcUserDetailsManager 并填充到 AuthenticationManagerBuilder 中。
- InMemoryUserDetailsManagerConfigurer:配置 InMemoryUserDetailsManager。

- DaoAuthenticationConfigurer：完成对 DaoAuthenticationProvider 的配置。

AbstractHttpConfigurer

AbstractHttpConfigurer 主要是为了给在 HttpSecurity 中使用的配置类添加一个方便的父类，提取出共同的操作。

```java
public abstract class AbstractHttpConfigurer<T extends
        AbstractHttpConfigurer<T, B>, B extends HttpSecurityBuilder<B>>
        extends SecurityConfigurerAdapter<DefaultSecurityFilterChain, B> {
    public B disable() {
        getBuilder().removeConfigurer(getClass());
        return getBuilder();
    }
    public T withObjectPostProcessor(ObjectPostProcessor<?>
                                                    objectPostProcessor) {
        addObjectPostProcessor(objectPostProcessor);
        return (T) this;
    }
}
```

可以看到，提取出来的方法其实就两个：一个 disable 表示禁用某一个配置（第 2 章中我们配置的.csrf().disable()），本质上就是从构建器的 configurers 集合中移除某一个配置类，这样在将来构建的时候就不存在该配置类，那么对应的功能也就不存在（被禁用）；另一个 withObjectPostProcessor 表示给某一个对象添加一个对象后置处理器，由于该方法的返回值是当前对象，所以该方法可以用在链式配置中。

AbstractHttpConfigurer 的实现类比较多，基本上都用来配置各种各样的过滤器，参见表 4-1。

表 4-1　AbstractHttpConfigurer 子类及其作用

配置类名称	作用
HttpBasicConfigurer	配置基于 Http Basic 认证的过滤器 BasicAuthenticationFilter
LogoutConfigurer	配置注销登录过滤器 LogoutFilter
RequestCacheConfigurer	配置请求缓存过滤器 RequestCacheAwareFilter
RememberMeConfigurer	配置记住我登录过滤器 RememberMeAuthenticationFilter
ServletApiConfigurer	配置包装原始请求过滤器 SecurityContextHolderAwareRequestFilter
DefaultLoginPageConfigurer	配置提供默认登录页面的过滤器 DefaultLoginPageGeneratingFilter 和默认注销页面的过滤器 DefaultLogoutPageGeneratingFilter
SessionManagementConfigurer	配置 Session 管理过滤器 SessionManagementFilter 和 ConcurrentSessionFilter
PortMapperConfigurer	配置一个共享的 PortMapper 实例，以便在 HTTP 和 HTTPS 之间重定向时确定端口
ExceptionHandlingConfigurer	配置异常处理过滤器 ExceptionTranslationFilter
HeadersConfigurer	配置安全相关的响应头信息
CsrfConfigurer	配置防范 CSRF 攻击过滤器 CsrfFilter

(续表)

配置类名称	作用
OAuth2ClientConfigurer	配置 OAuth2 相关的过滤器 OAuth2AuthorizationRequestRedirectFilter 和 OAuth2AuthorizationCodeGrantFilter
ImplicitGrantConfigurer	配置 OAuth2 认证请求重定向的过滤器 OAuth2AuthorizationRequestRedirectFilter
AnonymousConfigurer	配置匿名过滤器 AnonymousAuthenticationFilter
JeeConfigurer	配置 J2EE 身份预校验过滤器 J2eePreAuthenticatedProcessingFilter
ChannelSecurityConfigurer	配置请求协议处理过滤器 ChannelProcessingFilter
CorsConfigurer	配置处理跨域过滤器 CorsFilter
SecurityContextConfigurer	配置登录信息存储和恢复的过滤器 SecurityContextPersistenceFilter
OAuth2ResourceServerConfigurer	配置 OAuth2 身份请求认证过滤器 BearerTokenAuthenticationFilter
AbstractAuthenticationFilterConfigurer	身份认证配置类的父类
FormLoginConfigurer	配置身份认证过滤器 UsernamePasswordAuthenticationFilter 和默认登录页面的过滤器 DefaultLoginPageGeneratingFilter
OAuth2LoginConfigurer	配置 OAuth2 认证请求重定向的过滤器 OAuth2AuthorizationRequestRedirectFilter 和处理第三方回调过滤器 OAuth2LoginAuthenticationFilter
OpenIDLoginConfigurer	配置 OpenID 身份认证过滤器 OpenIDAuthenticationFilter
Saml2LoginConfigurer	配置 SAML2.0 身份认证过滤器 Saml2WebSsoAuthenticationFilter 和 Saml2WebSsoAuthenticationRequestFilter
X509Configurer	配置 X509 身份认证过滤器 X509AuthenticationFilter
AbstractInterceptUrlConfigurer	拦截器配置类的父类
UrlAuthorizationConfigurer	配置基于 URL 的权限认证拦截器 FilterSecurityInterceptor
ExpressionUrlAuthorizationConfigurer	配置基于 SpEL 表达式的 URL 权限认证拦截器 FilterSecurityInterceptor

GlobalAuthenticationConfigurerAdapter

GlobalAuthenticationConfigurerAdapter 主要用于配置全局 AuthenticationManagerBuilder，在 AuthenticationConfiguration 类中会自动使用 GlobalAuthenticationConfigurerAdapter 提供的 Bean 来配置全局 AuthenticationManagerBuilder。

在第 3 章介绍 ProviderManager 时曾经提到过，默认情况下 ProviderManager 有一个 parent，这个 parent 就是通过这里的全局 AuthenticationManagerBuilder 来构建的。

GlobalAuthenticationConfigurerAdapter 有四个不同的子类，如图 4-6 所示。

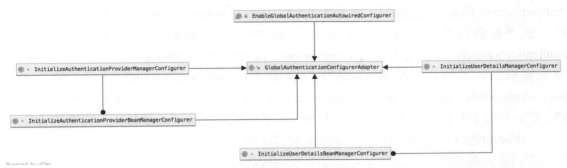

图 4-6　GlobalAuthenticationConfigurerAdapter 的子类

- InitializeAuthenticationProviderBeanManagerConfigurer：初始化全局的 AuthenticationProvider 对象。
- InitializeAuthenticationProviderManagerConfigurer：配置全局的 AuthenticationProvider 对象，配置过程就是从 Spring 容器中查找 AuthenticationProvider 并设置给全局的 AuthenticationManagerBuilder 对象。
- InitializeUserDetailsBeanManagerConfigurer：初始化全局的 UserDetailsService 对象。
- InitializeUserDetailsManagerConfigurer：配置全局的 UserDetailsService 对象，配置过程就是从 Spring 容器中查找 UserDetailsService，并设置给全局的 AuthenticationManagerBuilder 对象。
- EnableGlobalAuthenticationAutowiredConfigurer：从 Spring 容器中加载被@EnableGlobalAuthentication 注解标记的 Bean。

WebSecurityConfigurer

WebSecurityConfigurer 是一个空接口，我们可以通过它来自定义 WebSecurity。WebSecurityConfigurer 只有一个实现类就是 WebSecurityConfigurerAdapter，在大多数情况下，开发者通过继承 WebSecurityConfigurerAdapter 来实现对 WebSecurity 的自定义配置。

WebSecurityConfigurerAdapter

WebSecurityConfigurerAdapter 是一个可以方便创建 WebSecurityConfigurer 实例的基类，开发者可以通过覆盖 WebSecurityConfigurerAdapter 中的方法完成对 HttpSecurity 和 WebSecurity 的定制。在本书前面的章节中，我们所定制的 Spring Security 登录都是通过自定义类继承 WebSecurityConfigurerAdapter 来实现的。

在 WebSecurityConfigurerAdapter 中声明了两个 AuthenticationManagerBuilder 对象用来构建 AuthenticationManager：

```
private AuthenticationManagerBuilder authenticationBuilder;
private AuthenticationManagerBuilder localConfigureAuthenticationBldr;
```

其中，localConfigureAuthenticationBldr 对象负责构建全局的 AuthenticationManager，而 authenticationBuilder 则负责构建局部的 AuthenticationManager。局部的 AuthenticationManager 是和每一个 HttpSecurity 对象绑定的，而全局的 AuthenticationManager 对象则是所有局部

AuthenticationManager 的 parent。需要注意的是，localConfigureAuthenticationBldr 并非总是有用，在开发者没有重写 configure(AuthenticationManagerBuilder) 方法的情况下，全局的 AuthenticationManager 对象是由 AuthenticationConfiguration 类中的 getAuthenticationManager 方法提供的，如果用户重写了 configure(AuthenticationManagerBuilder) 方法，则全局的 AuthenticationManager 就由 localConfigureAuthenticationBldr 负责构建。这里可能会感觉有点绕，在后面的小节中，我们将通过实际的例子展示全局 AuthenticationManager 对象的构建。

WebSecurityConfigurerAdapter 类的初始化方法如下：

```java
public void init(final WebSecurity web) throws Exception {
    final HttpSecurity http = getHttp();
    web.addSecurityFilterChainBuilder(http).postBuildAction(() -> {
        FilterSecurityInterceptor securityInterceptor = http
                .getSharedObject(FilterSecurityInterceptor.class);
        web.securityInterceptor(securityInterceptor);
    });
}
protected final HttpSecurity getHttp() throws Exception {
    if (http != null) {
        return http;
    }
    AuthenticationEventPublisher eventPublisher =
                                    getAuthenticationEventPublisher();
    localConfigureAuthenticationBldr
                    .authenticationEventPublisher(eventPublisher);
    AuthenticationManager authenticationManager = authenticationManager();
    authenticationBuilder.parentAuthenticationManager(authenticationManager);
    Map<Class<?>, Object> sharedObjects = createSharedObjects();
    http = new HttpSecurity(objectPostProcessor, authenticationBuilder,
        sharedObjects);
    if (!disableDefaults) {
        http
            .csrf().and()
            .addFilter(new WebAsyncManagerIntegrationFilter())
            .exceptionHandling().and()
            .headers().and()
            .sessionManagement().and()
            .securityContext().and()
            .requestCache().and()
            .anonymous().and()
            .servletApi().and()
            .apply(new DefaultLoginPageConfigurer<>()).and()
            .logout();
        ClassLoader classLoader = this.context.getClassLoader();
        List<AbstractHttpConfigurer> defaultHttpConfigurers =
                SpringFactoriesLoader
                    .loadFactories(AbstractHttpConfigurer.class, classLoader);
        for (AbstractHttpConfigurer configurer : defaultHttpConfigurers) {
```

```
            http.apply(configurer);
        }
    }
    configure(http);
    return http;
}
protected void configure(HttpSecurity http) throws Exception {
    http
        .authorizeRequests()
        .anyRequest().authenticated()
        .and()
        .formLogin().and()
        .httpBasic();
}
protected void configure(AuthenticationManagerBuilder auth) throws Exception {
    this.disableLocalConfigureAuthenticationBldr = true;
}
protected AuthenticationManager authenticationManager() throws Exception {
    if (!authenticationManagerInitialized) {
        configure(localConfigureAuthenticationBldr);
        if (disableLocalConfigureAuthenticationBldr) {
            authenticationManager = authenticationConfiguration
                .getAuthenticationManager();
        }
        else {
            authenticationManager = localConfigureAuthenticationBldr.build();
        }
        authenticationManagerInitialized = true;
    }
    return authenticationManager;
}
```

（1）在 init 方法中，首先调用 getHttp 方法获取一个 HttpSecurity 实例，并将获取到的实例添加到 WebSecurity 对象中，再由 WebSecurity 对象进行构建。

（2）在 getHttp 方法中，如果 http 对象已经初始化，则直接返回，否则进行初始化操作。在初始化的过程中，给 localConfigureAuthenticationBldr 设置事件发布器，并调用 authenticationManager 方法获取全局的 AuthenticationManager 对象。

（3）在 authenticationManager 方法中，如果全局的 AuthenticationManager 对象还没有初始化，则先调用 configure 方法，该方法的逻辑很简单，就是将 disableLocalConfigure AuthenticationBldr 变量由 false 变为 true，接下来就会进入到 authenticationManager 方法的 if 分支中，通过调用 authenticationConfiguration.getAuthenticationManager() 方法获取全局的 AuthenticationManager 对象并返回。如果开发者自己重写了 configure(AuthenticationManager Builder)方法，则 disableLocalConfigureAuthenticationBldr 变量就一直是 false，没有机会变为 true，这样就会进入到 else 分支中，通过 localConfigureAuthenticationBldr 变量来构建 authenticationManager 对象。

（4）再次回到 getHttp 方法中，获取到全局的 authenticationManager 对象之后，设置给 authenticationBuilder，然后创建一个 HttpSecurity 实例出来，并为其配置上默认的过滤器。默认的配置完成后，调用 configure(HttpSecurity) 方法进行扩展配置，WebSecurityConfigurerAdapter 中对 configure(HttpSecurity) 方法提供了默认的实现，开发者也可以自定义该方法。

这就是 WebSecurityConfigurerAdapter 的初始化方法，其实就是创建并配置一个 HttpSecurity 实例，之后添加到 WebSecurity 中。

WebSecurityConfigurerAdapter 中的 configure 方法是一个空方法，可以用来配置 WebSecurity，代码如下：

```
public void configure(WebSecurity web) throws Exception {
}
```

一般来说，如果我们有一些静态资源不需要经过 Spring Security 过滤器，就可以通过重写该方法实现。

至此，在 Spring Security 初始化过程中，几个重要的组件都介绍完了，单纯的源码读者看起来可能会比较枯燥，在后面的小节中，我们会结合大量的应用案例，来帮助大家深入理解源码。

不过在讲解具体的案例之前，我们还是先来分析一遍 Spring Security 的初始化流程，将前面讲的这些知识点串起来。

4.1.6 初始化流程分析

在 Spring Boot 中使用 Spring Security，初始化就从 Spring Security 的自动化配置类中开始：

```
@Configuration(proxyBeanMethods = false)
@ConditionalOnClass(DefaultAuthenticationEventPublisher.class)
@EnableConfigurationProperties(SecurityProperties.class)
@Import({ SpringBootWebSecurityConfiguration.class,
        WebSecurityEnablerConfiguration.class,
        SecurityDataConfiguration.class })
public class SecurityAutoConfiguration {
   @Bean
   @ConditionalOnMissingBean(AuthenticationEventPublisher.class)
   public DefaultAuthenticationEventPublisher authenticationEventPublisher(
                        ApplicationEventPublisher publisher) {
      return new DefaultAuthenticationEventPublisher(publisher);
   }
}
```

可以看到，在自动化配置类 SecurityAutoConfiguration 中，最重要的就是导入了三个配置类，并且定义了一个默认的事件发布器。

导入的三个配置类中，SpringBootWebSecurityConfiguration 的主要作用是在开发者没有提供 WebSecurityConfigurerAdapter 实例的情况下，由其负责提供一个默认的 WebSecurity

ConfigurerAdapter 实例，代码如下：

```
@Configuration(proxyBeanMethods = false)
@ConditionalOnClass(WebSecurityConfigurerAdapter.class)
@ConditionalOnMissingBean(WebSecurityConfigurerAdapter.class)
@ConditionalOnWebApplication(type = Type.SERVLET)
public class SpringBootWebSecurityConfiguration {
    @Configuration(proxyBeanMethods = false)
    @Order(SecurityProperties.BASIC_AUTH_ORDER)
    static class DefaultConfigurerAdapter extends WebSecurityConfigurerAdapter {
    }
}
```

另一个导入的配置类 SecurityDataConfiguration 主要提供了一个 SecurityEvaluationContextExtension 实例，以便通过 SpEL 为经过身份验证的用户提供数据查询：

```
@Configuration(proxyBeanMethods = false)
@ConditionalOnClass(SecurityEvaluationContextExtension.class)
public class SecurityDataConfiguration {
    @Bean
    @ConditionalOnMissingBean
    public SecurityEvaluationContextExtension
                            securityEvaluationContextExtension() {
        return new SecurityEvaluationContextExtension();
    }
}
```

最后一个导入的配置类 WebSecurityEnablerConfiguration 则是我们分析的重点。

```
@Configuration(proxyBeanMethods = false)
@ConditionalOnBean(WebSecurityConfigurerAdapter.class)
@ConditionalOnMissingBean(name = BeanIds.SPRING_SECURITY_FILTER_CHAIN)
@ConditionalOnWebApplication(type = ConditionalOnWebApplication.Type.SERVLET)
@EnableWebSecurity
public class WebSecurityEnablerConfiguration {
}
```

WebSecurityEnablerConfiguration 配置类中添加了 @EnableWebSecurity 注解，而该注解的定义，引入了关键的配置类 WebSecurityConfiguration。

```
@Retention(value = java.lang.annotation.RetentionPolicy.RUNTIME)
@Target(value = { java.lang.annotation.ElementType.TYPE })
@Documented
@Import({ WebSecurityConfiguration.class,
        SpringWebMvcImportSelector.class,
        OAuth2ImportSelector.class })
@EnableGlobalAuthentication
@Configuration
public @interface EnableWebSecurity {
    boolean debug() default false;
```

}
```

可以看到，@EnableWebSecurity 是一个组合注解，首先导入了三个配置类：

- WebSecurityConfiguration：用来配置 WebSecurity（重点分析）。
- SpringWebMvcImportSelector：判断当前环境是否存在 Spring MVC，如果存在，则引入相关配置。
- OAuth2ImportSelector：判断当前环境是否存在 OAuth2，如果存在，则引入相关配置。

另外还有一个 @EnableGlobalAuthentication 注解，用来开启全局配置，代码如下：

```
@Retention(value = java.lang.annotation.RetentionPolicy.RUNTIME)
@Target(value = { java.lang.annotation.ElementType.TYPE })
@Documented
@Import(AuthenticationConfiguration.class)
@Configuration
public @interface EnableGlobalAuthentication {
}
```

可以看到，@EnableGlobalAuthentication 注解的主要功能是导入了配置类 AuthenticationConfiguration。

从上面的源码中我们可以看到，Spring Security 的自动化配置类主要导入了两个类：WebSecurityConfiguration 和 AuthenticationConfiguration。接下来我们就来分析这两个类。

4.1.6.1 WebSecurityConfiguration

WebSecurityConfiguration 配置类的功能，主要就是为了构建 Spring Security 过滤器链代理对象 FilterChainProxy。根据前面的分析，FilterChainProxy 是由 WebSecurity 来构建的，所以在 WebSecurityConfiguration 中会首先构建 WebSecurity 对象，再利用 WebSecurity 对象构建出 FilterChainProxy。

我们先来看一下 WebSecurityConfiguration 中定义的属性：

```
@Configuration(proxyBeanMethods = false)
public class WebSecurityConfiguration implements ImportAware,
                                                  BeanClassLoaderAware {
    private WebSecurity webSecurity;
    private Boolean debugEnabled;
    private List<SecurityConfigurer<Filter, WebSecurity>>
                                                  webSecurityConfigurers;
    private ClassLoader beanClassLoader;
    @Autowired(required = false)
    private ObjectPostProcessor<Object> objectObjectPostProcessor;
}
```

（1）WebSecurityConfiguration 类实现了 ImportAware 接口。ImportAware 接口一般是和 @Import 注解一起使用，实现了 ImportAware 接口的配置类可以方便地通过 setImportMetadata 方法获取到导入类中的数据配置。换句话说，WebSecurityConfiguration 实现了 ImportAware
```

接口，使用@Import 注解在@EnableWebSecurity 上导入 WebSecurityConfiguration 之后，在 WebSecurityConfiguration 的 setImportMetadata 方法中可以方便的获取到@EnableWebSecurity 注解中的属性值，这里主要是 debug 属性。另一方面，WebSecurityConfiguration 类通过实现 BeanClassLoaderAware 接口可以方便地获取到 ClassLoader 对象。

（2）webSecurity 对象是 WebSecurityConfiguration 中需要构建的 WebSecurity 对象。

（3）webSecurityConfigurers 集合中保存了所有的配置类，也就是 WebSecurityConfigurerAdapter 对象，一个 WebSecurityConfigurerAdapter 对象可以创建一个 HttpSecurity，进而构建出一条过滤器链，多个 WebSecurityConfigurerAdapter 对象就可以构建出多条过滤器链。

（4）beanClassLoader 是一个 ClassLoader。

（5）objectObjectPostProcessor 是一个对象后置处理器，注意这个对象是直接从 Spring 容器中注入的。下一小节会分析对象后置处理器是什么时候初始化并注册到 Spring 容器中去的。

这是 WebSecurityConfiguration 类中定义的属性。接下来，我们来看一下 setFilterChainProxySecurityConfigurer 方法，该方法主要用来构建一个 WebSecurity 对象，并且加载所有的配置类对象。

```
@Autowired(required = false)
public void setFilterChainProxySecurityConfigurer(
        ObjectPostProcessor<Object> objectPostProcessor,
@Value("#{@autowiredWebSecurityConfigurersIgnoreParents.getWebSecurityConfigurers()}") List<SecurityConfigurer<Filter, WebSecurity>> webSecurityConfigurers)
                                                            throws Exception {
    webSecurity = objectPostProcessor
                        .postProcess(new WebSecurity(objectPostProcessor));
    if (debugEnabled != null) {
        webSecurity.debug(debugEnabled);
    }
    webSecurityConfigurers.sort(AnnotationAwareOrderComparator.INSTANCE);
    Integer previousOrder = null;
    Object previousConfig = null;
    for (SecurityConfigurer<Filter, WebSecurity> config :
                                                webSecurityConfigurers) {
        Integer order = AnnotationAwareOrderComparator.lookupOrder(config);
        if (previousOrder != null && previousOrder.equals(order)) {
            throw new IllegalStateException("");
        }
        previousOrder = order;
        previousConfig = config;
    }
    for (SecurityConfigurer<Filter, WebSecurity> webSecurityConfigurer :
                                                webSecurityConfigurers) {
        webSecurity.apply(webSecurityConfigurer);
    }
    this.webSecurityConfigurers = webSecurityConfigurers;
}
```

```java
@Bean
public static AutowiredWebSecurityConfigurersIgnoreParents
                    autowiredWebSecurityConfigurersIgnoreParents(
    ConfigurableListableBeanFactory beanFactory) {
    return new AutowiredWebSecurityConfigurersIgnoreParents(beanFactory);
}
```

setFilterChainProxySecurityConfigurer 方法有两个参数，第一个参数 objectPostProcessor 是一个对象后置处理器，由于该方法有一个@Autowired 注解，会自动查找需要注入的参数，所以 objectPostProcessor 参数会自动注入进来。需要注意的是，@Autowired 注解的 required 属性为 false，所以在方法参数注入的时候，有就注入，没有则忽略。required 属性设置为 false 主要是针对第二个参数 webSecurityConfigurers，因为该参数的值是通过调用 autowiredWebSecurityConfigurersIgnoreParents 对象的 getWebSecurityConfigurers 方法获取的。autowiredWebSecurityConfigurersIgnoreParents 对象也是在当前类中注入到 Spring 容器中的，我们来看一下它的 getWebSecurityConfigurers 方法：

```java
public List<SecurityConfigurer<Filter, WebSecurity>>
                                        getWebSecurityConfigurers() {
    List<SecurityConfigurer<Filter, WebSecurity>> webSecurityConfigurers =
                                                    new ArrayList<>();
    Map<String, WebSecurityConfigurer> beansOfType = beanFactory
            .getBeansOfType(WebSecurityConfigurer.class);
    for (Entry<String, WebSecurityConfigurer> entry :
                                        beansOfType.entrySet()) {
        webSecurityConfigurers.add(entry.getValue());
    }
    return webSecurityConfigurers;
}
```

可以看到，在 getWebSecurityConfigurers 方法中主要是通过调用 beanFactory.getBeansOfType 方法来获取 Spring 容器中所有的 WebSecurityConfigurer 实例，也就是开发者自定义的各种各样继承自 WebSecurityConfigurerAdapter 的配置类。如果开发者没有自定义任何配置类，那么这里获取到的就是前面所讲的 SpringBootWebSecurityConfiguration 类中提供的默认配置类，将获取到的所有配置类实例放入 webSecurityConfigurers 集合中并返回。

返回 setFilterChainProxySecurityConfigurer 方法中，现在我们已经明白了第二个参数 webSecurityConfigurers 的含义了。在该方法中，首先创建一个 WebSecurity 实例，创建出来之后去对象后置处理器中走一圈，这样就将 webSecurity 对象注册到 Spring 容器中了。接下来，根据每一个配置类的@Order 注解对 webSecurityConfigurers 集合中的所有配置类进行排序，因为一个配置类对应一个过滤器链，当请求到来后，需要先和哪个过滤器链进行匹配，这里必然存在一个优先级问题，所以如果开发者自定义了多个配置类，则需要通过 @Order 注解标记多个配置类的优先级。排序完成后，进入到 for 循环中，检查是否存在优先级相等的配置类，如果存在，则直接抛出异常。最后再去遍历所有的配置类，调用 webSecurity.apply 方法将其添加到 webSecurity 父类中的 configurers 集合中（将来遍历该集合并分别调用配置类的 init 和 configure 方法完成配置类的初始化操作）。

这是 setFilterChainProxySecurityConfigurer 方法的执行逻辑，该方法主要用来初始化 WebSecurity 对象，同时收集到所有的自定义配置类。

有了 WebSecurity 对象和配置类，接下来就可以构建过滤器 FilterChainProxy 了。我们来看一下 springSecurityFilterChain 方法：

```java
@Bean(name = AbstractSecurityWebApplicationInitializer.DEFAULT_FILTER_NAME)
public Filter springSecurityFilterChain() throws Exception {
    boolean hasConfigurers = webSecurityConfigurers != null
            && !webSecurityConfigurers.isEmpty();
    if (!hasConfigurers) {
        WebSecurityConfigurerAdapter adapter = objectObjectPostProcessor
                .postProcess(new WebSecurityConfigurerAdapter() {
                });
        webSecurity.apply(adapter);
    }
    return webSecurity.build();
}
```

这里首先判断 webSecurityConfigurers 集合中是否存在配置类，如果不存在，则立马创建一个匿名的 WebSecurityConfigurerAdapter 对象并注册到 Spring 容器中，否则就直接调用 WebSecurity 的 build 方法进行构建。

根据前面小节的介绍，了解了 WebSecurity 对象的 build 方法执行后，首先会对所有的配置类即 WebSecurityConfigurerAdapter 实例进行构建，在 WebSecurityConfigurerAdapter 的 init 方法中，又会完成 HttpSecurity 的构建，而 HttpSecurity 的构建过程中，则会完成局部 AuthenticationManager 对象以及每一个具体的过滤器的构建。

这就是整个过滤器链的构建流程。

4.1.6.2 AuthenticationConfiguration

在 Spring Security 自动化配置类中导入的另外一个配置类是 AuthenticationConfiguration，该类的功能主要是做全局的配置，同时提供一个全局的 AuthenticationManager 实例。首先我们来看 AuthenticationConfiguration 类的定义：

```java
@Configuration(proxyBeanMethods = false)
@Import(ObjectPostProcessorConfiguration.class)
public class AuthenticationConfiguration {}
```

可以看到，AuthenticationConfiguration 类的定义中，导入了 ObjectPostProcessorConfiguration 配置，而 ObjectPostProcessorConfiguration 配置则提供了一个基本的对象后置处理器：

```java
@Configuration(proxyBeanMethods = false)
@Role(BeanDefinition.ROLE_INFRASTRUCTURE)
public class ObjectPostProcessorConfiguration {
```

```
    @Bean
    @Role(BeanDefinition.ROLE_INFRASTRUCTURE)
    public ObjectPostProcessor<Object> objectPostProcessor(
            AutowireCapableBeanFactory beanFactory) {
        return new AutowireBeanFactoryObjectPostProcessor(beanFactory);
    }
}
```

可以看到，ObjectPostProcessorConfiguration 类主要提供了一个 ObjectPostProcessor 实例，具体的实现类是 AutowireBeanFactoryObjectPostProcessor，根据 4.1.1 小节的介绍，该实现类主要用来将一个对象注册到 Spring 容器中去，我们在其他配置类中所见到的 ObjectPostProcessor 实例其实都是这里提供的。

这是 AuthenticationConfiguration 类的定义部分，AuthenticationConfiguration 类中的方法比较多，我们挑选出关键的部分分析一下：

```
@Bean
public AuthenticationManagerBuilder authenticationManagerBuilder(
        ObjectPostProcessor<Object> objectPostProcessor,
                                    ApplicationContext context) {
    LazyPasswordEncoder defaultPasswordEncoder =
                                new LazyPasswordEncoder(context);
    AuthenticationEventPublisher authenticationEventPublisher =
            getBeanOrNull(context, AuthenticationEventPublisher.class);
    DefaultPasswordEncoderAuthenticationManagerBuilder result = new
    DefaultPasswordEncoderAuthenticationManagerBuilder(objectPostProcessor,
                                            defaultPasswordEncoder);
    if (authenticationEventPublisher != null) {
        result.authenticationEventPublisher(authenticationEventPublisher);
    }
    return result;
}
@Bean
public static GlobalAuthenticationConfigurerAdapter
        enableGlobalAuthenticationAutowiredConfigurer(
                                    ApplicationContext context) {
    return new EnableGlobalAuthenticationAutowiredConfigurer(context);
}
@Bean
public static InitializeUserDetailsBeanManagerConfigurer
 initializeUserDetailsBeanManagerConfigurer(ApplicationContext context) {
    return new InitializeUserDetailsBeanManagerConfigurer(context);
}
@Bean
public static InitializeAuthenticationProviderBeanManagerConfigurer
 initializeAuthenticationProviderBeanManagerConfigurer(ApplicationContext
                                                    context) {
    return
```

```
            new InitializeAuthenticationProviderBeanManagerConfigurer(context);
    }
    public AuthenticationManager getAuthenticationManager() throws Exception {
        if (this.authenticationManagerInitialized) {
            return this.authenticationManager;
        }
        AuthenticationManagerBuilder authBuilder =
         this.applicationContext.getBean(AuthenticationManagerBuilder.class);
        if (this.buildingAuthenticationManager.getAndSet(true)) {
            return new AuthenticationManagerDelegator(authBuilder);
        }
        for (GlobalAuthenticationConfigurerAdapter config :
                                                globalAuthConfigurers) {
            authBuilder.apply(config);
        }
        authenticationManager = authBuilder.build();
        if (authenticationManager == null) {
            authenticationManager = getAuthenticationManagerBean();
        }
        this.authenticationManagerInitialized = true;
        return authenticationManager;
    }
```

（1）首先定义了一个 AuthenticationManagerBuilder 实例，目的是为了构建全局的 AuthenticationManager 对象，这个过程中会从 Spring 容器中查找 AuthenticationEventPublisher 实例设置给 AuthenticationManagerBuilder 对象。

（2）接下来构建了三个 Bean，这三个 Bean 的作用在 4.1.5 小节中已经介绍过了，这里就不再赘述了。

（3）getAuthenticationManager 方法则用来构建具体的 AuthenticationManager 对象，在该方法内部，会首先判断 AuthenticationManager 对象是否已经初始化，如果已经初始化，则直接返回 AuthenticationManager 对象，否则就先从 Spring 容器中获取到 AuthenticationManagerBuilder 对象。注意这里还多了一个 AuthenticationManagerDelegator 对象，这个主要是为了防止在初始化 AuthenticationManager 时进行无限递归。拿到 authBuilder 对象之后，接下来遍历 globalAuthConfigurers 配置类集合（也就是第二点中所说的三个配置类），将配置类分别添加到 authBuilder 对象中，然后进行构建，最终将构建结果返回。

这是全局 AuthenticationManager 的构建过程。

整体来说，AuthenticationConfiguration 的作用主要体现在两方面：第一就是导入了 ObjectPostProcessorConfiguration 配置类；第二则是提供了一个全局的 AuthenticationManager 对象。

如果开发者在自定义配置类中重写了 configure(AuthenticationManagerBuilder)方法，这里的全局 AuthenticationManager 对象将不会生效，而大部分情况下，开发者都会重写 configure(AuthenticationManagerBuilder)方法。

至此，Spring Security 初始化就讲解完了。然而这里的架构复杂，概念繁多，可能有读者看完之后还是理解不到位，因此，接下来我们将通过几个不同的案例，展示前面这些组件的不同用法，加深大家对 Spring Security 基础组件的理解。

4.2　ObjectPostProcessor 使用

前面介绍了 ObjectPostProcessor 的基本概念。相信读者已经明白，所有的过滤器都由对应的配置类来负责创建，配置类在将过滤器创建成功之后，会调用父类的 postProcess 方法，该方法最终会调用到 CompositeObjectPostProcessor 对象的 postProcess 方法，在该方法中，会遍历 CompositeObjectPostProcessor 对象所维护的 List 集合中存储的所有 ObjectPostProcessor 对象，并调用其 postProcess 方法对对象进行后置处理。默认情况下，CompositeObjectPostProcessor 对象中所维护的 List 集合中只有一个对象那就是 AutowireBeanFactoryObjectPostProcessor，调用 AutowireBeanFactoryObjectPostProcessor 的 postProcess 方法可以将对象注册到 Spring 容器中去。

开发者可以自定义 ObjectPostProcessor 对象，并添加到 CompositeObjectPostProcessor 所维护的 List 集合中，此时，当一个过滤器在创建成功之后，就会被两个对象后置处理器处理，第一个是默认的对象后置处理器，负责将对象注册到 Spring 容器中；第二个是我们自定义的对象后置处理器，可以完成一些个性化配置。

自定义 ObjectPostProcessor 对象比较典型的用法是动态权限配置（权限管理将在后续章节具体介绍），为了便于大家理解，笔者这里先通过一个大家熟悉的案例来展示 ObjectPostProcessor 的用法，后面在配置动态权限时，ObjectPostProcessor 的使用思路是一样的。

```java
@Configuration
public class SecurityConfig extends WebSecurityConfigurerAdapter {
    @Override
    protected void configure(HttpSecurity http) throws Exception {
        http.authorizeRequests()
                .anyRequest().authenticated()
                .and()
                .formLogin()
                .withObjectPostProcessor(new
            ObjectPostProcessor<UsernamePasswordAuthenticationFilter>() {
                @Override
                public <O extends UsernamePasswordAuthenticationFilter> O
                                                postProcess(O object) {
                    object.setUsernameParameter("name");
                    object.setPasswordParameter("passwd");
                    object.setAuthenticationSuccessHandler((req,resp,auth)->{
                        resp.getWriter().write("login success");
```

```
                });
                return object;
            }
        })
        .and()
        .csrf().disable();
    }
}
```

在这个案例中,调用 formLogin 方法之后,开启了 FormLoginConfigurer 的配置,FormLogin Configurer 的作用是为了配置 UsernamePasswordAuthenticationFilter 过滤器,在 formLogin 方法执行完毕后,我们调用 withObjectPostProcessor 方法对 UsernamePasswordAuthenticationFilter 过滤器进行二次处理,修改登录参数的 key,将登录用户名参数的 key 改为 name,将登录密码参数的 key 改为 passwd,同时配置一个登录成功的处理器。

方便起见,这里就不创建登录页面了,我们直接使用 POSTMAN 工具来进行登录接口测试,测试结果如图 4-7 所示。

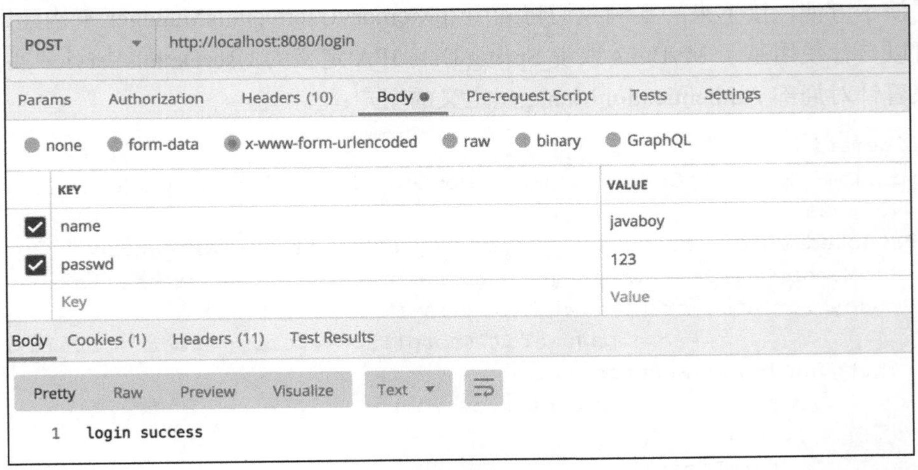

图 4-7 登录接口测试

> **提 示**
>
> 我们在第 2 章中介绍过 form 表单登录配置,实际项目中以第 2 章介绍的配置方式为准,这里主要是演示 ObjectPostProcessor 的用法。

4.3 多种用户定义方式

在前面的章节中,我们定义用户主要是两种方式:

(1)第一种方式是 2.4 节中使用的重写 configure(AuthenticationManagerBuilder)方法的方

式。

（2）第二种方式是 3.2 节中定义多个数据源时，我们直接向 Spring 容器中注入了 UserDetailsService 对象。

那么这两种用户定义方式有什么区别？

根据前面的源码分析可知，在 Spring Security 中存在两种类型的 AuthenticationManager，一种是全局的 AuthenticationManager，另一种则是局部的 AuthenticationManager。局部的 AuthenticationManager 由 HttpSecurity 进行配置，而全局的 AuthenticationManager 可以不用配置，系统会默认提供一个全局的 AuthenticationManager 对象，也可以通过重写 configure(AuthenticationManagerBuilder)方法进行全局配置。

当进行用户身份验证时，首先会通过局部的 AuthenticationManager 对象进行验证，如果验证失败，则会调用其 parent 也就是全局的 AuthenticationManager 再次进行验证。

所以开发者在定义用户时，也分为两种情况，一种是针对局部 AuthenticationManager 定义的用户，另一种则是针对全局 AuthenticationManager 定义的用户。

为了演示方便，接下来的案例我们将采用 InMemoryUserDetailsManager 来构建用户对象，读者也可以自行使用基于 MyBatis 或者 Spring Data JPA 定义的 UserDetailsService 实例。

先来看针对局部 AuthenticationManager 定义的用户：

```java
@Configuration
public class SecurityConfig extends WebSecurityConfigurerAdapter {
    @Override
    protected void configure(HttpSecurity http) throws Exception {
        InMemoryUserDetailsManager users = new InMemoryUserDetailsManager();
        users.createUser(User.withUsername("javaboy")
                        .password("{noop}123").roles("admin").build());
        http.authorizeRequests()
                .anyRequest().authenticated()
                .and()
                .formLogin()
                .permitAll()
                .and()
                .userDetailsService(users)
                .csrf().disable();
    }
}
```

在上面这段代码中，我们基于内存来管理用户，并向 users 中添加了一个用户，将配置好的 users 对象添加到 HttpSecurity 中，也就是配置到局部的 AuthenticationManager 中。

配置完成后，启动项目。项目启动成功后，我们就可以使用 javaboy/123 来登录系统了。

但是读者注意，当我们启动项目时，在 IDEA 控制台输出的日志中可以看到如下内容：

```
Using generated security password: cfc7f8b5-8346-492e-b25c-90c2c4501350
```

通过第 2 章的介绍可知，这个是系统自动生成的用户，那么我们是否可以使用系统自动生成的用户进行登录呢？答案是可以的，为什么呢？

回顾 2.1.2.1 小节，系统自动提供的用户对象实际上就是往 Spring 容器中注册了一个 InMemoryUserDetailsManager 对象。而在前面的代码中，我们没有重写 configure(AuthenticationManagerBuilder)方法，这意味着全局的 AuthenticationManager 是通过 AuthenticationConfiguration#getAuthenticationManager 方法自动生成的，在生成的过程中，会从 Spring 容器中查找对应的 UserDetailsService 实例进行配置（具体配置在 InitializeUserDetailsManagerConfigurer 类中）。所以系统自动提供的用户实际上相当于是全局 AuthenticationManager 对应的用户。

以上面的代码为例，当我们开始执行登录后，Spring Security 首先会调用局部 AuthenticationManager 去进行登录校验，如果登录的用户名/密码是 javaboy/123，那就直接登录成功，否则登录失败。当登录失败后，会继续调用局部 AuthenticationManager 的 parent 继续进行校验，此时如果登录的用户名/密码是 user/cfc7f8b5-8346-492e-b25c-90c2c4501350，则登录成功，否则登录失败。

这是针对局部 AuthenticationManager 定义的用户，我们也可以将定义的用户配置给全局的 AuthenticationManager，由于默认的全局 AuthenticationManager 在配置时会从 Spring 容器中查找 UserDetailsService 实例，所以我们如果针对全局 AuthenticationManager 配置用户，只需要往 Spring 容器中注入一个 UserDetailsService 实例即可，代码如下：

```java
@Configuration
public class SecurityConfig extends WebSecurityConfigurerAdapter {
    @Bean
    UserDetailsService us() {
        InMemoryUserDetailsManager users = new InMemoryUserDetailsManager();
        users.createUser(User.withUsername("江南一点雨")
                .password("{noop}123").roles("admin").build());
        return users;
    }
    @Override
    protected void configure(HttpSecurity http) throws Exception {
        InMemoryUserDetailsManager users = new InMemoryUserDetailsManager();
        users.createUser(User.withUsername("javaboy")
                .password("{noop}123").roles("admin").build());
        http.authorizeRequests()
                .anyRequest().authenticated()
                .and()
                .formLogin()
                .permitAll()
                .and()
                .userDetailsService(users)
                .csrf().disable();
    }
}
```

配置完成后，当我们启动项目时，全局的 AuthenticationManager 在配置时会去 Spring 容器中查找 UserDetailsService 实例，找到的就是我们自定义的 UserDetailsService 实例。当我们进行登录时，系统拿着我们输入的用户名/密码，首先和 javaboy/123 进行匹配，如果匹配不上的话，再去和江南一点雨/123 进行匹配。

当然，开发者也可以不使用 Spring Security 提供的默认的全局 AuthenticationManager 对象，而是通过重写 configure(AuthenticationManagerBuilder) 方法来自定义全局 Authentication Manager 对象：

```java
@Configuration
public class SecurityConfig extends WebSecurityConfigurerAdapter {
    @Override
    protected void configure(AuthenticationManagerBuilder auth)
                                                    throws Exception {
        auth.inMemoryAuthentication().withUser("javagirl")
                .password("{noop}123")
                .roles("admin");
    }
    @Override
    protected void configure(HttpSecurity http) throws Exception {
        InMemoryUserDetailsManager users = new InMemoryUserDetailsManager();
        users.createUser(User.withUsername("javaboy")
                .password("{noop}123").roles("admin").build());
        http.authorizeRequests()
                .anyRequest().authenticated()
                .and()
                .formLogin()
                .permitAll()
                .and()
                .userDetailsService(users)
                .csrf().disable();
    }
}
```

根据 4.1.5 小节中对 WebSecurityConfigurerAdapter 的源码分析可知，一旦我们重写了 configure(AuthenticationManagerBuilder)方法，则全局的 AuthenticationManager 对象将不再通过 AuthenticationConfiguration#getAuthenticationManager 方法来构建，而是通过 WebSecurityConfigurerAdapter 中的 localConfigureAuthenticationBldr 变量来构建，该变量也是我们重写的 configure(AuthenticationManagerBuilder)方法的参数。

配置完成后，当我们启动项目时，全局的 AuthenticationManager 在构建时会直接使用 configure(AuthenticationManagerBuilder)方法的 auth 变量去构建，使用的用户也是我们配置给 auth 变量的用户。当我们进行登录时，系统会将所输入的用户名/密码，首先和 javaboy/123 进行匹配，如果匹配不上的话，再去和 javagirl/123 进行匹配。

需要注意的是，一旦重写了 configure(AuthenticationManagerBuilder)方法，那么全局 AuthenticationManager 对象中使用的用户，将以 configure(AuthenticationManagerBuilder)方法中定义的用户为准。此时，如果我们还向 Spring 容器中注入了另外一个 UserDetailsService 实例，

那么该实例中定义的用户将不会生效（因为 AuthenticationConfiguration#getAuthenticationManager 方法没有被调用）。

这就是 Spring Security 中几种不同的用户定义方式，相信通过这几个案例，读者对于全局 AuthenticationManager 和局部 AuthenticationManager 对象会有更加深刻的理解。

4.4 定义多个过滤器链

在 Spring Security 中可以同时存在多个过滤器链，一个 WebSecurityConfigurerAdapter 的实例就可以配置一条过滤器链。

我们来看如下一个案例：

```java
@Configuration
public class SecurityConfig {
    @Bean
    UserDetailsService us() {
        InMemoryUserDetailsManager users = new InMemoryUserDetailsManager();
        users.createUser(User.withUsername("javaboy")
                        .password("{noop}123").roles("admin").build());
        return users;
    }
    @Configuration
    @Order(1)
    static class SecurityConfig01 extends WebSecurityConfigurerAdapter {
        @Override
        protected void configure(HttpSecurity http) throws Exception {
            InMemoryUserDetailsManager users =
                                    new InMemoryUserDetailsManager();
            users.createUser(User.withUsername("bar")
                            .password("{noop}123").roles("admin").build());
            http.antMatcher("/bar/**")
                    .authorizeRequests()
                    .anyRequest().authenticated()
                    .and()
                    .formLogin()
                    .loginProcessingUrl("/bar/login")
                    .successHandler((req, resp, auth) -> {
                      resp.setContentType("application/json;charset=utf-8");
                        String s = new ObjectMapper().writeValueAsString(auth);
                        resp.getWriter().write(s);
                    })
                    .permitAll()
                    .and()
                    .csrf().disable()
                    .userDetailsService(users);
```

```
        }
    }
    @Configuration
    @Order(2)
    static class SecurityConfig02 extends WebSecurityConfigurerAdapter {
        @Override
        protected void configure(AuthenticationManagerBuilder auth)
                                                    throws Exception {
            auth.inMemoryAuthentication().withUser("javagirl")
                    .password("{noop}123")
                    .roles("admin");
        }
        @Override
        protected void configure(HttpSecurity http) throws Exception {
            InMemoryUserDetailsManager users =
                                    new InMemoryUserDetailsManager();
            users.createUser(User.withUsername("foo")
                    .password("{noop}123").roles("admin").build());
            http.antMatcher("/foo/**")
                    .authorizeRequests()
                    .anyRequest().authenticated()
                    .and()
                    .formLogin()
                    .loginProcessingUrl("/foo/login")
                    .successHandler((req, resp, auth) -> {
                        resp.setContentType("application/json;charset=utf-8");
                        String s = new ObjectMapper().writeValueAsString(auth);
                        resp.getWriter().write(s);
                    })
                    .permitAll()
                    .and()
                    .csrf().disable()
                    .userDetailsService(users);
        }
    }
}
```

在 SecurityConfig 中分别定义两个静态内部类 SecurityConfig01 和 SecurityConfig02,两个配置类都继承自 WebSecurityConfigurerAdapter,可以分别配置一条过滤器链。

先来看 Security01。在 Security01 中,我们设置过滤器链的拦截规则是/bar/**,即如果请求路径是/bar/**格式的,则进入到 Security01 的过滤器链中进行处理。同时我们配置了局部 AuthenticationManager 对应的用户是 bar/123,由于没有重写 configure(AuthenticationManagerBuilder)方法,所以注册到 Spring 容器中的 UserDetailsService 将作为局部 AuthenticationManager 的 parent 对应的用户。换句话说,如果登录的路径是/bar/login,那么开发者可以使用 bar/123 和 javaboy/123 两个用户进行登录。登录效果如图 4-8 所示(注意登录路径是/bar/login)。

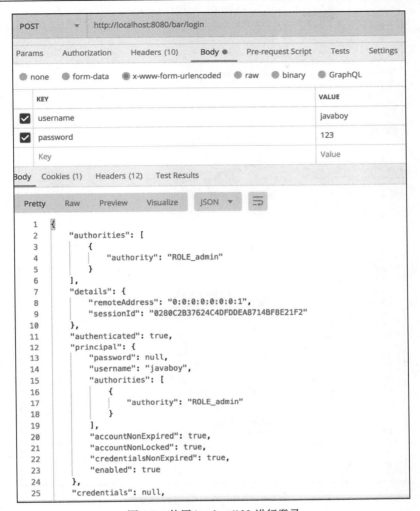

图 4-8　使用 javaboy/123 进行登录

再来看 SecurityConfig02。在 Security02 中，我们设置过滤器链的拦截规则是/foo/**，即如果请求路径是/foo/**格式的，则进入到 Security02 的过滤器链中进行处理，同时我们配置了局部 AuthenticationManager 对应的用户是 foo/123，由于重写了 configure(AuthenticationManagerBuilder)方法，在该方法中定义了局部 AuthenticationManager 的 parent 对应的用户，此时注册到 Spring 容器中的 UserDetailsService 实例对于/foo/**过滤器链不再生效。换句话说，如果登录路径是/foo/login，开发者可以使用 foo/123 和 javagirl/123 两个用户进行登录，而不可以使用 javaboy/123 进行登录。登录效果如图 4-9 所示（注意登录路径是/foo/login）。

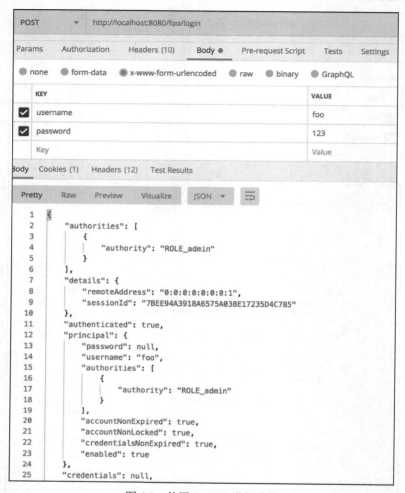

图 4-9　使用 foo/123 进行登录

需要注意的是，如果配置了多个过滤器链，需要使用 @Order 注解来标记不同配置的优先级（即不同过滤器链的优先级），数字越大优先级越低。当请求到来时，会按照过滤器链的优先级从高往低，依次进行匹配。

4.5　静态资源过滤

在一个实际项目中，并非所有的请求都需要经过 Spring Security 过滤器，有一些特殊的请求，例如静态资源等，一般来说并不需要经过 Spring Security 过滤器链，用户如果访问这些静态资源，直接返回对应的资源即可。

回顾 4.1.3 小节中关于 WebSecurity 的讲解，提到它里边维护了一个 ignoredRequests 变量，该变量记录的就是所有需要被忽略的请求，这些被忽略的请求将不再经过 Spring Security 过滤器。例如，笔者的静态资源目录结构如图 4-10 所示。

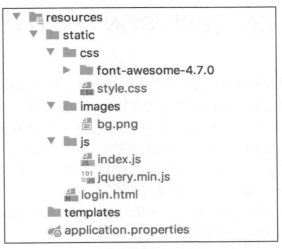

图 4-10　静态资源目录结构

现在这些静态资源的访问不需要经过 Spring Security 过滤器，具体配置方案如下：

```
@Configuration
public class SecurityConfig extends WebSecurityConfigurerAdapter {
    @Override
    public void configure(WebSecurity web) throws Exception {
        web.ignoring()
            .antMatchers("/login.html", "/css/**", "/js/**","/images/**");
    }
    @Override
    protected void configure(HttpSecurity http) throws Exception {
        http.authorizeRequests()
                .anyRequest().authenticated()
                .and()
                .formLogin()
                .and()
                .csrf().disable();
    }
}
```

重写 configure(WebSecurity)方法，并配置需要忽略的请求，这些需要忽略的地址，最终都会被添加到 ignoredRequests 集合中，并最终以过滤器链的形式呈现出来。换句话说，上面的配置中一共包含了五个过滤器链：configure(WebSecurity)方法中配置的四个以及 HttpSecurity 中配置的一个（即/**）。如果大家不能理解为什么会有五个过滤器链，可以回顾 4.1.3 小节中关于 WebSecurity 的分析以及 4.1.4 小节中关于 FilterChainProxy 的分析，这里不再赘述。

配置完成后，再次启动项目，此时不需要认证就可以访问/login.html 页面。

4.6 使用 JSON 格式登录

Spring Security 中默认的登录参数传递格式是 key/value 形式，也就是表单登录格式，在实际项目中，我们可能会通过 JSON 格式来传递登录参数，这就需要我们自定义登录过滤器来实现。

通过 3.1.4 小节的介绍，大家已经明白登录参数的提取是在 UsernamePasswordAuthenticationFilter 过滤器中完成的。如果我们要使用 JSON 格式登录，只需要模仿 UsernamePasswordAuthenticationFilter 过滤器定义自己的过滤器，再将自定义的过滤器放到 UsernamePasswordAuthenticationFilter 过滤器所在的位置即可。

思路理清了，我们来看代码实现。首先我们自定义一个 LoginFilter 继承自 UsernamePasswordAuthenticationFilter，代码如下：

```java
public class LoginFilter extends UsernamePasswordAuthenticationFilter {
    @Override
    public Authentication attemptAuthentication(HttpServletRequest request,
            HttpServletResponse response) throws AuthenticationException {
        if (!request.getMethod().equals("POST")) {
            throw new AuthenticationServiceException(
                    "Authentication method not supported: " +
                                                        request.getMethod());
        }
        if (request.getContentType()
                    .equalsIgnoreCase(MediaType.APPLICATION_JSON_VALUE)
            || request.getContentType()
                    .equalsIgnoreCase(MediaType.APPLICATION_JSON_UTF8_VALUE)) {
            Map<String, String> userInfo = new HashMap<>();
            try {
                userInfo =
        new ObjectMapper().readValue(request.getInputStream(), Map.class);
                String username = userInfo.get(getUsernameParameter());
                String password = userInfo.get(getPasswordParameter());
                UsernamePasswordAuthenticationToken authRequest =
                    new UsernamePasswordAuthenticationToken(username, password);
                setDetails(request, authRequest);
                return this.getAuthenticationManager()
                                            .authenticate(authRequest);
            } catch (IOException e) {
                e.printStackTrace();
            }
        }
        return super.attemptAuthentication(request, response);
    }
}
```

（1）首先确保进入该过滤器中的请求是 POST 请求。

（2）根据请求的 content-type 来判断参数是 JSON 格式的还是 key/value 格式的，如果是 JSON 格式的，则直接在当前方法中处理；如果是 key/value 格式的，那直接调用父类的 attemptAuthentication 方法处理即可。

（3）如果请求参数是 JSON 格式，则首先利用 jackson 提供的 ObjectMapper 工具，将输入流转为 Map 对象，然后从 Map 对象中分别提取出用户名/密码信息并构造出 UsernamePasswordAuthenticationToken 对象，然后调用 AuthenticationManager 的 authenticate 方法执行认证操作。

其实 LoginFilter 中，从请求中提取出 JSON 参数之后的认证逻辑和父类 UsernamePasswordAuthenticationFilter 中的认证逻辑是一致的，读者可以回顾第 3 章中关于 UsernamePasswordAuthenticationFilter 的分析。

LoginFilter 定义完成后，接下来我们将其添加到 Spring Security 过滤器链中，代码如下：

```java
@Configuration
public class SecurityConfig extends WebSecurityConfigurerAdapter {
    @Override
    protected void configure(AuthenticationManagerBuilder auth)
                                                    throws Exception {
        auth.inMemoryAuthentication()
                .withUser("javaboy")
                .password("{noop}123")
                .roles("admin");
    }
    @Override
    @Bean
    public AuthenticationManager authenticationManagerBean()
                                                    throws Exception {
        return super.authenticationManagerBean();
    }
    @Bean
    LoginFilter loginFilter() throws Exception {
        LoginFilter loginFilter = new LoginFilter();
        loginFilter.setAuthenticationManager(authenticationManagerBean());
        loginFilter.setAuthenticationSuccessHandler((req, resp, auth) -> {
            resp.setContentType("application/json;charset=utf-8");
            PrintWriter out = resp.getWriter();
            out.write(new ObjectMapper().writeValueAsString(auth));
        });
        return loginFilter;
    }
    @Override
    protected void configure(HttpSecurity http) throws Exception {
        http.authorizeRequests()
                .anyRequest().authenticated()
```

```
                .and()
                .formLogin()
                .and()
                .csrf().disable();
        http.addFilterAt(loginFilter(),
                        UsernamePasswordAuthenticationFilter.class);
    }
}
```

（1）首先重写 configure 方法来定义一个登录用户。

（2）重写父类的 authenticationManagerBean 方法来提供一个 AuthenticationManager 实例，一会将配置给 LoginFilter。

（3）配置 loginFilter 实例，同时将 AuthenticationManager 实例设置给 loginFilter，然后再设置登录成功回调。当然，我们也可以在 loginFilter 中配置用户名/密码的参数名或者登录失败的回调。

（4）最后在 HttpSecurity 中，调用 addFilterAt 方法将 loginFilter 过滤器添加到 UsernamePasswordAuthenticationFilter 过滤器所在的位置。

配置完成后，重启项目，此时我们就可以使用 JSON 格式的数据来进行登录操作了，如图 4-11 所示。

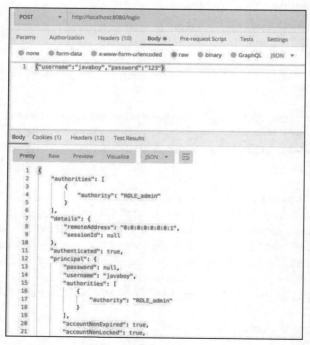

图 4-11　使用 JSON 格式登录

有读者可能会注意到，当我们想要获取一个 AuthenticationManager 实例时，有两种不同的方式，第一种方式是通过重写父类的 authenticationManager 方法获取，第二种则是通过重写父类的 authenticationManagerBean 方法获取。表面上看两种方式获取到的 AuthenticationManager 实例在这里都可以运行，但实际上是有区别的。区别在于第一种获取到的是全局的 AuthenticationManager 实例，而第二种获取到的是局部的 AuthenticationManager 实例，而 LoginFilter 作为过滤器链中的一环，显然应该配置局部的 AuthenticationManager 实例，因为如果将全局的 AuthenticationManager 实例配置给 LoginFilter，则局部 AuthenticationManager 实例所对应的用户就会失效，例如如下配置：

```java
@Configuration
public class SecurityConfig extends WebSecurityConfigurerAdapter {
    @Override
    protected void configure(AuthenticationManagerBuilder auth)
                                                    throws Exception {
        auth.inMemoryAuthentication()
                .withUser("javaboy")
                .password("{noop}123")
                .roles("admin");
    }
    @Override
    @Bean
    public AuthenticationManager authenticationManager() throws Exception {
        return super.authenticationManager();
    }
    @Bean
    LoginFilter loginFilter() throws Exception {
        LoginFilter loginFilter = new LoginFilter();
        loginFilter.setAuthenticationManager(authenticationManager());
        loginFilter.setAuthenticationSuccessHandler((req, resp, auth) -> {
            resp.setContentType("application/json;charset=utf-8");
            PrintWriter out = resp.getWriter();
            out.write(new ObjectMapper().writeValueAsString(auth));
        });
        return loginFilter;
    }

    @Override
    protected void configure(HttpSecurity http) throws Exception {
        InMemoryUserDetailsManager users = new InMemoryUserDetailsManager();
        users.createUser(User.withUsername("javagirl")
                            .password("{noop}123").roles("admin").build());
        http.authorizeRequests()
                .anyRequest().authenticated()
                .and()
                .formLogin()
                .and()
                .csrf().disable()
                .userDetailsService(users);
        http.addFilterAt(loginFilter(),
```

```
                    UsernamePasswordAuthenticationFilter.class);
        }
    }
```

在上面这段代码中,我们将无法使用 javagirl/123 进行登录,因为 LoginFilter 中指定了全局的 AuthenticationManager 来做验证,所以局部的 AuthenticationManager 实例失效了。

在实际应用中,如果需要自己配置一个 AuthenticationManager 实例,大部分情况下,我们都是通过重写 authenticationManagerBean 方法来获取。

4.7 添加登录验证码

在第 3 章中介绍了一种登录验证码的实现方案,我们通过学习了 Spring Security 的过滤器链之后,可能也会发现,使用过滤器链来实现登录验证码更加容易。这里就介绍一下使用过滤器实现验证码的方案。

验证码的生成方案依然和 3.3 节中的一致,这里不再赘述,主要介绍一下 LoginFilter 的定义以及配置。先来看 LoginFilter 的定义:

```
public class LoginFilter extends UsernamePasswordAuthenticationFilter {
    @Override
    public Authentication attemptAuthentication(HttpServletRequest request,
            HttpServletResponse response) throws AuthenticationException {
        if (!request.getMethod().equals("POST")) {
            throw new AuthenticationServiceException(
                "Authentication method not supported: "
                                            + request.getMethod());
        }
        String kaptcha = request.getParameter("kaptcha");
        String sessionKaptcha =
                (String) request.getSession().getAttribute("kaptcha");
        if (!StringUtils.isEmpty(kaptcha)
                    && !StringUtils.isEmpty(sessionKaptcha)
                    && kaptcha.equalsIgnoreCase(sessionKaptcha)) {
            return super.attemptAuthentication(request, response);
        }
        throw new AuthenticationServiceException("验证码输入错误");
    }
}
```

在 LoginFilter 中首先判断验证码是否正确,如果验证码输入错误,则直接抛出异常;如果验证码输入正确,则调用父类的 attemptAuthentication 方法进行登录校验。

接下来,在 SecurityConfig 中配置 LoginFilter:

```
@Configuration
public class SecurityConfig extends WebSecurityConfigurerAdapter {
```

```java
@Override
protected void configure(AuthenticationManagerBuilder auth)
        throws Exception {
    auth.inMemoryAuthentication()
            .withUser("javaboy")
            .password("{noop}123")
            .roles("admin");
}
@Override
@Bean
public AuthenticationManager authenticationManagerBean()
        throws Exception {
    return super.authenticationManagerBean();
}
@Bean
LoginFilter loginFilter() throws Exception {
    LoginFilter loginFilter = new LoginFilter();
    loginFilter.setFilterProcessesUrl("/doLogin");
    loginFilter.setAuthenticationManager(authenticationManagerBean());
    loginFilter.setAuthenticationSuccessHandler(new
                SimpleUrlAuthenticationSuccessHandler("/hello"));
    loginFilter.setAuthenticationFailureHandler(new
            SimpleUrlAuthenticationFailureHandler("/mylogin.html"));
    return loginFilter;
}
@Override
protected void configure(HttpSecurity http) throws Exception {
    http.authorizeRequests()
            .antMatchers("/vc.jpg").permitAll()
            .anyRequest().authenticated()
            .and()
            .formLogin()
            .loginPage("/mylogin.html")
            .permitAll()
            .and()
            .csrf().disable();
    http.addFilterAt(loginFilter(),
            UsernamePasswordAuthenticationFilter.class);
}
```

  这里的配置基本和 4.6 节中的配置一致。不同的是，我们修改了登录请求的处理地址，注意这个地址要在 LoginFilter 实例上配置。

  这里介绍的第二种添加登录验证码的方式，相比于第一种方式，这种验证码的添加方式更简单也更易于理解。

4.8 小　结

本章主要分析了 Spring Security 中的过滤器链以及初始化过程。Spring Security 初始化过程的理解对于开发者来说非常重要，这主要体现在两方面：第一，理解 Spring Security 初始化流程之后，开发者就可以随心所欲地根据自己的项目需求去定制 Spring Security；第二，如果在 Spring Security 使用的过程中出现了问题，理解了初始化流程之后，问题的排查也会变得非常容易。同时，为了帮助大家更好地理解 Spring Security 初始化流程，我们还通过六个案例演示了 Spring Security 中一些常见的个性化配置，以便读者更好地掌握 Spring Security。

第 5 章

密码加密

在前面的案例中，凡是涉及密码的地方，我们都采用明文存储，在实际项目中这肯定是不可取的，因为这会带来极高的安全风险。在企业级应用中，密码不仅需要加密，还需要加"盐"（盐的意思参见 5.2 节），最大程度地保证密码安全。本章我们就来详细谈一谈密码加密问题，学完本章之后，大家就会明白我们前面一直在使用的 {noop} 是什么意思了。

本章涉及的主要知识点有：

- 密码为什么要加密。
- 常见的密码加密方案。
- PasswordEncoder 详解。
- 加密方案自动升级。

5.1 密码为什么要加密

2011 年 12 月 21 日，有人在网络上公开了一个包含 600 万个 CSDN 用户资料的数据库，数据全部为明文储存，包含用户名、密码以及注册邮箱。事件发生后 CSDN 在微博、官方网站等渠道发出了声明，解释说此数据库系 2009 年备份所用，因不明原因泄漏，已经向警方报案，后又在官网发出了公开道歉信。在接下来的十多天里，金山、网易、京东、当当、新浪等多家公司被卷入到这次事件中。整个事件中最触目惊心的莫过于 CSDN 把用户密码明文存储，由于很多用户是多个网站共用一个密码，因此一个网站密码泄漏就会造成很大的安全隐患。由于有了这么多前车之鉴，我们现在做系统时，密码都要加密处理。

5.2 密码加密方案进化史

最早我们使用类似 SHA-256 这样的单向 Hash 算法。用户注册成功后，保存在数据库中的不再是用户的明文密码，而是经过 SHA-256 加密计算的一个字符串，当用户进行登录时，将用户输入的明文密码用 SHA-256 进行加密，加密完成之后，再和存储在数据库中的密码进行比对，进而确定用户登录信息是否有效。如果系统遭遇攻击，最多也只是存储在数据库中的密文被泄漏。

这样就绝对安全了吗？当然不是的。彩虹表是一个用于加密 Hash 函数逆运算的表，通常用于破解加密过的 Hash 字符串。为了降低彩虹表对系统安全性的影响，人们又发明了密码加"盐"，之前是直接将密码作为明文进行加密，现在再添加一个随机数（即盐）和密码明文混合在一起进行加密，这样即使密码明文相同，生成的加密字符串也是不同的。当然，这个随机数也需要以明文形式和密码一起存储在数据库中。当用户需要登录时，拿到用户输入的明文密码和存储在数据库中的盐一起进行 Hash 运算，再将运算结果和存储在数据库中的密文进行比较，进而确定用户的登录信息是否有效。

密码加盐之后，彩虹表的作用就大打折扣了，因为唯一的盐和明文密码总会生成唯一的 Hash 字符。

然而，随着计算机硬件的发展，每秒执行数十亿次 Hash 计算已经变得轻轻松松，这意味着即使给密码加密加盐也不再安全。

在 Spring Security 中，我们现在是用一种自适应单向函数（Adaptive One-way Functions）来处理密码问题，这种自适应单向函数在进行密码匹配时，会有意占用大量系统资源（例如 CPU、内存等），这样可以增加恶意用户攻击系统的难度。在 Spring Security 中，开发者可以通过 bcrypt、PBKDF2、scrypt 以及 argon2 来体验这种自适应单向函数加密。

由于自适应单向函数有意占用大量系统资源，因此每个登录认证请求都会大大降低应用程序的性能，但是 Spring Security 不会采取任何措施来提高密码验证速度，因为它正是通过这种方式来增强系统的安全性。当然，开发者也可以将用户名/密码这种长期凭证兑换为短期凭证，如会话、OAuth2 令牌等，这样既可以快速验证用户凭证信息，又不会损失系统的安全性。

5.3 PasswordEncoder 详解

Spring Security 中通过 PasswordEncoder 接口定义了密码加密和比对的相关操作：

```
public interface PasswordEncoder {
    String encode(CharSequence rawPassword);
    boolean matches(CharSequence rawPassword, String encodedPassword);
```

```
    default boolean upgradeEncoding(String encodedPassword) {
        return false;
    }
}
```

可以看到，PasswordEncoder 接口中一共有三个方法：

（1）encode：该方法用来对明文密码进行加密。
（2）matches：该方法用来进行密码比对。
（3）upgradeEncoding：该方法用来判断当前密码是否需要升级，默认返回 false 表示不需要升级。

针对密码的所有操作，PasswordEncoder 接口中都定义好了，不同的实现类将采用不同的密码加密方案对密码进行处理。

5.3.1 PasswordEncoder 常见实现类

BCryptPasswordEncoder

BCryptPasswordEncoder 使用 bcrypt 算法对密码进行加密，为了提高密码的安全性，bcrypt 算法故意降低运行速度，以增强密码破解的难度。同时 BCryptPasswordEncoder "为自己带盐"，开发者不需要额外维护一个"盐"字段，使用 BCryptPasswordEncoder 加密后的字符串就已经"带盐"了，即使相同的明文每次生成的加密字符串都不相同。

BCryptPasswordEncoder 的默认强度为 10，开发者可以根据自己的服务器性能进行调整，以确保密码验证时间约为 1 秒钟（官方建议密码验证时间为 1 秒钟，这样既可以提高系统安全性，又不会过多影响系统运行性能）。

Argon2PasswordEncoder

Argon2PasswordEncoder 使用 Argon2 算法对密码进行加密，Argon2 曾在 Password Hashing Competition 竞赛中获胜。为了解决在定制硬件上密码容易被破解的问题，Argon2 也是故意降低运算速度，同时需要大量内存，以确保系统的安全性。

Pbkdf2PasswordEncoder

Pbkdf2PasswordEncoder 使用 PBKDF2 算法对密码进行加密，和前面几种类似，PBKDF2 算法也是一种故意降低运算速度的算法，当需要 FIPS（Federal Information Processing Standard，美国联邦信息处理标准）认证时，PBKDF2 算法是一个很好的选择。

SCryptPasswordEncoder

SCryptPasswordEncoder 使用 scrypt 算法对密码进行加密，和前面的几种类似，scrypt 也是一种故意降低运算速度的算法，而且需要大量内存。

这四种就是我们前面所说的自适应单向函数加密。除了这几种，还有一些基于消息摘要算法的加密方案，这些方案都已经不再安全，但是出于兼容性考虑，Spring Security 并未移除相关类，主要有 LdapShaPasswordEncoder、MessageDigestPasswordEncoder、Md4PasswordEncoder、StandardPasswordEncoder 以及 NoOpPasswordEncoder（密码明文存储），这五种皆已废弃，这里对这些类也不做过多介绍。

除了上面介绍的这几种之外，还有一个非常重要的密码加密工具类，那就是 DelegatingPasswordEncoder。

5.3.2 DelegatingPasswordEncoder

根据前文的介绍，读者可能会认为 Spring Security 中默认的密码加密方案应该是四种自适应单向加密函数中的一种，其实不然，在 Spring Security 5.0 之后，默认的密码加密方案其实是 DelegatingPasswordEncoder。

从名字上来看，DelegatingPasswordEncoder 是一个代理类，而并非一种全新的密码加密方案。DelegatingPasswordEncoder 主要用来代理上面介绍的不同的密码加密方案。为什么采用 DelegatingPasswordEncoder 而不是某一个具体加密方式作为默认的密码加密方案呢？主要考虑了如下三方面的因素：

（1）兼容性：使用 DelegatingPasswordEncoder 可以帮助许多使用旧密码加密方式的系统顺利迁移到 Spring Security 中，它允许在同一个系统中同时存在多种不同的密码加密方案。

（2）便捷性：密码存储的最佳方案不可能一直不变，如果使用 DelegatingPasswordEncoder 作为默认的密码加密方案，当需要修改加密方案时，只需要修改很小一部分代码就可以实现。

（3）稳定性：作为一个框架，Spring Security 不能经常进行重大更改，而使用 DelegatingPasswordEncoder 可以方便地对密码进行升级（自动从一个加密方案升级到另外一个加密方案）。

那么 DelegatingPasswordEncoder 到底是如何代理其他密码加密方案的？又是如何对加密方案进行升级的？我们就从 PasswordEncoderFactories 类开始看起，因为正是由它里边的静态方法 createDelegatingPasswordEncoder 提供了默认的 DelegatingPasswordEncoder 实例：

```java
public class PasswordEncoderFactories {
    public static PasswordEncoder createDelegatingPasswordEncoder() {
        String encodingId = "bcrypt";
        Map<String, PasswordEncoder> encoders = new HashMap<>();
        encoders.put(encodingId, new BCryptPasswordEncoder());
        encoders.put("ldap", new org.springframework.security.crypto
                                    .password.LdapShaPasswordEncoder());
        encoders.put("MD4", new org.springframework.security.crypto
                                    .password.Md4PasswordEncoder());
        encoders.put("MD5", new org.springframework.security.crypto
                        .password.MessageDigestPasswordEncoder("MD5"));
        encoders.put("noop", org.springframework.security.crypto.password
                                    .NoOpPasswordEncoder.getInstance());
        encoders.put("pbkdf2", new Pbkdf2PasswordEncoder());
```

```
        encoders.put("scrypt", new SCryptPasswordEncoder());
        encoders.put("SHA-1", new org.springframework.security.crypto
                        .password.MessageDigestPasswordEncoder("SHA-1"));
        encoders.put("SHA-256", new org.springframework.security.crypto
                        .password.MessageDigestPasswordEncoder("SHA-256"));
        encoders.put("sha256", new org.springframework.security.crypto
                                .password.StandardPasswordEncoder());
        encoders.put("argon2", new Argon2PasswordEncoder());
        return new DelegatingPasswordEncoder(encodingId, encoders);
    }
    private PasswordEncoderFactories() {}
}
```

可以看到，在 createDelegatingPasswordEncoder 方法中，首先定义了 encoders 变量，encoders 中存储了每一种密码加密方案的 id 和所对应的加密类，例如 bcrypt 对应着 BcryptPasswordEncoder、argon2 对应着 Argon2PasswordEncoder、noop 对应着 NoOpPasswordEncoder。

encoders 创建完成后，最终新建一个 DelegatingPasswordEncoder 实例，并传入 encodingId 和 encoders 变量，其中 encodingId 默认值为 bcrypt，相当于代理类中默认使用的加密方案是 BCryptPasswordEncoder。

我们来分析一下 DelegatingPasswordEncoder 类的源码，由于源码比较长，我们就先从它的属性开始看起：

```
public class DelegatingPasswordEncoder implements PasswordEncoder {
    private static final String PREFIX = "{";
    private static final String SUFFIX = "}";
    private final String idForEncode;
    private final PasswordEncoder passwordEncoderForEncode;
    private final Map<String, PasswordEncoder> idToPasswordEncoder;
    private PasswordEncoder defaultPasswordEncoderForMatches =
                                    new UnmappedIdPasswordEncoder();
}
```

（1）首先定义了前缀 PREFIX 和后缀 SUFFIX，用来包裹将来生成的加密方案的 id。

（2）idForEncode 表示默认的加密方案 id。

（3）passwordEncoderForEncode 表示默认的加密方案（BCryptPasswordEncoder），它的值是根据 idForEncode 从 idToPasswordEncoder 集合中提取出来的。

（4）idToPasswordEncoder 用来保存 id 和加密方案之间的映射。

（5）defaultPasswordEncoderForMatches 是指默认的密码比对器，当根据密码加密方案的 id 无法找到对应的加密方案时，就会使用默认的密码比对器。defaultPasswordEncoderForMatches 的默认类型是 UnmappedIdPasswordEncoder，在 UnmappedIdPasswordEncoder 的 matches 方法中并不会做任何密码比对操作，直接抛出异常。

（6）最后看到的 DelegatingPasswordEncoder 也是 PasswordEncoder 接口的子类，所以接下来我们就来重点分析 PasswordEncoder 接口中三个方法在 DelegatingPasswordEncoder 中的具

体实现。

首先来看 encode 方法：

```
@Override
public String encode(CharSequence rawPassword) {
    return PREFIX + this.idForEncode + SUFFIX
                + this.passwordEncoderForEncode.encode(rawPassword);
}
```

encode 方法的实现逻辑很简单，具体的加密工作还是由加密类来完成，只不过在密码加密完成后，给加密后的字符串加上一个前缀{id}，用来描述所采用的具体加密方案。因此，encode 方法加密出来的字符串格式类似如下形式：

```
{bcrypt}$2a$10$dXJ3SW6G7P50lGmMkkmwe.20cQQubK3.HZWzG3YB1tlRy.fqvM/BG
{noop}123
{pbkdf2}23b44a6d129f3ddf3e3c8d29412723dcbde72445e8ef6bf3b508fbf17fa4ed4
```

不同的前缀代表了后面的字符串采用了不同的加密方案。

再来看密码比对方法 matches：

```
@Override
public boolean matches(CharSequence rawPassword,
                      String prefixEncodedPassword) {
    if (rawPassword == null && prefixEncodedPassword == null) {
        return true;
    }
    String id = extractId(prefixEncodedPassword);
    PasswordEncoder delegate = this.idToPasswordEncoder.get(id);
    if (delegate == null) {
        return this.defaultPasswordEncoderForMatches
            .matches(rawPassword, prefixEncodedPassword);
    }
    String encodedPassword = extractEncodedPassword(prefixEncodedPassword);
    return delegate.matches(rawPassword, encodedPassword);
}
private String extractId(String prefixEncodedPassword) {
    if (prefixEncodedPassword == null) {
        return null;
    }
    int start = prefixEncodedPassword.indexOf(PREFIX);
    if (start != 0) {
        return null;
    }
    int end = prefixEncodedPassword.indexOf(SUFFIX, start);
    if (end < 0) {
        return null;
    }
    return prefixEncodedPassword.substring(start + 1, end);
```

在 matches 方法中，首先调用 extractId 方法从加密字符串中提取出具体的加密方案 id，也就是 {} 中的字符，具体的提取方式就是字符串截取。拿到 id 之后，再去 idToPasswordEncoder 集合中获取对应的加密方案，如果获取到的为 null，说明不存在对应的加密实例，那么就会采用默认的密码匹配器 defaultPasswordEncoderForMatches；如果根据 id 获取到了对应的加密实例，则调用其 matches 方法完成密码校验。

可以看到，这里的 matches 方法非常灵活，可以根据加密字符串的前缀，去查找到不同的加密方案，进而完成密码校验。同一个系统中，加密字符串可以使用不同的前缀而互不影响。

最后，我们再来看一下 DelegatingPasswordEncoder 中的密码升级方法 upgradeEncoding：

```java
@Override
public boolean upgradeEncoding(String prefixEncodedPassword) {
    String id = extractId(prefixEncodedPassword);
    if (!this.idForEncode.equalsIgnoreCase(id)) {
        return true;
    }
    else {
        String encodedPassword = extractEncodedPassword(prefixEncodedPassword);
        return this.idToPasswordEncoder.get(id)
                                .upgradeEncoding(encodedPassword);
    }
}
```

可以看到，如果当前加密字符串所采用的加密方案不是默认的加密方案（BcryptPasswordEncoder），就会自动进行密码升级，否则就调用默认加密方案的 upgradeEncoding 方法判断密码是否需要升级。

至此，我们将 Spring Security 中的整个加密体系向读者简单介绍了一遍，接下来我们通过几个实际的案例来看一下加密方案要怎么用。

5.4 实 战

首先创建一个普通的 Spring Boot 工程，并引入 spring-boot-starter-security 依赖。

创建一个测试接口 /hello，代码如下：

```java
@RestController
public class HelloController {
    @GetMapping("/hello")
    public String hello() {
        return "hello";
    }
}
```

为了测试方便，我们首先在单元测试中执行如下代码，生成一段加密字符串（读者可以多次执行该方法，会发现相同的明文每次生成的密文都不相同，这也是 BCryptPasswordEncoder 的便捷之处）：

```
@Test
void contextLoads() {
    BCryptPasswordEncoder encoder = new BCryptPasswordEncoder();
    System.out.println(encoder.encode("123"));
}
```

接下来自定义 SecurityConfig 类：

```
@Configuration
public class SecurityConfig extends WebSecurityConfigurerAdapter {
    @Bean
    PasswordEncoder passwordEncoder() {
        return new BCryptPasswordEncoder();
    }
    @Override
    protected void configure(AuthenticationManagerBuilder auth)
                                                    throws Exception {
        auth.inMemoryAuthentication()
                .withUser("javaboy")
                .password("$2a$10$XtBXprcqjT/sGPEOY5y1eurS.V.9U7/M5RD1i32k1uAhXQHK4//U6")
                .roles("admin");
    }
    @Override
    protected void configure(HttpSecurity http) throws Exception {
        http.authorizeRequests()
                .anyRequest().authenticated()
                .and()
                .formLogin()
                .and()
                .csrf().disable();
    }
}
```

（1）首先我们将一个 BCryptPasswordEncoder 实例注册到 Spring 容器中，这将代替默认的 DelegatingPasswordEncoder。

（2）在定义用户时，设置的密码字符串就是前面单元测试方法执行生成的加密字符串。

配置完成后，启动项目。项目启动成功后，我们就可以使用 javaboy/123 进行登录了。

另一方面，由于默认使用的是 DelegatingPasswordEncoder，所以也可以不配置 PasswordEncoder 实例，只在密码前加上前缀：

```
@Configuration
public class SecurityConfig extends WebSecurityConfigurerAdapter {
```

```
    @Override
    protected void configure(AuthenticationManagerBuilder auth)
                                                    throws Exception {
        auth.inMemoryAuthentication()
                .withUser("javaboy")
                .password("{bcrypt}$2a$10$XtBXprcqjT/sGPEOY5y1eurS.V.9U7/M5RD1i32k1uAhXQHK4//U6")
                .roles("admin")
                .and()
                .withUser("江南一点雨")
                .password("{noop}123")
                .roles("user");
    }
    @Override
    protected void configure(HttpSecurity http) throws Exception {
        http.authorizeRequests()
                .anyRequest().authenticated()
                .and()
                .formLogin()
                .and()
                .csrf().disable();
    }
}
```

如果我们不提供 PasswordEncoder 实例的话，默认使用的就是 DelegatingPasswordEncoder，所以需要给加密字符串加上前缀{bcrypt}，同时再添加一个新用户，新用户的密码是{noop}123，这个表示密码明文存储，密码就是 123。

配置完成后，再次重启项目，此时我们就可以使用 javaboy/123 和江南一点雨/123 两个用户进行登录了。

5.5　加密方案自动升级

使用 DelegatingPasswordEncoder 的另外一个好处就是会自动进行密码加密方案升级，这个功能在整合一些"老破旧"系统时非常有用。接下来，我们通过一个案例展示如何进行加密方案升级。

我们在 5.4 节的案例上继续完善。

首先创建一个 security05 数据库，向数据库中添加一张 user 表，并添加一条用户数据，在添加的用户数据中，用户密码是{noop}123：

```
DROP TABLE IF EXISTS `user`;

CREATE TABLE `user` (
  `id` int(11) unsigned NOT NULL AUTO_INCREMENT,
```

```
  `username` varchar(255) COLLATE utf8mb4_unicode_ci DEFAULT NULL,
  `password` varchar(255) COLLATE utf8mb4_unicode_ci DEFAULT NULL,
  PRIMARY KEY (`id`)
) ENGINE=InnoDB DEFAULT CHARSET=utf8mb4 COLLATE=utf8mb4_unicode_ci;

INSERT INTO `user` (`id`, `username`, `password`)
VALUES (1,'javaboy','{noop}123');
```

接下来，在项目中添加 MyBatis 依赖和 MySQL 数据库驱动依赖：

```
<dependency>
    <groupId>org.mybatis.spring.boot</groupId>
    <artifactId>mybatis-spring-boot-starter</artifactId>
    <version>2.1.3</version>
</dependency>
<dependency>
    <groupId>mysql</groupId>
    <artifactId>mysql-connector-java</artifactId>
</dependency>
```

然后在 application.properties 中配置数据库连接信息：

```
spring.datasource.username=root
spring.datasource.password=123
spring.datasource.url=jdbc:mysql:///security05?useUnicode=true&characterEncoding=UTF-8&serverTimezone=Asia/Shanghai
```

接下来创建 User 实体类，为了方便，这里只创建三个属性 id、username 以及 password，其他方法默认都返回 true 即可，代码如下：

```java
public class User implements UserDetails {
    private Long id;
    private String username;
    private String password;
    @Override
    public Collection<? extends GrantedAuthority> getAuthorities() {
        return null;
    }
    @Override
    public String getPassword() {
        return password;
    }
    @Override
    public String getUsername() {
        return username;
    }
    @Override
    public boolean isAccountNonExpired() {
        return true;
    }
```

```
    @Override
    public boolean isAccountNonLocked() {
        return true;
    }
    @Override
    public boolean isCredentialsNonExpired() {
        return true;
    }
    @Override
    public boolean isEnabled() {
        return true;
    }
    //省略其他 get/set
}
```

创建 UserService，代码如下：

```
@Configuration
public class UserService implements UserDetailsService,
                                    UserDetailsPasswordService {
    @Autowired
    UserMapper userMapper;
    @Override
    public UserDetails updatePassword(UserDetails user, String newPassword) {
        Integer result = userMapper
                    .updatePassword(user.getUsername(), newPassword);
        if (result == 1) {
            ((User) user).setPassword(newPassword);
        }
        return user;
    }
    @Override
    public UserDetails loadUserByUsername(String username)
                            throws UsernameNotFoundException {
        return userMapper.loadUserByUsername(username);
    }
}
```

和前面第 2 章中定义的 UserService 不同，这里的 UserService 多实现了一个接口 UserDetailsPasswordService，并实现了该接口中的 updatePassword 方法。当系统判断密码加密方案需要升级的时候，就会自动调用 updatePassword 方法去修改数据库中的密码。当数据库中的密码修改成功后，修改 user 对象中的 password 属性，并将 user 对象返回（回顾 3.1.2 小节中关于 DaoAuthenticationProvider 的讲解，在其 createSuccessAuthentication 方法中触发了密码加密方案自动升级）。

接下来在 UserMapper 中定义相关方法，代码如下：

```
@Mapper
public interface UserMapper {
```

```
    User loadUserByUsername(String username);
    Integer updatePassword(@Param("username") String username,
                           @Param("newPassword") String newPassword);
}
```

UserMapper.xml 定义如下:

```xml
<!DOCTYPE mapper
        PUBLIC "-//mybatis.org//DTD Mapper 3.0//EN"
        "http://mybatis.org/dtd/mybatis-3-mapper.dtd">
<mapper namespace="org.javaboy.passwordencoder.mapper.UserMapper">
    <select id="loadUserByUsername"
                    resultType="org.javaboy.passwordencoder.model.User">
        select * from user where username=#{username};
    </select>
    <update id="updatePassword">
        update user set password = #{newPassword} where username=#{username};
    </update>
</mapper>
```

最后,在 SecurityConfig 中配置 UserService 实例:

```java
@Configuration
public class SecurityConfig extends WebSecurityConfigurerAdapter {
    @Autowired
    UserService userService;
    @Override
    protected void configure(AuthenticationManagerBuilder auth)
                                                    throws Exception {
        auth.userDetailsService(userService);
    }
    @Override
    protected void configure(HttpSecurity http) throws Exception {
        http.authorizeRequests()
                .anyRequest().authenticated()
                .and()
                .formLogin()
                .and()
                .csrf().disable();
    }
}
```

配置完成后,我们启动项目。

在登录之前,数据库中用户信息如图 5-1 所示。

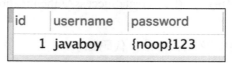

图 5-1 登录之前用户信息

接下来访问 http://localhost:8080/login 进行登录，登录成功之后，再去查看数据库，此时用户密码已经自动更新了，如图 5-2 所示。

图 5-2　自动更新后的用户信息

如果开发者使用了 DelegatingPasswordEncoder，只要数据库中存储的密码加密方案不是 DelegatingPasswordEncoder 中默认的 BCryptPasswordEncoder，在登录成功之后，都会自动升级为 BCryptPasswordEncoder 加密。

这就是加密方案的升级。

在同一种密码加密方案中，也有可能存在升级的情况。例如，开发者在创建 BcryptPasswordEncoder 实例时有一个强度参数 strength，该参数取值在 4~31 之间，默认值为 10。在图 5-2 中大家看到的加密字符串就是 strength 为 10 时生成的加密字符串。我们可以自己来修改 strength 参数，配置如下：

```java
@Configuration
public class SecurityConfig extends WebSecurityConfigurerAdapter {
    @Autowired
    UserService userService;
    @Bean
    PasswordEncoder passwordEncoder() {
        String encodingId = "bcrypt";
        Map<String, PasswordEncoder> encoders = new HashMap<>();
        encoders.put(encodingId, new BCryptPasswordEncoder(31));
        encoders.put("ldap", new LdapShaPasswordEncoder());
        encoders.put("MD4", new Md4PasswordEncoder());
        encoders.put("MD5", new MessageDigestPasswordEncoder("MD5"));
        encoders.put("noop", NoOpPasswordEncoder.getInstance());
        encoders.put("pbkdf2", new Pbkdf2PasswordEncoder());
        encoders.put("scrypt", new SCryptPasswordEncoder());
        encoders.put("SHA-1", new MessageDigestPasswordEncoder("SHA-1"));
        encoders.put("SHA-256", new MessageDigestPasswordEncoder("SHA-256"));
        encoders.put("sha256", new StandardPasswordEncoder());
        encoders.put("argon2", new Argon2PasswordEncoder());
        return new DelegatingPasswordEncoder(encodingId, encoders);
    }
    @Override
    protected void configure(AuthenticationManagerBuilder auth)
                                                    throws Exception {
        auth.userDetailsService(userService);
    }
    @Override
    protected void configure(HttpSecurity http) throws Exception {
        http.authorizeRequests()
```

```
                .anyRequest().authenticated()
                .and()
                .formLogin()
                .and()
                .csrf().disable();
    }
}
```

这里我们自己来提供一个 DelegatingPasswordEncoder 实例，同时在构建 BcryptPasswordEncoder 实例时，传入一个 strength 参数为 31，配置完成后，重启项目。

项目启动成功之后，再次进行登录操作，登录成功后，我们发现数据库中保存的用户密码从图 5-2 变为图 5-3，完成了升级操作。

id	username	password
1	javaboy	{bcrypt}$2a$31$g3w1T6s7Mp86oP7MEYHXCuDsEtT9gxcxVjfaCYCeqKA2.fzuAa.2u

图 5-3　修改密码加密强度 strength 参数之后，密码再次进行加密升级

这就是 Spring Security 中提供的密码升级功能。在升级一些"老破旧"系统时，这个功能非常好用。

5.6　是谁的 PasswordEncoder

根据前面 3.1.2 小节的介绍，大家知道 PasswordEncoder 做密码校验主要是在 DaoAuthenticationProvider 中完成的；根据 3.1.3 小节的介绍，DaoAuthenticationProvider 是被某一个 ProviderManager 管理的；而根据 4.1 节的介绍，AuthenticationManager（即 ProviderManager）实例有全局和局部之分，那么如果开发者配置了 PasswordEncoder 实例，是在全局的 AuthenticationManager 中使用，还是在局部的 AuthenticationManager 中使用呢？

我们先来看 DaoAuthenticationProvider 中的构造方法：

```
public DaoAuthenticationProvider() {
    setPasswordEncoder(PasswordEncoderFactories
                                .createDelegatingPasswordEncoder());
}
public void setPasswordEncoder(PasswordEncoder passwordEncoder) {
    this.passwordEncoder = passwordEncoder;
    this.userNotFoundEncodedPassword = null;
}
```

可以看到，当系统创建一个 DaoAuthenticationProvider 实例的时候，就会自动调用 setPasswordEncoder 方法来指定一个默认的 PasswordEncoder，默认的 PasswordEncoder 实例就是 DelegatingPasswordEncoder。

在全局的 AuthenticationManager 创建过程中，在 InitializeUserDetailsManagerConfigurer#

configure 方法中，有如下一段代码：

```
PasswordEncoder passwordEncoder = getBeanOrNull(PasswordEncoder.class);
UserDetailsPasswordService passwordManager
                    = getBeanOrNull(UserDetailsPasswordService.class);
DaoAuthenticationProvider provider = new DaoAuthenticationProvider();
provider.setUserDetailsService(userDetailsService);
if (passwordEncoder != null) {
    provider.setPasswordEncoder(passwordEncoder);
}
```

首先调用 getBeanOrNull 方法，从 Spring 容器中获取一个 PasswordEncoder 实例；然后创建一个 DaoAuthenticationProvider 实例，如果 passwordEncoder 不为 null，就设置给 provider 实例。从这段代码中可以看到，在上一小节中我们注册到 Spring 容器的 PasswordEncoder 实例，可以在这里被获取到并设置给 provider。如果我们没有向 Spring 容器中注册 PasswordEncoder 实例，则 provider 中使用默认的 DelegatingPasswordEncoder。

根据前面 4.1.5 小节的介绍可知，全局 AuthenticationManager 也有可能是通过 WebSecurityConfigurerAdapter 中的 localConfigureAuthenticationBldr 变量来构建的，localConfigureAuthenticationBldr 变量在构建 AuthenticationManager 实例时，使用的是 LazyPasswordEncoder，就是一个懒加载的 PasswordEncoder 实例，代码如下：

```
static class LazyPasswordEncoder implements PasswordEncoder {
    private ApplicationContext applicationContext;
    private PasswordEncoder passwordEncoder;
    LazyPasswordEncoder(ApplicationContext applicationContext) {
        this.applicationContext = applicationContext;
    }
    @Override
    public String encode(CharSequence rawPassword) {
        return getPasswordEncoder().encode(rawPassword);
    }
    @Override
    public boolean matches(CharSequence rawPassword,
        String encodedPassword) {
        return getPasswordEncoder().matches(rawPassword, encodedPassword);
    }
    @Override
    public boolean upgradeEncoding(String encodedPassword) {
        return getPasswordEncoder().upgradeEncoding(encodedPassword);
    }
    private PasswordEncoder getPasswordEncoder() {
        if (this.passwordEncoder != null) {
            return this.passwordEncoder;
        }
        PasswordEncoder passwordEncoder = getBeanOrNull(PasswordEncoder.class);
        if (passwordEncoder == null) {
            passwordEncoder
```

```
            = PasswordEncoderFactories.createDelegatingPasswordEncoder();
    }
    this.passwordEncoder = passwordEncoder;
    return passwordEncoder;
}
private <T> T getBeanOrNull(Class<T> type) {
    try {
        return this.applicationContext.getBean(type);
    } catch(NoSuchBeanDefinitionException notFound) {
        return null;
    }
}
```

可以看到，在 LazyPasswordEncoder 中，使用 getPasswordEncoder 方法去获取一个 Password Encoder 实例，具体的获取过程就是去 Spring 容器中查找，找到了就直接使用，没找到的话，则调用 PasswordEncoderFactories.createDelegatingPasswordEncoder()方法生成默认的 Delegating PasswordEncoder。

在 WebSecurityConfigurerAdapter 中，用来构建局部 AuthenticationManager 实例的 authenticationBuilder 变量也用的是 LazyPasswordEncoder，这里不再赘述。

经过上面的分析可知，如果开发者向 Spring 容器中注册了一个 PasswordEncoder 实例，那么无论是全局的 AuthenticationManager 还是局部的 AuthenticationManager，都将使用该 PasswordEncoder 实例；如果开发者没有提供任何 PasswordEncoder 实例，那么无论是全局的 AuthenticationManager 还是局部的 AuthenticationManager，都将使用 DelegatingPassword Encoder 实例。

5.7 小 结

本章主要讲解了 Spring Security 中的密码加密问题，探讨了为什么密码要加密、密码加密方案的演变历史、Spring Security 中对于密码加密方案的支持、DelegatingPasswordEncoder 的详细用法以及如何进行密码加密方案升级。通过本章的学习，读者对于密码加密方案已经有了一个基本认知，在前 4 章中我们反复使用的密码{noop}123，相信读者也已经理解其含义了。

第 6 章

RememberMe

可能有的人没有留意过，RememberMe 其实是我们日常开发中一个非常常见的需求。一般来说，我们依赖于会话 Session 来保持登录状态，而 Session 依赖于 Cookie，如果浏览器关闭后重新打开，会话就会丢失，用户需要重新登录。然而，在日常访问一些网站如购物网站、博客网站等，我们会发现有的网站即使你关闭了浏览器再重新打开，登录状态还依然有效！这个功能的实现，就依赖于我们本章要讲的 RememberMe。

本章涉及的主要知识点有：

- RememberMe 简介。
- RememberMe 基本用法。
- RememberMe 持久化令牌。
- 二次校验。
- RememberMe 原理分析。

6.1 RememberMe 简介

RememberMe 这个功能非常常见，图 6-1 所示就是 QQ 邮箱登录时的"记住我"选项。

提到 RememberMe，一些初学者往往会有一些误解，认为 RememberMe 功能就是把用户名/密码用 Cookie 保存在浏览器中，下次登录时不用再次输入用户名/密码。这个理解显然是不对的。

我们这里所说的 RememberMe 是一种服务器端的行为。传统的登录方式基于 Session 会话，一旦用户关闭浏览器重新打开，就要再次登录，这样太过于烦琐。如果能有一种机制，让用户关闭并重新打开浏览器之后，还能继续保持认证状态，就会方便很多，RememberMe 就是为了解决这一需求而生的。

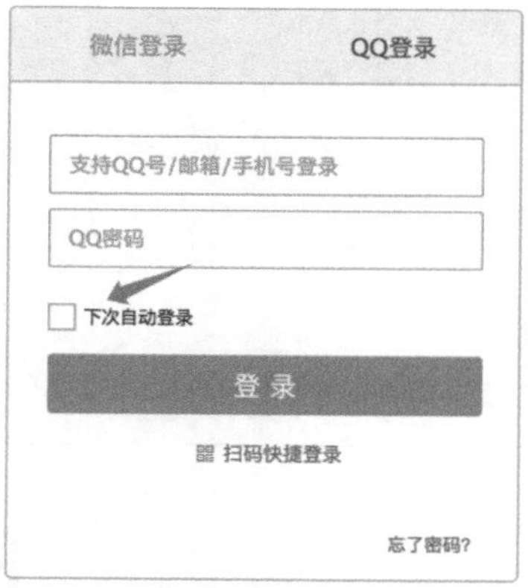

图 6-1　QQ 邮箱登录时的记住我选项

具体的实现思路就是通过 Cookie 来记录当前用户身份。当用户登录成功之后，会通过一定的算法，将用户信息、时间戳等进行加密，加密完成后，通过响应头带回前端存储在 Cookie 中，当浏览器关闭之后重新打开，如果再次访问该网站，会自动将 Cookie 中的信息发送给服务器，服务器对 Cookie 中的信息进行校验分析，进而确定出用户的身份，Cookie 中所保存的用户信息也是有时效的，例如三天、一周等。

敏锐的读者可能已经发现这种方式是存在安全隐患的。所谓鱼与熊掌不可兼得，要想使用便利，就要牺牲一定的安全性，不过在本章中，我们将会介绍通过持久化令牌以及二次校验来降低使用 RememberMe 所带来的安全风险。

6.2　RememberMe 基本用法

我们先来看一种最简单的用法。

首先创建一个 Spring Boot 工程，引入 spring-boot-starter-security 依赖。工程创建成功后，添加一个 HelloController 并创建一个测试接口，代码如下：

```
@RestController
public class HelloController {
    @GetMapping("/hello")
    public String hello() {
        return "hello";
    }
}
```

然后创建 SecurityConfig 配置文件：

```
@Configuration
public class SecurityConfig extends WebSecurityConfigurerAdapter {
```

```java
@Bean
PasswordEncoder passwordEncoder() {
    return NoOpPasswordEncoder.getInstance();
}
@Override
protected void configure(AuthenticationManagerBuilder auth)
                                                throws Exception {
    auth.inMemoryAuthentication()
            .withUser("javaboy")
            .password("123")
            .roles("admin");
}
@Override
protected void configure(HttpSecurity http) throws Exception {
    http.authorizeRequests()
            .anyRequest().authenticated()
            .and()
            .formLogin()
            .and()
            .rememberMe()
            .key("javaboy")
            .and()
            .csrf().disable();
}
```

这里我们主要是调用了 HttpSecurity 中的 rememberMe 方法并配置了一个 key，该方法最终会向过滤器链中添加 RememberMeAuthenticationFilter 过滤器。

配置完成后，启动项目，当我们访问/hello 接口时，会自动重定向到登录页面，如图 6-2 所示。

图 6-2　添加了 RememberMe 功能后的默认登录页面

可以看到，此时的默认登录页面多了一个 RememberMe 选项，勾选上 RememberMe，登录成功之后，我们就可以访问/hello 接口了。访问完成后，关闭浏览器再重新打开，此时不需要登录就可以直接访问/hello 接口；同时，如果关闭掉服务端重新打开，再去访问/hello 接口，发现此时也不需要登录了。

那么这一切是怎么实现的呢？打开浏览器控制台，我们来分析整个登录过程。

首先，当我们单击登录按钮时，多了一个请求参数 remember-me，如图 6-3 所示。

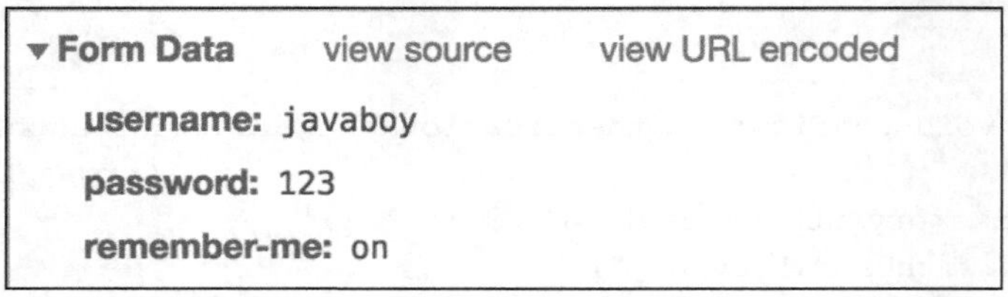

图 6-3　开启 RememberMe 功能后的请求参数

很明显，remember-me 参数就是用来告诉服务端是否开启 RememberMe 功能，如果开发者自定义登录页面，那么默认情况下，是否开启 RememberMe 的参数就是 remember-me。

当请求成功后，在响应头中多出了一个 Set-Cookie，如图 6-4 所示。

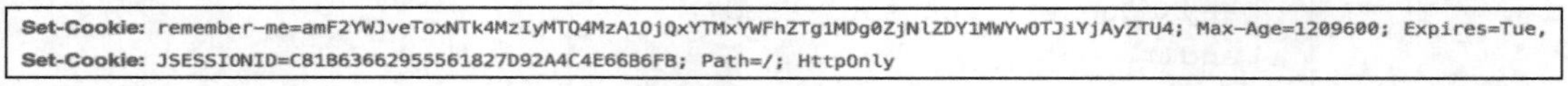

图 6-4　响应头中多出一个 Set-Cookie 字段

在响应头中给出了一个 remember-me 字符串。以后所有请求的请求头 Cookie 字段，都会自动携带上这个令牌，服务端利用该令牌可以校验用户身份是否合法。

大致的流程就是这样，但是大家发现这种方式安全隐患很大，一旦 remember-me 令牌泄漏，恶意用户就可以拿着这个令牌去随意访问系统资源。持久化令牌和二次校验可以在一定程度上降低该问题带来的风险。

6.3　持久化令牌

使用持久化令牌实现 RememberMe 的体验和使用普通令牌的登录体验是一样的，不同的是服务端所做的事情变了。持久化令牌在普通令牌的基础上，新增了 series 和 token 两个校验参数，当使用用户名/密码的方式登录时，series 才会自动更新；而一旦有了新的会话，token 就会重新生成。所以，如果令牌被人盗用，一旦对方基于 RememberMe 登录成功后，就会生成新的 token，你自己的登录令牌就会失效，这样就能及时发现账户泄漏并作出处理，比如清除自动登录令牌、通知用户账户泄漏等。

Spring Security 中对于持久化令牌提供了两种实现：JdbcTokenRepositoryImpl 和 InMemoryTokenRepositoryImpl，前者是基于 JdbcTemplate 来操作数据库，后者则是操作存储在内存中的数据。由于 InMemoryTokenRepositoryImpl 的使用场景很少，因此这里主要介绍基于 JdbcTokenRepositoryImpl 的配置。

我们在 6.2 小节的案例上继续完善。

首先创建一个 security06 数据库，然后我们需要一张表来记录令牌信息，创建表的 SQL

脚本在在 JdbcTokenRepositoryImpl 类中的 CREATE_TABLE_SQL 变量上已经定义好了，代码如下：

```
public static final String CREATE_TABLE_SQL = "create table persistent_logins
(username varchar(64) not null, series varchar(64) primary key, "
+ "token varchar(64) not null, last_used timestamp not null)";
```

我们直接将变量中定义的 SQL 脚本拷贝出来到数据库中执行，生成一张 persistent_logins 表用来记录令牌信息。persistent_logins 表一共就四个字段：username 表示登录用户名、series 表示生成的 series 字符串、token 表示生成的 token 字符串、last_used 则表示上次使用时间。

接下来，在项目中引入 JdbcTemplate 依赖和 MySQL 数据库驱动依赖：

```xml
<dependency>
    <groupId>org.springframework.boot</groupId>
    <artifactId>spring-boot-starter-jdbc</artifactId>
</dependency>
<dependency>
    <groupId>mysql</groupId>
    <artifactId>mysql-connector-java</artifactId>
</dependency>
```

然后在 application.properties 中配置数据库连接信息：

```
spring.datasource.url=jdbc:mysql:///security06?useUnicode=true&characterEncoding=UTF-8&serverTimezone=Asia/Shanghai
spring.datasource.username=root
spring.datasource.password=123
```

最后修改 SecurityConfig：

```java
@Configuration
public class SecurityConfig extends WebSecurityConfigurerAdapter {
    @Autowired
    DataSource dataSource;
    @Bean
    JdbcTokenRepositoryImpl jdbcTokenRepository() {
        JdbcTokenRepositoryImpl jdbcTokenRepository =
                                    new JdbcTokenRepositoryImpl();
        jdbcTokenRepository.setDataSource(dataSource);
        return jdbcTokenRepository;
    }
    @Bean
    PasswordEncoder passwordEncoder() {
        return NoOpPasswordEncoder.getInstance();
    }
    @Override
    protected void configure(AuthenticationManagerBuilder auth)
                                                throws Exception {
        auth.inMemoryAuthentication()
```

```
                .withUser("javaboy")
                .password("123")
                .roles("admin");
    }
    @Override
    protected void configure(HttpSecurity http) throws Exception {
        http.authorizeRequests()
                .anyRequest().authenticated()
                .and()
                .formLogin()
                .and()
                .rememberMe()
                .tokenRepository(jdbcTokenRepository())
                .and()
                .csrf().disable();
    }
}
```

在配置中我们提供了一个 JdbcTokenRepositoryImpl 实例，并为其配置了数据源，最后在配置 RememberMe 时通过 tokenRepository 方法指定 JdbcTokenRepositoryImpl 实例。

配置完成后，启动项目并进行登录测试。登录成功后，我们发现数据库表中多了一条记录，如图 6-5 所示。

username	series	token	last_used
javaboy	gBZ3bOxzoUojjFD7ElbPJg==	Z/s6ZCRIKU1YE/1RT/B2sQ==	2020-08-11 23:59:12

图 6-5　登录成功后，数据库中自动存储的登录令牌信息

此时如果关闭浏览器重新打开，再去访问/hello 接口，访问时并不需要登录，但是访问成功之后，数据库中的 token 字段会发生变化。同时，如果服务端重启之后，浏览器再去访问/hello 接口，依然不需要登录，但是 token 字段也会更新，因为这两种情况中都有新会话的建立，所以 token 会更新，而 series 则不会更新。当然，如果用户注销登录，则数据库中和该用户相关的登录记录会自动清除。

可以看到，持久化令牌比前面的普通令牌安全系数提高了不少，但是依然存在风险。安全问题和用户的使用便捷性就像一个悖论，想要用户使用方便，不可避免地要牺牲一点安全性。对于开发者而言，要做的就是如何将系统存在的安全风险降到最低。

接下来我们就来看二次校验，继续提高 RememberMe 登录的安全性。

6.4　二次校验

二次校验就是将系统中的资源分为敏感的和不敏感的，如果用户使用了 RememberMe 的方式登录，则访问敏感资源时会自动跳转到登录页面，要求用户重新登录；如果使用了用户名

/密码的方式登录，则可以访问所有资源。这种方式相当于牺牲了一点用户体验，但是却换取了系统安全。

配置方式也很简单，首先我们提供三个测试接口，代码如下：

```
@RestController
public class HelloController {
    @GetMapping("/hello")
    public String hello() {
        return "hello";
    }
    @GetMapping("/admin")
    public String admin() {
        return "admin";
    }
    @GetMapping("/rememberme")
    public String rememberme() {
        return "rememberme";
    }
}
```

现在假设：

（1）/hello 接口：认证后才可以访问，无论通过何种认证方式。

（2）/admin 接口：认证后才可以访问，但是必须是通过用户名/密码的方式认证。

（3）/rememberme 接口：认证后才可以访问，但是必须是通过 RememberMe 的方式认证。

在 SecurityConfig 中进行配置（在 6.3 节的 SecurityConfig 基础上修改 HttpSecurity 配置即可）：

```
@Configuration
public class SecurityConfig extends WebSecurityConfigurerAdapter {
    //省略其他配置
    @Override
    protected void configure(HttpSecurity http) throws Exception {
        http.authorizeRequests()
                .antMatchers("/admin").fullyAuthenticated()
                .antMatchers("/rememberme").rememberMe()
                .anyRequest().authenticated()
                .and()
                .formLogin()
                .and()
                .rememberMe()
                .key("javaboy")
                .tokenRepository(jdbcTokenRepository())
                .and()
                .csrf().disable();
    }
}
```

配置完成后，重启项目进行测试。使用用户名/密码登录成功后，访问/hello 接口和/admin 接口是没有问题的。如果访问/rememberme 接口，则会出现如图 6-6 所示的错误提示，表示没有权限访问。

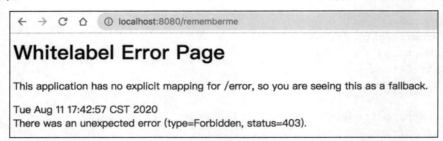

图 6-6　基于用户名/密码登录成功后没有权限访问 /rememberme 接口

此时关闭浏览器再重新打开，就可以访问/rememberme 接口了，但是如果去访问/admin 接口，则系统会自动跳转到登录页面，要求用户重新登录。

至此，RememberMe 的使用基本上就和大家介绍完了，接下来我们通过源码，对其原理进行深入分析。

6.5　原理分析

从 RememberMeServices 接口开始介绍。

RememberMeServices 接口定义如下：

```
public interface RememberMeServices {
    Authentication autoLogin(HttpServletRequest request,
                             HttpServletResponse response);
    void loginFail(HttpServletRequest request, HttpServletResponse response);
    void loginSuccess(HttpServletRequest request,
                      HttpServletResponse response,
                      Authentication successfulAuthentication);
}
```

这里一共定义了三个方法：

（1）autoLogin 方法可以从请求中提取出需要的参数，完成自动登录功能。

（2）loginFail 方法是自动登录失败的回调。

（3）loginSuccess 方法是自动登录成功的回调。

RememberMeServices 接口的继承关系如图 6-7 所示。

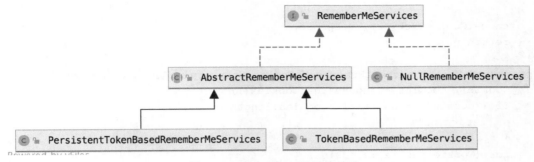

图 6-7　RememberMe 继承关系图

NullRememberMeServices 是一个空的实现，这里不做讨论，我们来重点分析另外三个实现类。

AbstractRememberMeServices

AbstractRememberMeServices 对于 RememberMeServices 接口中定义的方法提供了基本的实现，这里就以接口中定义的方法为思路，分析 AbstractRememberMeServices 中的具体实现。

首先我们来看 autoLogin 及其相关方法：

```
public final Authentication autoLogin(HttpServletRequest request,
                                      HttpServletResponse response) {
    String rememberMeCookie = extractRememberMeCookie(request);
    if (rememberMeCookie == null) {
        return null;
    }
    if (rememberMeCookie.length() == 0) {
        cancelCookie(request, response);
        return null;
    }
    UserDetails user = null;
    try {
        String[] cookieTokens = decodeCookie(rememberMeCookie);
        user = processAutoLoginCookie(cookieTokens, request, response);
        userDetailsChecker.check(user);
        return createSuccessfulAuthentication(request, user);
    }
    catch (CookieTheftException cte) {
        cancelCookie(request, response);
        throw cte;
    }
    catch (UsernameNotFoundException noUser) {
    }
    catch (InvalidCookieException invalidCookie) {
    }
    catch (AccountStatusException statusInvalid) {
    }
    catch (RememberMeAuthenticationException e) {
```

```java
        }
        cancelCookie(request, response);
        return null;
    }
    protected String extractRememberMeCookie(HttpServletRequest request) {
        Cookie[] cookies = request.getCookies();
        if ((cookies == null) || (cookies.length == 0)) {
            return null;
        }
        for (Cookie cookie : cookies) {
            if (cookieName.equals(cookie.getName())) {
                return cookie.getValue();
            }
        }
        return null;
    }
    protected String[] decodeCookie(String cookieValue)
                                                throws InvalidCookieException {
        for (int j = 0; j < cookieValue.length() % 4; j++) {
            cookieValue = cookieValue + "=";
        }
        try {
            Base64.getDecoder().decode(cookieValue.getBytes());
        }
        catch (IllegalArgumentException e) {
            throw new InvalidCookieException(
                    "Cookie token was not Base64 encoded; value was '" + cookieValue
                    + "'");
        }
        String cookieAsPlainText =
                new String(Base64.getDecoder().decode(cookieValue.getBytes()));
        String[] tokens =
         StringUtils.delimitedListToStringArray(cookieAsPlainText,DELIMITER);
        for (int i = 0; i < tokens.length; i++){
            try
            {
                tokens[i] = URLDecoder
                        .decode(tokens[i], StandardCharsets.UTF_8.toString());
            }
            catch (UnsupportedEncodingException e)
            {
            }
        }
        return tokens;
    }
    protected void cancelCookie(HttpServletRequest request,
                            HttpServletResponse response) {
        Cookie cookie = new Cookie(cookieName, null);
```

```
    cookie.setMaxAge(0);
    cookie.setPath(getCookiePath(request));
    if (cookieDomain != null) {
        cookie.setDomain(cookieDomain);
    }
    if (useSecureCookie == null) {
        cookie.setSecure(request.isSecure());
    }
    else {
        cookie.setSecure(useSecureCookie);
    }
    response.addCookie(cookie);
}
protected Authentication createSuccessfulAuthentication(
                            HttpServletRequest request,
                            UserDetails user) {
    RememberMeAuthenticationToken auth =
            new RememberMeAuthenticationToken(key, user,
                authoritiesMapper.mapAuthorities(user.getAuthorities()));
    auth.setDetails(authenticationDetailsSource.buildDetails(request));
    return auth;
}
```

autoLogin 方法主要功能就是从当前请求中提取出令牌信息，根据令牌信息完成自动登录功能，登录成功之后会返回一个认证后的 Authentication 对象，我们来看一下该方法的具体实现：

（1）首先调用 extractRememberMeCookie 方法从当前请求中提取出需要的 Cookie 信息，即 remember-me 对应的值。如果这个值为 null，表示本次请求携带的 Cookie 中没有 remember-me，这次不需要自动登录，直接返回 null 即可。如果 remember-me 对应的值长度为 0，则在返回 null 之前，执行一下 cancelCookie 函数，将 Cookie 中 remember-me 的值置为 null。

（2）接下来调用 decodeCookie 方法对获取到的令牌进行解析。具体方式是，先用 Base64 对令牌进行还原（如果令牌字符串长度不是 4 的倍数，则在令牌末尾补上一个或者多个 "="，以使其长度变为 4 的倍数，之所以要是 4 的倍数，这和 Base64 编解码的原理有关，感兴趣的读者可以自行学习 Base64 编解码的原理，并不难），还原之后的字符串分为三部分，三部分之间用 ":" 隔开，第一部分是当前登录用户名，第二部分是时间戳，第三部分是一个签名。也就是说，我们一开始在浏览器中看到的 remember-me 令牌，其实是一个 Base64 编码后的字符串，解码后的信息包含三部分，读者可以根据 decodeCookie 中的方法自行尝试对令牌进行解码。最后将这三部分分别提取出来组成一个数组返回。

（3）调用 processAutoLoginCookie 方法对 Cookie 进行验证，如果验证通过，则返回登录用户对象，然后对用户状态进行检验（账户是否可用、账户是否锁定等）。processAutoLoginCookie 方法是一个抽象方法，具体实现在 AbstractRememberMeServices 的子类中。

（4）最后调用 createSuccessfulAuthentication 方法创建登录成功的用户对象，不同于使用用户名/密码登录，本次登录成功后创建的用户对象类型是 RememberMeAuthenticationToken。

接下来我们再来看一下自动登录成功和自动登录失败的回调：

```java
public final void loginFail(HttpServletRequest request,
                            HttpServletResponse response) {
    cancelCookie(request, response);
    onLoginFail(request, response);
}
protected void onLoginFail(HttpServletRequest request,
                           HttpServletResponse response) {
}
public final void loginSuccess(HttpServletRequest request,
                               HttpServletResponse response,
                               Authentication successfulAuthentication) {
    if (!rememberMeRequested(request, parameter)) {
        return;
    }
    onLoginSuccess(request, response, successfulAuthentication);
}
protected abstract void onLoginSuccess(HttpServletRequest request,
                                       HttpServletResponse response,
                                       Authentication successfulAuthentication);
protected boolean rememberMeRequested(HttpServletRequest request,
                                      String parameter) {
    if (alwaysRemember) {
        return true;
    }
    String paramValue = request.getParameter(parameter);
    if (paramValue != null) {
        if (paramValue.equalsIgnoreCase("true")
            || paramValue.equalsIgnoreCase("on")
            || paramValue.equalsIgnoreCase("yes")
            || paramValue.equals("1")) {
            return true;
        }
    }
    return false;
}
```

（1）登录失败时，首先取消 Cookie 的设置，然后调用 onLoginFail 方法完成失败处理，onLoginFail 方法是一个空方法，如果有需要，开发者可以自行重写该方法，一般来说不需要重写。

（2）登录成功时，会首先调用 rememberMeRequested 方法，判断当前请求是否开启了自动登录。开发者可以在服务端配置 alwaysRemember，这样无论前端参数是什么，都会开启自动登录，如果开发者没有配置 alwaysRemember，则根据前端传来的 remember-me 参数进行判断，remember-me 参数的值如果是 true、on（默认）、yes 或者 1，表示开启自动登录。如果开启了自动登录，则调用 onLoginSuccess 方法进行登录成功的处理。onLoginSuccess 是一个抽象方法，具体实现在 AbstractRememberMeServices 的子类中。

最后再来看 AbstractRememberMeServices 中一个比较重要的方法 setCookie，在自动登录成功后，将调用该方法把令牌信息放入响应头中并最终返回到前端：

```java
protected void setCookie(String[] tokens, int maxAge,
                    HttpServletRequest request,
                    HttpServletResponse response) {
    String cookieValue = encodeCookie(tokens);
    Cookie cookie = new Cookie(cookieName, cookieValue);
    cookie.setMaxAge(maxAge);
    cookie.setPath(getCookiePath(request));
    if (cookieDomain != null) {
        cookie.setDomain(cookieDomain);
    }
    if (maxAge < 1) {
        cookie.setVersion(1);
    }
    if (useSecureCookie == null) {
        cookie.setSecure(request.isSecure());
    }
    else {
        cookie.setSecure(useSecureCookie);
    }
    cookie.setHttpOnly(true);
    response.addCookie(cookie);
}
protected String encodeCookie(String[] cookieTokens) {
    StringBuilder sb = new StringBuilder();
    for (int i = 0; i < cookieTokens.length; i++) {
        try
        {
            sb.append(URLEncoder.encode(cookieTokens[i],
                            StandardCharsets.UTF_8.toString()));
        }
        catch (UnsupportedEncodingException e)
        {
        }
        if (i < cookieTokens.length - 1) {
            sb.append(DELIMITER);
        }
    }
    String value = sb.toString();
    sb = new StringBuilder(new String(
                    Base64.getEncoder().encode(value.getBytes())));
    while (sb.charAt(sb.length() - 1) == '=') {
        sb.deleteCharAt(sb.length() - 1);
```

```
        }
        return sb.toString();
}
```

（1）首先调用 encodeCookie 方法对要返回到前端的数据进行 Base64 编码，具体方式是将数组中的数据拼接成一个字符串并用":"隔开，然后对其进行 Base64 编码。

（2）将编码后的字符串放入 Cookie 中，并配置 Cookie 的过期时间、path、domain、secure、httpOnly 等属性，最终将配置好的 Cookie 对象放入响应头中。

这便是 AbstractRememberMeServices 中的几个主要方法，还有其他一些辅助的方法都比较简单，读者可以自行研究。

TokenBasedRememberMeServices

TokenBasedRememberMeServices 是 AbstractRememberMeServices 的实现类之一，在 6.2 节中，我们讲解 RememberMe 的基本用法时，最终起作用的就是 TokenBasedRememberMeServices。作为 AbstractRememberMeServices 的子类，TokenBasedRememberMeServices 中最重要的方法就是对 AbstractRememberMeServices 中所定义的两个抽象方法 processAutoLoginCookie 和 onLoginSuccess 的实现。

我们先来看 processAutoLoginCookie 方法：

```java
protected UserDetails processAutoLoginCookie(String[] cookieTokens,
        HttpServletRequest request, HttpServletResponse response) {
    if (cookieTokens.length != 3) {
        throw new InvalidCookieException("Cookie token did not contain 3"
            + " tokens, but contained '" + Arrays.asList(cookieTokens) + "'");
    }
    long tokenExpiryTime;
    try {
        tokenExpiryTime = new Long(cookieTokens[1]);
    }
    catch (NumberFormatException nfe) {
        throw new InvalidCookieException(
            "Cookie token[1] did not contain a valid number (contained '"
                + cookieTokens[1] + "')");
    }
    if (isTokenExpired(tokenExpiryTime)) {
        throw new InvalidCookieException("Cookie token[1] has expired"
                + " (expired on '" + new Date(tokenExpiryTime)
                + "'; current time is '" + new Date() + "')");
    }
    UserDetails userDetails = getUserDetailsService().loadUserByUsername(
            cookieTokens[0]);
    String expectedTokenSignature = makeTokenSignature(tokenExpiryTime,
            userDetails.getUsername(), userDetails.getPassword());
    if (!equals(expectedTokenSignature, cookieTokens[2])) {
        throw new InvalidCookieException("Cookie token[2] contained
```

```
                        signature '" + cookieTokens[2] + "' but expected '"
                                            + expectedTokenSignature + "'");
    }
    return userDetails;
}
protected String makeTokenSignature(long tokenExpiryTime, String username,
        String password) {
    String data = username + ":" + tokenExpiryTime + ":" + password
                                            + ":" + getKey();
    MessageDigest digest;
    try {
        digest = MessageDigest.getInstance("MD5");
    }
    catch (NoSuchAlgorithmException e) {
        throw new IllegalStateException("No MD5 algorithm available!");
    }
    return new String(Hex.encode(digest.digest(data.getBytes())));
}
```

processAutoLoginCookie 方法主要用来验证 Cookie 中的令牌信息是否合法：

（1）首先判断 cookieTokens 长度是否为 3，不为 3 说明格式不对，则直接抛出异常。

（2）从 cookieTokens 数组中提取出第 1 项，也就是过期时间，判断令牌是否过期，如果已经过期，则抛出异常。

（3）根据用户名（cookieTokens 数组的第 0 项）查询出当前用户对象。

（4）调用 makeTokenSignature 方法生成一个签名，签名的生成过程如下：首先将用户名、令牌过期时间、用户密码以及 key 组成一个字符串，中间用 ":" 隔开，然后通过 MD5 消息摘要算法对该字符串进行加密，并将加密结果转为一个字符串返回。

（5）判断第 4 步生成的签名和通过 Cookie 传来的签名是否相等（即 cookieTokens 数组的第 2 项），如果相等，表示令牌合法，则直接返回用户对象，否则抛出异常。

再来看登录成功的回调函数 onLoginSuccess：

```
public void onLoginSuccess(HttpServletRequest request,
                    HttpServletResponse response,
                    Authentication successfulAuthentication) {
    String username = retrieveUserName(successfulAuthentication);
    String password = retrievePassword(successfulAuthentication);
    if (!StringUtils.hasLength(username)) {
      return;
    }
    if (!StringUtils.hasLength(password)) {
      UserDetails user = getUserDetailsService().loadUserByUsername(username);
      password = user.getPassword();
      if (!StringUtils.hasLength(password)) {
          return;
      }
```

```
    }
    int tokenLifetime =
            calculateLoginLifetime(request, successfulAuthentication);
    long expiryTime = System.currentTimeMillis();
    expiryTime += 1000L * (tokenLifetime < 0 ? TWO_WEEKS_S : tokenLifetime);
    String signatureValue = makeTokenSignature(expiryTime, username, password);
    setCookie(new String[] { username,
                        Long.toString(expiryTime), signatureValue },
            tokenLifetime, request, response);
}
```

（1）在这个回调中，首先获取用户名和密码信息，如果用户密码在用户登录成功后已经从 successfulAuthentication 对象中擦除了，则从数据库中重新加载出用户密码。

（2）计算出令牌的过期时间，令牌默认有效期是两周。

（3）根据令牌的过期时间、用户名以及用户密码，计算出一个签名。

（4）调用 setCookie 方法设置 Cookie，第一个参数是一个数组，数组中一共包含三项：用户名、过期时间以及签名，在 setCookie 方法中会将数组转为字符串，并进行 Base64 编码后响应给前端。

看完 processAutoLoginCookie 和 onLoginSuccess 两个方法的实现，相信读者对于令牌的生成和校验已经非常清楚了，这里再总结一下：

当用户通过用户名/密码的形式登录成功后，系统会根据用户的用户名、密码以及令牌的过期时间计算出一个签名，这个签名使用 MD5 消息摘要算法生成，是不可逆的。然后再将用户名、令牌过期时间以及签名拼接成一个字符串，中间用 ":" 隔开，对拼接好的字符串进行 Base64 编码，然后将编码后的结果返回到前端，也就是我们在浏览器中看到的令牌。当用户关闭浏览器再次打开，访问系统资源时会自动携带上 Cookie 中的令牌，服务端拿到 Cookie 中的令牌后，先进行 Base64 解码，解码后分别提取出令牌中的三项数据；接着根据令牌中的数据判断令牌是否已经过期，如果没有过期，则根据令牌中的用户名查询出用户信息；接着再计算出一个签名和令牌中的签名进行对比，如果一致，表示令牌是合法令牌，自动登录成功，否则自动登录失败。

PersistentTokenBasedRememberMeServices

PersistentTokenBasedRememberMeServices 类作为 AbstractRememberMeServices 的另一个实现类，在 6.3 节的案例中，使用的就是 PersistentTokenBasedRememberMeServices。

在持久化令牌中，存储在数据库中的数据被封装成了一个对象 PersistentRememberMeToken，其定义如下：

```
public class PersistentRememberMeToken {
    private final String username;
    private final String series;
    private final String tokenValue;
    private final Date date;
    //省略getter/setter
```

}

username 表示登录用户名，series 和 tokenValue 则是自动生成的，date 表示上次使用时间。

PersistentTokenBasedRememberMeServices 里边重要的方法也是 processAutoLoginCookie 和 onLoginSuccess，我们分别来看一下。

先来看 processAutoLoginCookie 方法：

```
protected UserDetails processAutoLoginCookie(String[] cookieTokens,
        HttpServletRequest request, HttpServletResponse response) {
    if (cookieTokens.length != 2) {
        throw new InvalidCookieException("Cookie token did not contain " + 2
            + " tokens, but contained '" + Arrays.asList(cookieTokens) + "'");
    }
    final String presentedSeries = cookieTokens[0];
    final String presentedToken = cookieTokens[1];
    PersistentRememberMeToken token = tokenRepository
                                    .getTokenForSeries(presentedSeries);
    if (token == null) {
        throw new RememberMeAuthenticationException(
            "No persistent token found for series id: " + presentedSeries);
    }
    if (!presentedToken.equals(token.getTokenValue())) {
        tokenRepository.removeUserTokens(token.getUsername());
        throw new CookieTheftException(messages.getMessage(
                "PersistentTokenBasedRememberMeServices.cookieStolen",
                    "Invalid remember-me token (Series/token) mismatch.
                        Implies previous cookie theft attack."));
    }
    if (token.getDate().getTime() + getTokenValiditySeconds() * 1000L
                                    < System.currentTimeMillis()) {
        throw new RememberMeAuthenticationException("Remember-me login has
                                    expired");
    }
    PersistentRememberMeToken newToken = new PersistentRememberMeToken(
                                    token.getUsername(),
                                    token.getSeries(),
                                    generateTokenData(), new Date());
    try {
        tokenRepository.updateToken(newToken.getSeries(),
                        newToken.getTokenValue(),newToken.getDate());
        addCookie(newToken, request, response);
    }
    catch (Exception e) {
        throw new RememberMeAuthenticationException(
            "Autologin failed due to data access problem");
    }
    return getUserDetailsService().loadUserByUsername(token.getUsername());
}
```

```java
private void addCookie(PersistentRememberMeToken token,
                      HttpServletRequest request,
                      HttpServletResponse response) {
    setCookie(new String[] { token.getSeries(), token.getTokenValue() },
        getTokenValiditySeconds(), request, response);
}
```

（1）不同于 TokenBasedRememberMeServices 中的 processAutoLoginCookie 方法，这里 cookieTokens 数组的长度为 2，第一项是 series，第二项是 token。

（2）从 cookieTokens 数组中分别提取出 series 和 token，然后根据 series 去数据库中查询出一个 PersistentRememberMeToken 对象。如果查询出来的对象为 null，表示数据库中并没有 series 对应的值，本次自动登录失败；如果查询出来的 token 和从 cookieTokens 中解析出来的 token 不相同，说明自动登录令牌已经泄漏（恶意用户利用令牌登录后，数据库中的 token 变了），此时移除当前用户的所有自动登录记录并抛出异常。

（3）根据数据库中查询出来的结果判断令牌是否过期，如果过期就抛出异常。

（4）生成一个新的 PersistentRememberMeToken 对象，用户名和 series 不变，token 重新生成，date 也使用当前时间。newToken 生成后，根据 series 去修改数据库中的 token 和 date （即每次自动登录后都会产生新的 token 和 date）。

（5）调用 addCookie 方法添加 Cookie，在 addCookie 方法中，会调用到我们前面所说的 setCookie 方法，但是要注意第一个数组参数中只有两项：series 和 token（即返回到前端的令牌是通过对 series 和 token 进行 Base64 编码得到的）。

（6）最后将根据用户名查询用户对象并返回。

再来看登录成功的回调函数 onLoginSuccess：

```java
protected void onLoginSuccess(HttpServletRequest request,
                              HttpServletResponse response,
                              Authentication successfulAuthentication) {
    String username = successfulAuthentication.getName();
    PersistentRememberMeToken persistentToken = new PersistentRememberMeToken(
        username, generateSeriesData(), generateTokenData(), new Date());
    try {
        tokenRepository.createNewToken(persistentToken);
        addCookie(persistentToken, request, response);
    }
    catch (Exception e) {
    }
}
```

登录成功后，构建一个 PersistentRememberMeToken 对象，对象中的 series 和 token 参数都是随机生成的，然后将生成的对象存入数据库中，再调用 addCookie 方法添加相关的 Cookie 信息。

PersistentTokenBasedRememberMeServices 和 TokenBasedRememberMeServices 还是有一些明显的区别的：前者返回给前端的令牌是将 series 和 token 组成的字符串进行 Base64 编码后

返回给前端；后者返回给前端的令牌则是将用户名、过期时间以及签名组成的字符串进行 Base64 编码后返回给前端。

那么 RememberMeServices 是在何时被调用的？这就要回到我们一开始的配置中了。

当开发者配置 .rememberMe().key("javaboy") 时，实际上是引入了配置类 RememberMeConfigurer，根据第 4 章的介绍，我们知道对于 RememberMeConfigurer 而言最重要的就是 init 和 configure 方法，我们先来看其 init 方法：

```
public void init(H http) throws Exception {
    validateInput();
    String key = getKey();
    RememberMeServices rememberMeServices = getRememberMeServices(http, key);
    http.setSharedObject(RememberMeServices.class, rememberMeServices);
    LogoutConfigurer<H> logoutConfigurer
                            = http.getConfigurer(LogoutConfigurer.class);
    if (logoutConfigurer != null && this.logoutHandler != null) {
        logoutConfigurer.addLogoutHandler(this.logoutHandler);
    }
    RememberMeAuthenticationProvider authenticationProvider =
                        new RememberMeAuthenticationProvider(key);
    authenticationProvider = postProcess(authenticationProvider);
    http.authenticationProvider(authenticationProvider);
    initDefaultLoginFilter(http);
}
```

在这里首先获取了一个 key，这个 key 就是开发者一开始配置的 key，如果没有配置，则会自动生成一个 UUID 字符串。如果开发者使用普通的 RememberMe，即没有使用持久化令牌，则建议开发者自行配置该 key，因为使用默认的 UUID 字符串，系统每次重启都会生成新的 key，会导致之前下发的 remember-me 失效。

有了 key 之后，接下来再去获取 RememberMeServices 实例，如果开发者配置了 tokenRepository，则获取到的 RememberMeServices 实例是 PersistentTokenBasedRememberMeServices，否则获取到 TokenBasedRememberMeServices，即系统通过有没有配置 tokenRepository 来确定使用哪种类型的 RememberMeServices。

同时，init 方法中还配置了一个 RememberMeAuthenticationProvider，该实例主要用来校验 key。

再来看 RememberMeConfigurer 的 configure 方法：

```
public void configure(H http) {
    RememberMeAuthenticationFilter rememberMeFilter =
     new RememberMeAuthenticationFilter(
        http.getSharedObject(AuthenticationManager.class),
        this.rememberMeServices);
    if (this.authenticationSuccessHandler != null) {
        rememberMeFilter
         .setAuthenticationSuccessHandler(this.authenticationSuccessHandler);
```

```
        }
        rememberMeFilter = postProcess(rememberMeFilter);
        http.addFilter(rememberMeFilter);
    }
```

configure 方法中主要创建了一个 RememberMeAuthenticationFilter，创建时传入 RememberMeServices 实例，最后将创建好的 RememberMeAuthenticationFilter 加入到过滤器链中。最后我们再来看一下 RememberMeAuthenticationFilter 中的 doFilter 是如何"运筹帷幄"的：

```
public void doFilter(ServletRequest req, ServletResponse res,
            FilterChain chain) throws IOException, ServletException {
    HttpServletRequest request = (HttpServletRequest) req;
    HttpServletResponse response = (HttpServletResponse) res;
    if (SecurityContextHolder.getContext().getAuthentication() == null) {
        Authentication rememberMeAuth = rememberMeServices.autoLogin(request,
                                                            response);
        if (rememberMeAuth != null) {
            try {
                rememberMeAuth =
                        authenticationManager.authenticate(rememberMeAuth);
                SecurityContextHolder.getContext()
                                    .setAuthentication(rememberMeAuth);
                onSuccessfulAuthentication(request, response,rememberMeAuth);
                if (this.eventPublisher != null) {
                    eventPublisher
                            .publishEvent(new
InteractiveAuthenticationSuccessEvent(SecurityContextHolder.getContext()
                                        .getAuthentication(),
                                            this.getClass()));
                }
                if (successHandler != null) {
                    successHandler.onAuthenticationSuccess(request, response,
                            rememberMeAuth);
                    return;
                }
            }
            catch (AuthenticationException authenticationException) {
                rememberMeServices.loginFail(request, response);
                onUnsuccessfulAuthentication(request, response,
                        authenticationException);
            }
        }
        chain.doFilter(request, response);
    }
    else {
        chain.doFilter(request, response);
    }
}
```

（1）请求到达过滤器之后，首先判断 SecurityContextHolder 中是否有值，没值的话表示用户尚未登录，此时调用 autoLogin 方法进行自动登录。

（2）当自动登录成功后返回的 rememberMeAuth 不为 null 时，表示自动登录成功，此时调用 authenticate 方法对 key 进行校验，并且将登录成功的用户信息保存到 SecurityContextHolder 对象中，然后调用登录成功回调，并发布登录成功事件。需要注意的是，登录成功的回调并不包含 RememberMeServices 中的 loginSuccess 方法。

（3）如果自动登录失败，则调用 rememberMeServices.loginFail 方法处理登录失败回调。onUnsuccessfulAuthentication 和 onSuccessfulAuthentication 都是该过滤器中定义的空方法，并没有任何实现。

这就是 RememberMeAuthenticationFilter 过滤器所做的事情，成功将 RememberMeServices 的服务集成进来。

最后再额外说一下 RememberMeServices#loginSuccess 方法的调用位置。该方法是在 AbstractAuthenticationProcessingFilter#successfulAuthentication 中触发的，也就是说，无论你是否开启了 RememberMe 功能，该方法都会被调用。只不过在 RememberMeServices#loginSuccess 方法的具体实现中，会去判断是否开启了 RememberMe，进而决定是否在响应中添加对应的 Cookie。

至此，整个 RememberMe 的用法还有原理就介绍完了。

6.6 小　结

本章主要介绍了 RememberMe 功能的基本用法，以及为了提高 RememberMe 安全性所采取的持久化令牌方案和二次校验方案。RememberMe 可以避免让用户频繁登录，进而提高用户的使用体验，但是也在一定程度上牺牲了系统的安全性，而持久化令牌和二次校验则可以将这种安全风险降至最低。

第 7 章

会话管理

当用户通过浏览器登录成功之后，用户和系统之间就会保持一个会话（Session），通过这个会话，系统可以确定出访问用户的身份。Spring Security 中和会话相关的功能由 SessionManagementFilter 和 SessionAuthenticationStrategy 接口的组合来处理，过滤器委托该接口对会话进行处理，比较典型的用法有防止会话固定攻击、配置会话并发数等。本章我们就来学习 Spring Security 中关于会话的处理。

本章涉及的主要知识点有：

- 会话简介。
- 会话并发管理。
- 会话固定攻击与防御。
- Session 共享。

7.1 会话简介

当浏览器调用登录接口登录成功后，服务端会和浏览器之间建立一个会话（Session），浏览器在每次发送请求时都会携带一个 SessionId，服务端则根据这个 SessionId 来判断用户身份。当浏览器关闭后，服务端的 Session 并不会自动销毁，需要开发者手动在服务端调用 Session 销毁方法，或者等 Session 过期时间到了自动销毁。

在 Spring Security 中，与 HttpSession 相关的功能由 SessionManagementFilter 和 SessionAuthenticationStrategy 接口来处理，SessionManagementFilter 过滤器将 Session 相关操作委托给 SessionAuthenticationStrategy 接口去完成。

7.2 会话并发管理

会话并发管理就是指在当前系统中，同一个用户可以同时创建多少个会话，如果一台设备对应一个会话，那么也可以简单理解为同一个用户可以同时在多少台设备上进行登录。默认情况下，同一用户在多少台设备上登录并没有限制，不过开发者可以在 Spring Security 中对此进行配置。

7.2.1 实战

首先创建一个 Spring Boot 项目，引入 Spring Security 依赖 spring-boot-starter-security。添加配置类，内容如下：

```
@Configuration
public class SecurityConfig extends WebSecurityConfigurerAdapter {
    @Override
    protected void configure(AuthenticationManagerBuilder auth)
                                                    throws Exception {
        auth.inMemoryAuthentication()
                .withUser("javaboy")
                .password("{noop}123")
                .roles("admin");
    }
    @Override
    protected void configure(HttpSecurity http) throws Exception {
        http.authorizeRequests()
                .anyRequest().authenticated()
                .and()
                .formLogin()
                .and()
                .csrf()
                .disable()
                .sessionManagement()
                .maximumSessions(1);
    }
    @Bean
    HttpSessionEventPublisher httpSessionEventPublisher() {
        return new HttpSessionEventPublisher();
    }
}
```

和前面章节的配置类相比，这里主要有两点变化：

（1）在 configure(HttpSecurity)方法中通过 sessionManagement()方法开启会话配置，并设

置会话并发数为 1。

（2）提供一个 httpSessionEventPublisher 实例。Spring Security 中通过一个 Map 集合来维护当前的 HttpSession 记录，进而实现会话的并发管理。当用户登录成功时，就向集合中添加一条 HttpSession 记录；当会话销毁时，就从集合中移除一条 HttpSession 记录。HttpSessionEventPublisher 实现了 HttpSessionListener 接口，可以监听到 HttpSession 的创建和销毁事件，并将 HttpSession 的创建/销毁事件发布出去，这样，当有 HttpSession 销毁时，Spring Security 就可以感知到该事件了。

接下来再提供一个测试 Controller：

```
@RestController
public class HelloController {
    @GetMapping("/")
    public String hello() {
        return "hello";
    }
}
```

配置完成后，启动项目。这次测试我们需要两个浏览器，如果读者使用了 Chrome 浏览器，那么也可以使用 Chrome 浏览器中的多用户方式（相当于两个浏览器）。

先在第一个浏览器中输入 http://localhost:8080，此时会自动跳转到登录页面，完成登录操作，就可以访问到数据了；接下来在第二个浏览器中也输入 http://localhost:8080，也需要登录，完成登录操作；当第二个浏览器登录成功后，再回到第一个浏览器，刷新页面，结果如图 7-1 所示。

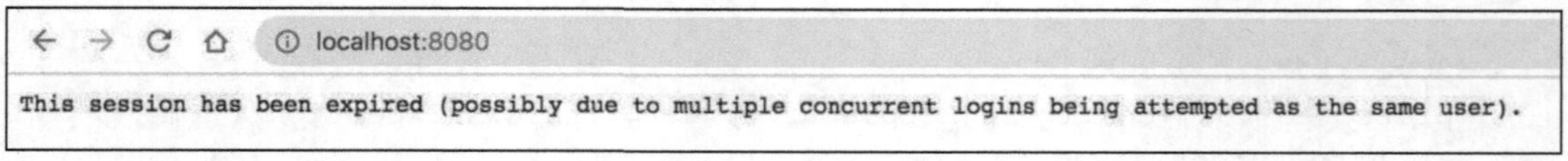

图 7-1　被"挤下线"之后的提示

从提示信息中可以看到，由于使用同一用户身份进行并发登录，所以当前会话已经失效。如果有需要，开发者也可以自定义会话销毁后的行为，代码如下：

```
http.authorizeRequests()
    .anyRequest().authenticated()
    .and()
    .formLogin()
    .and()
    .csrf()
    .disable()
    .sessionManagement()
    .maximumSessions(1)
    .expiredUrl("/login");
```

最后的 expiredUrl 方法配置了当会话失效后（即被人"挤下线"后），自动重定向到/login

页面。如果是前后端分离的项目，就不需要页面跳转了，直接返回一段 JSON 提示即可，配置如下：

```
http.authorizeRequests()
    .anyRequest().authenticated()
    .and()
    .formLogin()
    .and()
    .csrf()
    .disable()
    .sessionManagement()
    .maximumSessions(1)
    .expiredSessionStrategy(event -> {
        HttpServletResponse response = event.getResponse();
        response.setContentType("application/json;charset=utf-8");
        Map<String, Object> result = new HashMap<>();
        result.put("status", 500);
        result.put("msg", "当前会话已经失效，请重新登录");
        String s = new ObjectMapper().writeValueAsString(result);
        response.getWriter().print(s);
        response.flushBuffer();
    });
```

此时，当被人挤下线之后，服务端就会返回一段 JSON 响应。

这是一种被"挤下线"的效果，后面登录的用户会把前面登录的用户"挤下线"。还有一种是禁止后来者登录，即一旦当前用户登录成功，后来者无法再次使用相同的用户登录，直到当前用户主动注销登录，配置如下：

```
@Configuration
public class SecurityConfig extends WebSecurityConfigurerAdapter {
    //省略用户配置
    @Override
    protected void configure(HttpSecurity http) throws Exception {
        http.authorizeRequests()
                .anyRequest().authenticated()
                .and()
                .formLogin()
                .and()
                .csrf()
                .disable()
                .sessionManagement()
                .maximumSessions(1)
                .maxSessionsPreventsLogin(true);
    }
    @Bean
    HttpSessionEventPublisher httpSessionEventPublisher() {
        return new HttpSessionEventPublisher();
    }
```

}
```

这里主要是调用 maxSessionsPreventsLogin()方法,通过设置参数为 true,来禁止后来者登录。

配置完成后,重启项目。首先在第一个浏览器上进行登录,登录成功后再去第二个浏览器上进行登录,此时就会登录失败,结果如图 7-2 所示。当第一个浏览器主动执行注销登录后,第二个浏览器就可以登录了。

图 7-2　超过了当前用户的最大会话数,登录失败

### 7.2.2　原理分析

接下来我们来分析上面的效果是怎么实现的。这里涉及了比较多的类,我们逐个来看。

#### 7.2.2.1　SessionInformation

SessionInformation 主要用作 Spring Security 框架内的会话记录,代码如下:

```
public class SessionInformation implements Serializable {
    private Date lastRequest;
    private final Object principal;
    private final String sessionId;
    private boolean expired = false;
    public void refreshLastRequest() {
        this.lastRequest = new Date();
    }
    // 省略 getter/setter
}
```

这里定义了四个属性:

(1) lastRequest:最近一次请求的时间。

(2) principal:会话对应的主体(用户)。

(3) sessionId:会话 Id。
```

(4) expired：会话是否过期。

(5) refreshLastRequest()：该方法用来更新最近一次请求的时间。

7.2.2.2 SessionRegistry

SessionRegistry 是一个接口，主要用来维护 SessionInformation 实例，该接口只有一个实现类 SessionRegistryImpl，所以这里我们就不看接口了，直接来看实现类 SessionRegistryImpl，SessionRegistryImpl 类的定义比较长，我们拆开来看，先来看属性的定义：

```java
public class SessionRegistryImpl implements SessionRegistry,
    ApplicationListener<SessionDestroyedEvent> {
  private final ConcurrentMap<Object, Set<String>> principals;
  private final Map<String, SessionInformation> sessionIds;
  public SessionRegistryImpl() {
    this.principals = new ConcurrentHashMap<>();
    this.sessionIds = new ConcurrentHashMap<>();
  }
  public SessionRegistryImpl(ConcurrentMap<Object, Set<String>> principals,
                    Map<String, SessionInformation> sessionIds) {
    this.principals=principals;
    this.sessionIds=sessionIds;
  }
  public void onApplicationEvent(SessionDestroyedEvent event) {
    String sessionId = event.getId();
    removeSessionInformation(sessionId);
  }
}
```

SessionRegistryImpl 实现了 SessionRegistry 和 ApplicationListener 两个接口，实现了 ApplicationListener 接口，并通过重写其 onApplicationEvent 方法，就可以接收到 HttpSession 的销毁事件，进而移除掉 HttpSession 的记录。

SessionRegistryImpl 中一共定义了两个属性：

(1) principals：该变量用来保存当前登录主体（用户）和 SessionId 之间的关系，key 就是当前登录主体（即当前登录用户对象），value 则是当前登录主体所对应的会话 Id 的集合。

(2) sessionIds：该变量用来保存 sessionId 和 SessionInformation 之间的映射关系，key 是 sessionId，value 则是 SessionInformation。

> **注 意**
>
> 由于 principals 集合中采用当前登录用户对象做 key，将对象作为集合中的 key，需要重写其 equals 方法和 hashCode 方法。在前面的案例中，由于我们使用了系统默认定义的 User 类，该类已经重写了 equals 方法和 hashCode 方法。如果开发者自定义用户类，记得重写其 equals 方法和 hashCode 方法，否则会话并发管理会失效。

继续来看 SessionRegistryImpl 中的其他方法：

```java
public List<Object> getAllPrincipals() {
    return new ArrayList<>(principals.keySet());
}
public List<SessionInformation> getAllSessions(Object principal,
                                    boolean includeExpiredSessions) {
    final Set<String> sessionsUsedByPrincipal = principals.get(principal);
    if (sessionsUsedByPrincipal == null) {
        return Collections.emptyList();
    }
    List<SessionInformation> list = new ArrayList<>(
            sessionsUsedByPrincipal.size());
    for (String sessionId : sessionsUsedByPrincipal) {
        SessionInformation sessionInformation =
                                    getSessionInformation(sessionId);
        if (sessionInformation == null) {
            continue;
        }
        if (includeExpiredSessions || !sessionInformation.isExpired()) {
            list.add(sessionInformation);
        }
    }
    return list;
}
public SessionInformation getSessionInformation(String sessionId) {
    return sessionIds.get(sessionId);
}
public void refreshLastRequest(String sessionId) {
    SessionInformation info = getSessionInformation(sessionId);
    if (info != null) {
        info.refreshLastRequest();
    }
}
```

（1）getAllPrincipals：该方法返回所有的登录用户对象。

（2）getAllSessions：该方法返回某一个用户所对应的所有 SessionInformation。方法第一个参数就是用户对象，第二个参数表示是否包含已经过期的 Session。具体操作就是从 principals 变量中获取该用户对应的所有 sessionId，然后调用 getSessionInformation 方法从 sessionIds 变量中获取每一个 sessionId 所对应的 SessionInformation，最终将获取到的 SessionInformation 存入集合中返回。

（3）getSessionInformation：该方法主要是根据 sessionId 从 sessionIds 集合中获取对应的 SessionInformation。

（4）refreshLastRequest：根据传入的 sessionId 找到对应的 SessionInformation，并调用其 refreshLastRequest 方法刷新最后一次请求的时间。

再来看会话的保存操作：

```java
public void registerNewSession(String sessionId, Object principal) {
    if (getSessionInformation(sessionId) != null) {
        removeSessionInformation(sessionId);
    }
    sessionIds.put(sessionId,
            new SessionInformation(principal, sessionId, new Date()));
    principals.compute(principal, (key, sessionsUsedByPrincipal) -> {
        if (sessionsUsedByPrincipal == null) {
            sessionsUsedByPrincipal = new CopyOnWriteArraySet<>();
        }
        sessionsUsedByPrincipal.add(sessionId);
        return sessionsUsedByPrincipal;
    });
}
```

当用户登录成功后，会执行会话保存操作，传入当前请求的 sessionId 和当前登录主体 principal 对象。如果 sessionId 已经存在，则先将其移除，然后先往 sessionIds 中保存，key 是 sessionId，value 则是一个新创建的 SessionInformation 对象。

在向 principals 集合中保存时使用了 compute 方法（如果读者对 Java8 中的 compute 方法还不太熟悉，可以自行学习，这里不做过多介绍），第一个参数就是当前登录主体，第二个参数则进行了计算。如果当前登录主体在 principals 中已经有对应的 value，则在 value 的基础上继续添加一个 sessionId。如果当前登录主体在 principals 中没有对应的 value，则新建一个 sessionsUsedByPrincipal 对象，然后再将 sessionId 添加进去。

最后我们再来看会话的移除操作：

```java
public void removeSessionInformation(String sessionId) {
    SessionInformation info = getSessionInformation(sessionId);
    if (info == null) {
        return;
    }
    sessionIds.remove(sessionId);
    principals.computeIfPresent(info.getPrincipal(),
                                (key, sessionsUsedByPrincipal) -> {
        sessionsUsedByPrincipal.remove(sessionId);
        if (sessionsUsedByPrincipal.isEmpty()) {
            sessionsUsedByPrincipal = null;
        }
        return sessionsUsedByPrincipal;
    });
}
```

移除也是两方面的工作，一方面就是从 sessionIds 变量中移除，这个直接调用 remove 方法即可；另一方面就是从 principals 变量中移除，principals 中 key 是当前登录的用户对象，value 则是一个集合，里边保存着当前用户对应的所有 sessionId，这里主要是移除 value 中对应的 sessionId。

7.2.2.3 SessionAuthenticationStrategy

SessionAuthenticationStrategy 是一个接口，主要在用户登录成功后，对 HttpSession 进行处理。它里边只有一个 onAuthentication 方法，用来处理和 HttpSession 相关的事情：

```java
public interface SessionAuthenticationStrategy {
    void onAuthentication(Authentication authentication,
                          HttpServletRequest request,
                          HttpServletResponse response)
                          throws SessionAuthenticationException;
}
```

SessionAuthenticationStrategy 有如下一些实现类：

- CsrfAuthenticationStrategy：CsrfAuthenticationStrategy 和 CSRF 攻击有关，该类主要负责在身份验证后删除旧的 CsrfToken 并生成一个新的 CsrfToken。
- ConcurrentSessionControlAuthenticationStrategy：该类主要用来处理 Session 并发问题。前面案例中 Session 并发的控制，实际上就是通过该类来完成的。
- RegisterSessionAuthenticationStrategy：该类用于在认证成功后将 HttpSession 信息记录到 SessionRegistry 中。
- CompositeSessionAuthenticationStrategy：这是一个复合策略，它里边维护了一个集合，集合中保存了多个不同的 SessionAuthenticationStrategy 对象，相当于该类代理了多个 SessionAuthenticationStrategy 对象，大部分情况下，在 Spring Security 框架中直接使用的也是该类的实例。
- NullAuthenticatedSessionStrategy：这是一个空的实现，未做任何处理。
- AbstractSessionFixationProtectionStrategy：处理会话固定攻击的基类。
- ChangeSessionIdAuthenticationStrategy：通过修改 sessionId 来防止会话固定攻击。
- SessionFixationProtectionStrategy：通过创建一个新的会话来防止会话固定攻击。

ConcurrentSessionControlAuthenticationStrategy

在前面的案例中，起主要作用的是 ConcurrentSessionControlAuthenticationStrategy，因此这里先对该类进行重点分析，先来看它里边的 onAuthentication 方法：

```java
public void onAuthentication(Authentication authentication,
        HttpServletRequest request, HttpServletResponse response) {
    final List<SessionInformation> sessions =
    sessionRegistry.getAllSessions(authentication.getPrincipal(), false);
    int sessionCount = sessions.size();
    int allowedSessions = getMaximumSessionsForThisUser(authentication);
    if (sessionCount < allowedSessions) {
        return;
    }
    if (allowedSessions == -1) {
        return;
    }
    if (sessionCount == allowedSessions) {
```

```
        HttpSession session = request.getSession(false);
        if (session != null) {
            for (SessionInformation si : sessions) {
                if (si.getSessionId().equals(session.getId())) {
                    return;
                }
            }
        }
    }
    allowableSessionsExceeded(sessions, allowedSessions, sessionRegistry);
}
```

在该方法中，首先从 sessionRegistry 中获取当前用户的所有未失效的 SessionInformation 实例，然后获取到当前项目允许的最大 session 数。如果获取到的 SessionInformation 实例数小于当前项目允许的最大 session 数，说明当前登录没问题，直接 return 即可。如果允许的最大 session 数量为-1，则表示应用并不限制登录并发数，当前登录也没有问题，直接返回即可。如果获取到的 SessionInformation 实例等于当前项目允许的最大 session 数，则去判断当前登录的 sessionId 是否存在于获取到的 SessionInformation 实例中，如果存在，说明登录也没问题，直接返回即可。

如果在前面的判断中没有 return，说明当前用户登录的并发数已经超过允许的并发数了，进入到 allowableSessionsExceeded 方法中进行处理，代码如下：

```
protected void allowableSessionsExceeded(
                    List<SessionInformation> sessions,
                    int allowableSessions, SessionRegistry registry)
                    throws SessionAuthenticationException {
    if (exceptionIfMaximumExceeded || (sessions == null)) {
        throw new SessionAuthenticationException(messages.getMessage(
        "ConcurrentSessionControlAuthenticationStrategy.exceededAllowed",
            new Object[] {allowableSessions},
            "Maximum sessions of {0} for this principal exceeded"));
    }
    sessions.sort(Comparator.comparing(SessionInformation::getLastRequest));
    int maximumSessionsExceededBy = sessions.size() - allowableSessions + 1;
    List<SessionInformation> sessionsToBeExpired =
                        sessions.subList(0, maximumSessionsExceededBy);
    for (SessionInformation session: sessionsToBeExpired) {
        session.expireNow();
    }
}
```

如果 exceptionIfMaximumExceeded 属性为 true，则直接抛出异常，该属性的值也就是我们在 SecurityConfig 中通过 maxSessionsPreventsLogin 方法配置的值，即禁止后来者登录，抛出异常后，本次登录失败。否则说明不禁止后来者登录，此时对查询出来的当前用户所有登录会话按照最后一次请求时间进行排序，然后计算出需要过期的 session 数量，从 sessions 集合中取出来进行遍历，依次调用其 expireNow 方法使之过期。

这便是 ConcurrentSessionControlAuthenticationStrategy 类的实现逻辑。

RegisterSessionAuthenticationStrategy

在前面的案例中，默认也用到了 RegisterSessionAuthenticationStrategy，该类的作用主要是向 SessionRegistry 中记录 HttpSession 信息，我们来看一下它的 onAuthentication 方法：

```
public void onAuthentication(Authentication authentication,
    HttpServletRequest request, HttpServletResponse response) {
  sessionRegistry.registerNewSession(request.getSession().getId(),
      authentication.getPrincipal());
}
```

可以看到，这里的 onAuthentication 方法非常简单，就是调用 registerNewSession 方法向 sessionRegistry 中添加一条登录会话信息。

CompositeSessionAuthenticationStrategy

CompositeSessionAuthenticationStrategy 类相当于一个代理类，默认使用的其实就是该类的实例，我们来看一下该类的 onAuthentication 方法：

```
public void onAuthentication(Authentication authentication,
    HttpServletRequest request, HttpServletResponse response)
        throws SessionAuthenticationException {
  for (SessionAuthenticationStrategy delegate : this.delegateStrategies) {
    delegate.onAuthentication(authentication, request, response);
  }
}
```

可以看到，这里就是遍历它所维护的 SessionAuthenticationStrategy 集合，然后分别调用其 onAuthentication 方法。

在前面的案例中，主要涉及这些 SessionAuthenticationStrategy 实例，还有其他一些 SessionAuthenticationStrategy 实例，我们将在接下来的小节中详细介绍，这里不再赘述。

7.2.2.4　SessionManagementFilter

和会话并发管理相关的过滤器主要有两个，先来看第一个 SessionManagementFilter。

SessionManagementFilter 主要用来处理 RememberMe 登录时的会话管理：即如果用户使用了 RememberMe 的方式进行认证，则认证成功后需要进行会话管理，相关的管理操作通过 SessionManagementFilter 过滤器触发。我们来看一下该过滤器的 doFilter 方法：

```
public void doFilter(ServletRequest req, ServletResponse res,
                FilterChain chain)throws IOException, ServletException {
  HttpServletRequest request = (HttpServletRequest) req;
  HttpServletResponse response = (HttpServletResponse) res;
  if (request.getAttribute(FILTER_APPLIED) != null) {
    chain.doFilter(request, response);
    return;
  }
```

```
        request.setAttribute(FILTER_APPLIED, Boolean.TRUE);
        if (!securityContextRepository.containsContext(request)) {
            Authentication authentication = SecurityContextHolder.getContext()
                    .getAuthentication();
            if (authentication != null
                            && !trustResolver.isAnonymous(authentication)) {
                try {
                sessionAuthenticationStrategy.onAuthentication(authentication,
                        request, response);
                }
                catch (SessionAuthenticationException e) {
                    SecurityContextHolder.clearContext();
                    failureHandler.onAuthenticationFailure(request, response, e);
                    return;
                }
                securityContextRepository
                 .saveContext(SecurityContextHolder.getContext(),request, response);
            }
            else {
                if (request.getRequestedSessionId() != null
                        && !request.isRequestedSessionIdValid()) {
                    if (invalidSessionStrategy != null) {
                        invalidSessionStrategy
                                .onInvalidSessionDetected(request, response);
                        return;
                    }
                }
            }
        }
        chain.doFilter(request, response);
    }
```

在该过滤器中，通过 containsContext 方法去判断当前会话中是否存在 SPRING_SECURITY_CONTEXT 变量。如果是正常的认证流程，则 SPRING_SECURITY_CONTEXT 变量是存在于当前会话中的（本书 2.3 节关于 containsContext 方法的详细阐述）。那么什么时候不存在呢？有两种情况：

（1）用户使用了 RememberMe 方式进行认证。

（2）用户匿名访问某一个接口。

对于第一种情况，SecurityContextHolder 中获取到的当前用户实例是 RememberMe AuthenticationToken；对于第二种情况，SecurityContextHolder 中获取到的当前用户实例是 AnonymousAuthenticationToken。所以，接下来就是对这两种情况进行区分，如果是第一种情况，则调用 SessionAuthenticationStrategy 中的 onAuthentication 方法进行会话管理；如果是第二种情况，则进行会话失效处理。

7.2.2.5　ConcurrentSessionFilter

ConcurrentSessionFilter 过滤器是一个处理会话并发管理的过滤器，我们来看一下它的 doFilter 方法：

```java
public void doFilter(ServletRequest req, ServletResponse res,
                FilterChain chain)throws IOException, ServletException {
    HttpServletRequest request = (HttpServletRequest) req;
    HttpServletResponse response = (HttpServletResponse) res;
    HttpSession session = request.getSession(false);
    if (session != null) {
        SessionInformation info =
                sessionRegistry.getSessionInformation(session.getId());
        if (info != null) {
            if (info.isExpired()) {
                doLogout(request, response);
                this.sessionInformationExpiredStrategy.onExpiredSessionDetected(
                  new SessionInformationExpiredEvent(info, request, response));
                    return;
            }
            else {
                sessionRegistry.refreshLastRequest(info.getSessionId());
            }
        }
    }
    chain.doFilter(request, response);
}
```

从 doFilter 方法中可以看到，当请求通过时，首先获取当前会话，如果当前会话不为 null，则进而获取当前会话所对应的 SessionInformation 实例；如果 SessionInformation 实例已经过期，则调用 doLogout 方法执行注销操作，同时调用会话过期的回调；如果 SessionInformation 实例没有过期，则刷新当前会话的最后一次请求时间。

7.2.2.6　Session 创建时机

在讲解配置类之前，需要先了解一下 Spring Security 中 Session 的创建时机问题。在 Spring Security 中，HttpSession 的创建策略一共分为四种：

- ALWAYS：如果 HttpSession 不存在，就创建。
- NEVER：从不创建 HttpSession，但是如果 HttpSession 已经存在了，则会使用它。
- IF_REQUIRED：当有需要时，会创建 HttpSession，默认即此。
- STATELESS：从不创建 HttpSession，也不使用 HttpSession。

需要注意的是，这四种策略仅仅是指 Spring Security 中 HttpSession 的创建策略，而并非整个应用程序中 HttpSession 的创建策略。前三种策略都好理解，第四种策略完全不使用 HttpSession，有读者可能会有疑惑，完全不使用 HttpSession，那么 Spring Security 还能发挥作用吗？当然是可以的！如果系统使用了无状态认证方式，就可以使用 STATELESS 策略，这就意味着服务端不会创建 HttpSession，客户端的每一个请求都需要携带认证信息，同时，一些和 HttpSession 相关的过滤器也将失效，如 SessionManagementFilter、ConcurrentSessionFilter 等。

一般来说，我们使用默认的 IF_REQUIRED 即可，如果读者需要配置，可以通过如下方式进行：

```
@Override
protected void configure(HttpSecurity http) throws Exception {
    http.sessionManagement()
            .sessionCreationPolicy(SessionCreationPolicy.IF_REQUIRED);
}
```

7.2.2.7 SessionManagementConfigurer

最后我们再来看看 SessionManagementConfigurer，正是在该配置类中，完成了上面两个过滤器的配置。

作为一个配置类，我们主要看 SessionManagementConfigurer 的 init 方法和 configure 方法，先来看 init 方法：

```
public void init(H http) {
    SecurityContextRepository securityContextRepository = http
            .getSharedObject(SecurityContextRepository.class);
    boolean stateless = isStateless();
    if (securityContextRepository == null) {
        if (stateless) {
            http.setSharedObject(SecurityContextRepository.class,
                    new NullSecurityContextRepository());
        }
        else {
            HttpSessionSecurityContextRepository httpSecurityRepository =
                            new HttpSessionSecurityContextRepository();
            httpSecurityRepository
                    .setDisableUrlRewriting(!this.enableSessionUrlRewriting);
            httpSecurityRepository
                    .setAllowSessionCreation(isAllowSessionCreation());
            AuthenticationTrustResolver trustResolver = http
                    .getSharedObject(AuthenticationTrustResolver.class);
            if (trustResolver != null) {
                httpSecurityRepository.setTrustResolver(trustResolver);
            }
            http.setSharedObject(SecurityContextRepository.class,
                    httpSecurityRepository);
        }
    }
    RequestCache requestCache = http.getSharedObject(RequestCache.class);
    if (requestCache == null) {
        if (stateless) {
            http.setSharedObject(RequestCache.class, new NullRequestCache());
```

```
        }
    }
    http.setSharedObject(SessionAuthenticationStrategy.class,
            getSessionAuthenticationStrategy(http));
    http.setSharedObject(InvalidSessionStrategy.class,
                                        getInvalidSessionStrategy());
}
```

在该方法中，首先从 HttpSecurity 中获取 SecurityContextRepository 实例，如果没有获取到，则进行创建。创建的时候分两种情况，如果 Spring Security 中的 HttpSession 创建策略是 STATELESS，则使用 NullSecurityContextRepository 来保存 SecurityContext（相当于不保存，参见本书 2.3 节）；如果 Spring Security 中的 HttpSession 创建策略不是 STATELESS，则构建 HttpSessionSecurityContextRepository 对象，并最终存入 HttpSecurity 的共享对象中以备使用。

如果 HttpSession 创建策略是 STATELESS，还需要将保存在 HttpSecurity 共享对象中的请求缓存对象替换为 NullRequestCache 的实例。

最后则是分别构建 SessionAuthenticationStrategy 实例和 InvalidSessionStrategy 实例存入 HttpSecurity 共享对象中，其中 SessionAuthenticationStrategy 实例是通过 getSessionAuthenticationStrategy 方法来获取的，在该方法中，一共构建了三个 SessionAuthenticationStrategy 实例，分别是 ConcurrentSessionControlAuthenticationStrategy、ChangeSessionIdAuthenticationStrategy 以及 RegisterSessionAuthenticationStrategy，并将这三个实例由 CompositeSessionAuthenticationStrategy 进行代理，所以 getSessionAuthenticationStrategy 方法最终返回的是 CompositeSessionAuthenticationStrategy 类的实例。

再来看 configure 方法的定义：

```
public void configure(H http) {
    SecurityContextRepository securityContextRepository = http
            .getSharedObject(SecurityContextRepository.class);
    SessionManagementFilter sessionManagementFilter =
    new SessionManagementFilter(securityContextRepository,
                        getSessionAuthenticationStrategy(http));
    if (this.sessionAuthenticationErrorUrl != null) {
        sessionManagementFilter.setAuthenticationFailureHandler(
            new SimpleUrlAuthenticationFailureHandler(
                this.sessionAuthenticationErrorUrl));
    }
    InvalidSessionStrategy strategy = getInvalidSessionStrategy();
    if (strategy != null) {
        sessionManagementFilter.setInvalidSessionStrategy(strategy);
    }
    AuthenticationFailureHandler failureHandler =
                        getSessionAuthenticationFailureHandler();
    if (failureHandler != null) {
      sessionManagementFilter.setAuthenticationFailureHandler(failureHandler);
    }
    AuthenticationTrustResolver trustResolver = http
```

```
            .getSharedObject(AuthenticationTrustResolver.class);
    if (trustResolver != null) {
        sessionManagementFilter.setTrustResolver(trustResolver);
    }
    sessionManagementFilter = postProcess(sessionManagementFilter);
    http.addFilter(sessionManagementFilter);
    if (isConcurrentSessionControlEnabled()) {
        ConcurrentSessionFilter concurrentSessionFilter =
                                            createConcurrencyFilter(http);
        concurrentSessionFilter = postProcess(concurrentSessionFilter);
        http.addFilter(concurrentSessionFilter);
    }
}
```

configure 方法中主要是构建了两个过滤器 SessionManagementFilter 和 ConcurrentSessionFilter。SessionManagementFilter 过滤器在创建时，也是通过 getSessionAuthenticationStrategy 方法获取 SessionAuthenticationStrategy 实例并传入 sessionManagementFilter 实例中，然后为其配置各种回调函数，最终将创建好的 SessionManagementFilter 加入 HttpSecurity 过滤器链中。

如果配置了会话并发控制（只要用户调用.maximumSessions()方法配置了会话最大并发数，就算开启了会话并发控制），就再创建一个 ConcurrentSessionFilter 过滤器链并加入 HttpSecurity 中。

这就是 SessionManagementConfigurer 的主要功能。

7.2.2.8 AbstractAuthenticationFilterConfigurer

看完前面的分析，读者可能还是有疑问，登录成功后，Session 并发管理到底是在哪里触发的？虽然经过前面的分析，大家知道有两个过滤器存在：SessionManagementFilter 和 ConcurrentSessionFilter，但是前者在用户使用 RememberMe 认证时，才会触发 Session 并发管理，后者则根本不会触发 Session 并发管理，那么用户登录成功后，到底是在哪里触发 Session 并发管理的呢？

这里我们可以回到登录过滤器 AbstractAuthenticationProcessingFilter 的 doFilter 方法中去看一看了。我们发现，在其 doFilter 方法中有如下一段代码：

```
public void doFilter(ServletRequest req, ServletResponse res,
            FilterChain chain)throws IOException, ServletException {
    //省略
    try {
        authResult = attemptAuthentication(request, response);
        if (authResult == null) {
            return;
        }
        sessionStrategy.onAuthentication(authResult, request, response);
    }
    //省略
}
```

可以看到，在调用 attemptAuthentication 方法进行登录认证之后，接下来就调用了 sessionStrategy.onAuthentication 方法触发 Session 并发管理。

这里的 sessionStrategy 对象则是在 AbstractAuthenticationFilterConfigurer 类的 configure 方法中进行配置的，代码如下：

```
public void configure(B http) throws Exception {
    //省略
    SessionAuthenticationStrategy sessionAuthenticationStrategy = http
        .getSharedObject(SessionAuthenticationStrategy.class);
    if (sessionAuthenticationStrategy != null) {
      authFilter
        .setSessionAuthenticationStrategy(sessionAuthenticationStrategy);
    }
    //省略
}
```

可以看到，这里从 HttpSecurity 的共享对象中获取到 SessionAuthenticationStrategy 实例（在 SessionManagementConfigurer#init 方法中存入 HttpSecurity 共享对象），并设置到 authFilter 过滤器中。

我们再来梳理一下：

用户通过用户名/密码发起一个认证请求，当认证成功后，在 AbstractAuthenticationProcessingFilter#doFilter 方法中触发了 Session 并发管理。默认的 sessionStrategy 是 CompositeSessionAuthenticationStrategy，它一共代理了三个 SessionAuthenticationStrategy，分别是 ConcurrentSessionControlAuthenticationStrategy、ChangeSessionIdAuthenticationStrategy 以及 RegisterSessionAuthenticationStrategy。当前请求在这三个 SessionAuthenticationStrategy 中分别走一圈，第一个用来判断当前登录用户的 Session 数是否已经超出限制，如果超出限制就根据配置好的规则作出处理；第二个用来修改 sessionId（防止会话固定攻击）；第三个用来将当前 Session 注册到 SessionRegistry 中。使用用户名/密码的方式完成认证，将不会涉及 ConcurrentSessionFilter 和 SessionManagementFilter 两个过滤器。如果用户使用了 RememberMe 的方式来进行身份认证，则会通过 SessionManagementFilter#doFilter 方法触发 Session 并发管理。当用户认证成功后，以后的每一次请求都会经过 ConcurrentSessionFilter，在该过滤器中，判断当前会话是否已经过期，如果过期就执行注销登录流程；如果没有过期，则更新最近一次请求时间。

7.3 会话固定攻击与防御

7.3.1 什么是会话固定攻击

会话固定攻击（Session fixation attacks）是一种潜在的风险，恶意攻击者有可能通过访问当前应用程序来创建会话，然后诱导用户以相同的会话 ID 登录（通常是将会话 ID 作为参数放在请求链接中，然后诱导用户去单击），进而获取用户的登录身份。

举一个简单的会话固定攻击的例子（以 www.javaboy.org 为例，此网站为笔者所有）：

（1）攻击者自己可以正常访问 javaboy 网站，在访问的过程中，网站给攻击者分配了一个 sessionid。

（2）攻击者利用自己拿到的 sessionid 构造一个 javaboy 网站的链接，并把该链接发送给受害者。

（3）受害者使用该链接登录 javaboy 网站（该链接中含有 sessionid），登录成功后，一个合法的会话就成功建立了。

（4）攻击者利用手里的 sessionid 冒充受害者。

在这个过程中，如果 javaboy 网站支持 URL 重写，那么攻击还会变得更加容易。

什么是 URL 重写？用户如果在浏览器中禁用了 cookie，那么 sessionid 自然也用不了，所以有的服务端就支持把 sessionid 放在请求地址中，类似下面这样：

```
http://www.javaboy.org;jsessionid=xxxxxx
```

如果服务端支持这种 URL 重写，那么对于攻击者来说，按照上面的攻击流程，构造一个这样的地址就太容易了。

7.3.2 会话固定攻击防御策略

Spring Security 中从三方面入手防范会话固定攻击：

（1）Spring Security 中默认自带了 Http 防火墙，如果 sessionid 放在地址栏中，这个请求就会直接被拦截下来（本书第 8 章会详细分析 Http 防火墙）。

（2）在 Http 响应的 Set-Cookie 字段中有 httpOnly 属性，这样避免了通过 XSS 攻击来获取 Cookie 中的会话信息，进而达成会话固定攻击（在第 6 章的 RememberMe 认证中，服务端返回令牌信息时也设置了 httpOnly 属性为 true）。

（3）既然会话固定攻击是由于 SessionId 不变导致的，那么其中一个解决办法就是在用户登录成功后，改变 SessionId，Spring Security 中默认实现了该种方案，实现类就是 7.2 节所讲的 ChangeSessionIdAuthenticationStrategy。

前两种都是默认行为，一般来说不需要做更改。第三种方案，Spring Security 中有几种不同的配置策略，我们先来看一下配置方式：

```
http.sessionManagement().sessionFixation().changeSessionId();
```

通过 sessionFixation() 方法开启会话固定攻击防御的配置，一共有四种不同的策略，不同策略对应了不同的 SessionAuthenticationStrategy：

（1）changeSessionId()：用户登录成功后，直接修改 HttpSession 的 SessionId 即可，默认

方案即此，对应的处理类是 ChangeSessionIdAuthenticationStrategy。

（2）none()：用户登录成功后，HttpSession 不做任何变化，对应的处理类是 NullAuthenticatedSessionStrategy。

（3）migrateSession()：用户登录成功后，创建一个新的 HttpSession 对象，并将旧的 HttpSession 中的数据拷贝到新的 HttpSession 中，对应的处理类是 SessionFixationProtectionStrategy。

（4）newSession()：用户登录成功后，创建一个新的 HttpSession 对象，对应的处理类也是 SessionFixationProtectionStrategy，只不过将其里边的 migrateSessionAttributes 属性设置为 false。需要注意的是，该方法并非所有的属性都不拷贝，一些 Spring Security 使用的属性，如请求缓存，还是会从旧的 HttpSession 上复制到新的 HttpSession。

这四种策略，无论使用哪种，都相当于配置了一个 SessionAuthenticationStrategy，在 7.2.2 小节的分析中大家已经知道，默认使用的是 ChangeSessionIdAuthenticationStrategy，如果开发者在这里配置了其他类型的 SessionAuthenticationStrategy，就会替代掉默认使用的 ChangeSessionIdAuthenticationStrategy。

一般来说，我们使用 Spring Security 默认的方案就可以防御会话固定攻击了，即什么都不做，就可以防御会话固定攻击，这也是 Spring Security 强大之处！

7.4 Session 共享

7.4.1 集群会话方案

前面所讲的会话管理都是单机上的会话管理，如果当前是集群环境，前面所讲的会话管理方案就会失效。需要注意的是，我们这里讨论的范畴是有状态登录，如果用户采用无状态的认证方案，那么就不涉及会话，也就不存在接下来要讨论的问题。

我们先来看一幅简单的集群架构图，如图 7-3 所示。

如图 7-3 所示，如果项目是集群化部署，我们可以采用 Nginx 做反向代理服务器，所有到达 Nginx 上的请求被转发到不同的 Tomcat 实例上，每个 Tomcat 各自保存自己的会话信息。根据前面的讲解，Spring Security 中通过维护一张会话注册表来实现会话的并发管理，现在每个 Tomcat 上都有一张会话注册表，所以如果还按照之前的方式去配置会话并发管理，那必然是不生效的。

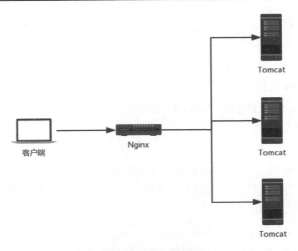

图 7-3　简化版的集群架构图

为了解决集群环境下的会话问题，我们有三种方案：

（1）Session 复制：多个服务之间互相复制 Session 信息，这样每个服务中都包含有所有的 Session 信息了，Tomcat 通过 IP 组播对这种方案提供支持。但是这种方案占用带宽、有时延，服务数量越多效率越低，所以这种方案使用较少。

（2）Session 粘滞：也叫会话保持，就是在 Nginx 上通过一致性 Hash，将 Hash 结果相同的请求总是分发到一个服务上去。这种方案可以解决一部分集群会话带来的问题，但是无法解决集群中的会话并发管理问题。

（3）Session 共享：Session 共享就是将不同服务的会话统一放在一个地方，所有的服务共享一个会话。一般使用一些 Key-Value 数据库来存储 Session，例如 Memcached 或者 Redis 等，比较常见的方案是使用 Redis 存储，Session 共享方案由于其简便性与稳定性，是目前使用较多的方案。Session 共享架构图如图 7-4 所示。

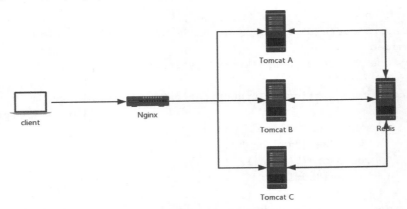

图 7-4　简化版的 Session 共享架构图

Session 共享目前使用比较多的是 spring-session，利用 spring-session 可以方便地实现 Session 的管理。

7.4.2 实战

首先启动一个 Redis 实例。

新建 Spring Boot 工程，分别引入 Web、Redis、Spring Security 以及 Spring Session 依赖，代码如下：

```xml
<dependency>
    <groupId>org.springframework.boot</groupId>
    <artifactId>spring-boot-starter-data-redis</artifactId>
</dependency>
<dependency>
    <groupId>org.springframework.boot</groupId>
    <artifactId>spring-boot-starter-security</artifactId>
</dependency>
<dependency>
    <groupId>org.springframework.boot</groupId>
    <artifactId>spring-boot-starter-web</artifactId>
</dependency>
<dependency>
    <groupId>org.springframework.session</groupId>
    <artifactId>spring-session-data-redis</artifactId>
</dependency>
```

接下来在 application.properties 中配置 Redis 连接信息：

```
spring.redis.password=123
spring.redis.host=127.0.0.1
spring.redis.port=6379
```

再来提供一个 SecurityConfig，代码如下：

```java
@Configuration
public class SecurityConfig extends WebSecurityConfigurerAdapter {
    @Autowired
    FindByIndexNameSessionRepository sessionRepository;
    @Override
    protected void configure(AuthenticationManagerBuilder auth)
                                                    throws Exception {
        auth.inMemoryAuthentication()
                .withUser("javaboy")
                .password("{noop}123")
                .roles("admin");
    }
    @Override
    protected void configure(HttpSecurity http) throws Exception {
        http.authorizeRequests()
                .anyRequest().authenticated()
```

```
                .and()
                .formLogin()
                .and()
                .csrf()
                .disable()
                .sessionManagement()
                .maximumSessions(1)
                .sessionRegistry(sessionRegistry());
    }
    @Bean
    SpringSessionBackedSessionRegistry sessionRegistry() {
        return new SpringSessionBackedSessionRegistry(sessionRepository);
    }
}
```

在这段配置中，我们首先注入了一个 FindByIndexNameSessionRepository 对象，这是一个会话的存储和加载工具。在前面的案例中，会话信息是保存在内存中的，现在会话信息保存在 Redis 中，具体的保存和加载过程则是由 FindByIndexNameSessionRepository 接口的实现类来完成，默认是 RedisIndexedSessionRepository，即我们一开始注入的实际上是一个 RedisIndexedSessionRepository 类型的对象。

接下来我们还配置了一个 SpringSessionBackedSessionRegistry 实例，构建时传入了 sessionRepository。SpringSessionBackedSessionRegistry 继承自 SessionRegistry，用来维护会话信息注册表。

最后在 HttpSecurity 中配置 sessionRegistry 即可，相当于 spring-session 提供的 SpringSessionBackedSessionRegistry 接管了会话信息注册表的维护工作。

需要注意的是，引入了 spring-session 之后，不再需要配置 HttpSessionEventPublisher 实例，因为 spring-session 中通过 SessionRepositoryFilter 将请求对象重新封装为 SessionRepositoryRequestWrapper，并重写了 getSession 方法。在重写的 getSession 方法中，最终返回的是 HttpSessionWrapper 实例，而在 HttpSessionWrapper 定义时，就重写了 invalidate 方法。当调用会话的 invalidate 方法去销毁会话时，就会调用 RedisIndexedSessionRepository 中的方法，从 Redis 中移除对应的会话信息，所以不再需要 HttpSessionEventPublisher 实例。

最后再配置一个测试 Controller：

```
@RestController
public class HelloController {
    @GetMapping("/")
    public String hello(HttpSession session) {
        return session.getClass().toString();
    }
}
```

在测试接口中返回 HttpSession 的类型以验证我们前面的讲解。

配置完成后，我们对项目进行打包，单击 IntelliJ IDEA 右侧的 Maven→Lifecycle→package

进行打包，如图 7-5 所示。

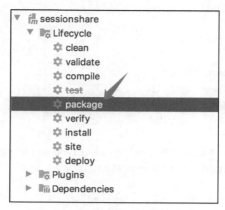

图 7-5 单击按钮对项目进行打包

打包完成后，进入 target 目录下，会有一个 jar，执行如下命令，分别启动两个实例：

```
java -jar sessionshare-0.0.1-SNAPSHOT.jar --server.port=8080
java -jar sessionshare-0.0.1-SNAPSHOT.jar --server.port=8081
```

两个实例启动完成后，这两个实例实际上共用了一个会话。接下来准备两个浏览器，先用浏览器 1 访问 8080 端口的项目，并完成登录操作；然后再用浏览器 2 访问 8081 端口的项目并完成登录操作。当浏览器 2 登录成功后，我们再去刷新浏览器 1，此时发现会话已经过期，说明集群环境下的会话管理已经生效。

7.5 小　结

本章主要介绍了 Spring Security 中的会话管理，从会话简介、会话并发管理、会话固定攻击与防御以及集群会话方案四个方面进行介绍，并对 Spring Security 中会话管理的相关源码进行了深入分析。在传统的 Web 开发中，会话的重要性不言而喻，而当下手机 App、小程序等的兴起，一定程度上降低了对服务端会话的需求，无状态登录也成为一个重要的备选方案。不过，即使这样，读者还是很有必要掌握 Web 中的会话管理，因为即使是无状态登录，实现思路还是与会话管理的思路高度相似。

第 8 章

HttpFirewall

HttpFirewall 是 Spring Security 提供的 Http 防火墙，它可以用于拒绝潜在的危险请求或者包装这些请求进而控制其行为。通过 HttpFirewall 可以对各种非法请求提前进行拦截并处理，降低损失。代码层面，HttpFirewall 被注入到 FilterChainProxy 中，并在 Spring Security 过滤器链执行之前被触发。

本章涉及的主要知识点有：

- HttpFirewall 简介。
- HttpFirewall 严格模式。
- HttpFirewall 普通模式。

8.1 HttpFirewall 简介

在 Servlet 容器规范中，为 HttpServletRequest 定义了一些属性，如 contextPath、servletPath、pathInfo、queryString 等，这些属性都可以通过 get 方法获取。

然而，在 Servlet 容器规范中并没有定义这些属性可以包含哪些值，例如在 servletPath 和 pathInfo 中都可以包含 RFC2396 规范（https://www.ietf.org/rfc/rfc2396.txt）中定义的参数，不同容器对此处理方案也不同，有的容器会对此进行预处理，有的容器则不会。这种比较混乱的处理方式有可能会造成安全隐患，因此 Spring Security 中通过 HttpFirewall 来检查请求路径以及参数是否合法，如果合法，才会进入到过滤器链中进行处理。

HttpFirewall 是一个接口，它只有两个方法：

```
public interface HttpFirewall {
```

```
    FirewalledRequest getFirewalledRequest(HttpServletRequest request)
            throws RequestRejectedException;
    HttpServletResponse getFirewalledResponse(HttpServletResponse response);
}
```

getFirewalledRequest 方法用来对请求对象进行校验并封装，getFirewalledResponse 方法则对响应对象进行封装。

FirewalledRequest 是封装后的请求类，但实际上该类只是在 HttpServletRequestWrapper 的基础上增加了 reset 方法。当 Spring Security 过滤器链执行完毕时，由 FilterChainProxy 负责调用该 reset 方法，以便重置全部或者部分属性。

FirewalledResponse 是封装后的响应类，该类主要重写了 sendRedirect、setHeader、addHeader 以及 addCookie 四个方法，在每一个方法中都对其参数进行校验，以确保参数中不含有\r 和\n。

HttpFirewall 一共有两个实现类，如图 8-1 所示。

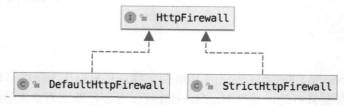

图 8-1　HttpFirewall 的两个实现类

- DefaultHttpFirewall：虽然名字中包含 Default，但这并不是框架默认使用的 Http 防火墙，它只是一个检查相对宽松的防火墙。
- StrictHttpFirewall：这是一个检查严格的 Http 防火墙，也是框架默认使用的 Http 防火墙。

HttpFirewall 中对请求的合法性校验在 FilterChainProxy#doFilterInternal 方法中触发，具体可以参考 4.1.4 小节，这里不再赘述。

需要注意的是 HttpFirewall 的配置位置，在 Spring Security 框架中有两个地方涉及了 HttpFirewall 实例的获取：

（1）在 FilterChainProxy 属性定义中，默认创建的 HttpFirewall 实例就是 StrictHttpFirewall。

（2）FilterChainProxy 是在 WebSecurity#performBuild 方法中构建的，而 WebSecurity 实现了 ApplicationContextAware 接口，并实现了接口中的 setApplicationContext 方法，在该方法中，从 Spring 容器中查找到 HttpFirewall 对象并赋值给 httpFirewall 属性。最终在 performBuild 方法中，将 FilterChainProxy 对象构建成功后，如果 httpFirewall 不为 null，就把 httpFirewall 配置给 FilterChainProxy 对象。

根据以上两点可以得出：如果 Spring 容器中存在 HttpFirewall 实例，则最终使用 Spring 容器提供的 HttpFirewall 实例；如果 Spring 容器中不存在 HttpFirewall 实例，则使用 FilterChainProxy 中默认定义的 StrictHttpFirewall。进而可知，如果开发者不想使用默认的 StrictHttpFirewall 实例，则只需要自己提供一个 HttpFirewall 并注册到 Spring 容器中即可。

8.2 HttpFirewall 严格模式

HttpFirewall 严格模式就是使用 StrictHttpFirewall，默认即此。本节我们将对严格模式中的规则逐一进行分析。

在 FilterChainProxy#doFilterInternal 中触发请求校验的方法如下：

```java
private void doFilterInternal(ServletRequest request,
                ServletResponse response,
        FilterChain chain) throws IOException, ServletException {
    FirewalledRequest fwRequest = firewall
        .getFirewalledRequest((HttpServletRequest) request);
    HttpServletResponse fwResponse = firewall
        .getFirewalledResponse((HttpServletResponse) response);
    //省略其他
    vfc.doFilter(fwRequest, fwResponse);
}
```

可以看到，请求的校验主要是在 getFirewalledRequest 方法中完成的。在进入 Spring Security 过滤器链之前，请求对象和响应对象都分别换成 FirewalledRequest 和 FirewalledResponse 了。如前面所述，FirewalledResponse 主要对响应头参数进行校验，比较简单，这里不再赘述。不过需要注意的是，无论是 FirewalledRequest 还是 FirewalledResponse，在经过 Spring Security 过滤器链的时候，还会通过装饰器模式增强其功能，所以开发者最终在接口中拿到的 HttpServletRequest 和 HttpServletResponse 对象，并不是这里的 FirewalledRequest 和 FirewalledResponse。

我们来重点分析 getFirewalledRequest 方法中所做的校验。

StrictHttpFirewall#getFirewalledRequest 源码如下：

```java
public FirewalledRequest getFirewalledRequest(HttpServletRequest request)
                            throws RequestRejectedException {
    rejectForbiddenHttpMethod(request);
    rejectedBlacklistedUrls(request);
    rejectedUntrustedHosts(request);
    if (!isNormalized(request)) {
        throw new RequestRejectedException("The request was rejected because the
                            URL was not normalized.");
    }
    String requestUri = request.getRequestURI();
    if (!containsOnlyPrintableAsciiCharacters(requestUri)) {
        throw new RequestRejectedException("The requestURI was rejected because
                it can only contain printable ASCII characters.");
    }
    return new FirewalledRequest(request) {
```

```
            @Override
            public void reset() {
            }
        };
}
```

可以看到，在返回对象之前，一共做了五个校验：

（1）rejectForbiddenHttpMethod：校验请求方法是否合法。
（2）rejectedBlacklistedUrls：校验请求中的非法字符。
（3）rejectedUntrustedHosts：检验主机信息。
（4）isNormalized：判断参数格式是否合法。
（5）containsOnlyPrintableAsciiCharacters：判断请求字符是否合法。

下面，我们来逐一分析这五个校验方法。

8.2.1　rejectForbiddenHttpMethod

rejectForbiddenHttpMethod 方法主要用来判断请求方法是否合法：

```java
private void rejectForbiddenHttpMethod(HttpServletRequest request) {
    if (this.allowedHttpMethods == ALLOW_ANY_HTTP_METHOD) {
        return;
    }
    if (!this.allowedHttpMethods.contains(request.getMethod())) {
        throw new RequestRejectedException("The request was rejected because the
            HTTP method \"" + request.getMethod() +
            "\" was not included within the whitelist " +
            this.allowedHttpMethods);
    }
}
```

allowedHttpMethods 是一个 Set 集合，默认情况下该集合中包含七个常见的方法：DELETE、GET、HEAD、OPTIONS、PATCH、POST、PUT，ALLOW_ANY_HTTP_METHOD 变量默认情况下则是一个空的 Set 集合。根据 rejectForbiddenHttpMethod 方法中的定义，只要你的请求方法是这七个中的任意一个，请求都是可以通过的，不会被拦截。当然开发者也可以根据实际需求修改 allowedHttpMethods 变量的值，进而调整允许的请求方法。

第一种修改方式如下：

```java
@Configuration
public class SecurityConfig extends WebSecurityConfigurerAdapter {
    @Bean
    HttpFirewall httpFirewall() {
        StrictHttpFirewall strictHttpFirewall = new StrictHttpFirewall();
        Set<String> allowedHttpMethods = new HashSet<>();
        allowedHttpMethods.add(HttpMethod.POST.name());
```

```
            strictHttpFirewall.setAllowedHttpMethods(allowedHttpMethods);
            return strictHttpFirewall;
        }
        //省略其他
    }
```

由开发者自己提供一个 HttpFirewall 实例，并调用 setAllowedHttpMethods 方法来传入一个 Set 集合，集合中保存着允许通过的请求方法，这个集合最终会被赋值给 allowedHttpMethods 变量。配置完成后，重启项目，此时再去访问，就只有 POST 请求可以被处理了，如果发送 GET 请求，那服务端将抛出异常，代码如下：

```
org.springframework.security.web.firewall.RequestRejectedException: The
request was rejected because the HTTP method "GET" was not included within
the whitelist [POST]
```

第二种修改方式如下：

```
@Bean
HttpFirewall httpFirewall() {
    StrictHttpFirewall strictHttpFirewall = new StrictHttpFirewall();
    strictHttpFirewall.setUnsafeAllowAnyHttpMethod(true);
    return strictHttpFirewall;
}
```

这种方式是直接调用 setUnsafeAllowAnyHttpMethod 方法并设置参数为 true，表示允许所有的请求通过。该方法会设置让 allowedHttpMethods 等于 ALLOW_ANY_HTTP_METHOD，这样会导致在 rejectForbiddenHttpMethod 方法的第一个 if 分支中直接返回，进而达到允许所有请求通过的目的。

8.2.2　rejectedBlacklistedUrls

rejectedBlacklistedUrls 主要用来校验请求 URL 是否规范，对于不规范的请求将会直接拒绝掉。什么样的请求算是不规范的请求呢？

（1）如果请求 URL 地址中在编码之前或者编码之后，包含了分号，即 ;、%3b、%3B，则该请求会被拒绝。可以通过 setAllowSemicolon 方法开启或者关闭这一规则。

（2）如果请求 URL 地址中在编码之前或者编码之后，包含了斜杠，即 %2f、%2F，则该请求会被拒绝。可以通过 setAllowUrlEncodedSlash 方法开启或者关闭这一规则。

（3）如果请求 URL 地址中在编码之前或者编码之后，包含了反斜杠，即\\、%5c、%5C，则该请求会被拒绝。可以通过 setAllowBackSlash 方法开启或者关闭这一规则。

（4）如果请求 URL 在编码之后包含了%25，亦或者在编码之前包含了%，则该请求会被拒绝。可以通过 setAllowUrlEncodedPercent 方法开启或者关闭这一规则。

（5）如果请求 URL 在 URL 编码后包含了英文句号%2e 或者%2E，则该请求会被拒绝。可以通过 setAllowUrlEncodedPeriod 方法开启或者关闭这一规则。

```java
private void rejectedBlacklistedUrls(HttpServletRequest request) {
    for (String forbidden : this.encodedUrlBlacklist) {
        if (encodedUrlContains(request, forbidden)) {
            throw new RequestRejectedException("The request was rejected
              because the URL contained a potentially malicious String \""
                                                + forbidden + "\"");
        }
    }
    for (String forbidden : this.decodedUrlBlacklist) {
        if (decodedUrlContains(request, forbidden)) {
            throw new RequestRejectedException("The request was rejected
              because the URL contained a potentially malicious String \""
                                                + forbidden + "\"");
        }
    }
}
```

这里一共包含两个 for 循环。第一个校验编码后的请求地址，第二个校验解码后的请求地址。

在 encodedUrlContains 方法中我们可以看到，这里主要是校验了 contextPath 和 requestURI 两个属性，这两个属性是客户端传递来的字符串，未做任何更改。

而在 decodedUrlContains 方法中，主要校验了 servletPath、pathInfo 两个属性，读者可能会觉得这不是重复校验了吗？前面的 requestURI 已经包含所有了！

这里需要注意，requestURI 是客户端发来的请求，是原封不动的，而 servletPath 和 pathInfo 是经过解码的请求地址，所以两者是不一样的。例如客户端发送的请求是 http://localhost:8080/get%3baaa，那么 requestURI 的值就是 http://localhost:8080/get%3baaa，而 servletPath 的值则是 /get;aaa（假设 contextPath 为空），即在 servletPath 中，将%3b 还原为分号了。

如果请求地址中含有不规范字符，例如请求 http://localhost:8080/get%3baaa 地址，则控制台报错如下：

```
org.springframework.security.web.firewall.RequestRejectedException: The
request was rejected because the URL contained a potentially malicious String
"%3b"
```

8.2.3　rejectedUntrustedHosts

rejectedUntrustedHosts 方法主要用来校验 Host 是否受信任：

```java
private void rejectedUntrustedHosts(HttpServletRequest request) {
    String serverName = request.getServerName();
    if (serverName != null && !this.allowedHostnames.test(serverName)) {
        throw new RequestRejectedException("The request was rejected because the
                            domain " + serverName + " is untrusted.");
    }
}
```

从这里可以看出主要是对 serverName 的校验，allowedHostnames 默认总是返回 true，即默认信任所有的 Host，开发者可以根据实际需求对此进行配置，代码如下：

```
@Bean
HttpFirewall httpFirewall() {
    StrictHttpFirewall strictHttpFirewall = new StrictHttpFirewall();
    strictHttpFirewall.setAllowedHostnames(
            (hostname) -> hostname.equalsIgnoreCase("local.javaboy.org"));
    return strictHttpFirewall;
}
```

这段配置表示 Host 必须是 local.javaboy.org，其他 Host 将不被信任。配置完成后，重启项目，此时如果访问 http://localhost:8080/get，控制台将会报错，代码如下：

```
org.springframework.security.web.firewall.RequestRejectedException:
The request was rejected because the domain localhost is untrusted.
```

使用 http://local.javaboy.org:8080/get 地址则可以正常访问。

8.2.4　isNormalized

isNormalized 方法主要用来检查请求地址是否规范，什么样的地址就算规范呢？即不包含 "./"、"/../" 以及 "/." 三种字符。

```
private static boolean isNormalized(HttpServletRequest request) {
    if (!isNormalized(request.getRequestURI())) {
        return false;
    }
    if (!isNormalized(request.getContextPath())) {
        return false;
    }
    if (!isNormalized(request.getServletPath())) {
        return false;
    }
    if (!isNormalized(request.getPathInfo())) {
        return false;
    }
    return true;
}
```

可以看到，该方法对 requestURI、contextPath、servletPath 以及 pathInfo 分别进行了校验。如果开发者请求 http://local.javaboy.org:8080/get/../ 地址，则控制台报错如下：

```
org.springframework.security.web.firewall.RequestRejectedException: The
request was rejected because the URL was not normalized.
```

8.2.5　containsOnlyPrintableAsciiCharacters

containsOnlyPrintableAsciiCharacters 方法用来校验请求地址中是否包含不可打印的 ASCII 字符：

```java
private static boolean containsOnlyPrintableAsciiCharacters(String uri) {
    int length = uri.length();
    for (int i = 0; i < length; i++) {
        char c = uri.charAt(i);
        if (c < '\u0020' || c > '\u007e') {
            return false;
        }
    }
    return true;
}
```

可打印的 ASCII 字符范围在'\u0020'到'\u007e'之间，对应的十进制就是 32~126 之间，在此范围之外的，属于不可打印的 ASCII 字符。

这就是 StrictHttpFirewall 中的所有校验规则了。其中前三种，开发者可以通过相关方法调整参数进而调整校验行为，后面两种则不可调整。

8.3　HttpFirewall 普通模式

HttpFirewall 普通模式就是使用 DefaultHttpFirewall，该类的校验规则就要简单很多，我们也来看一下其 getFirewalledRequest 方法：

```java
public FirewalledRequest getFirewalledRequest(HttpServletRequest request)
                                throws RequestRejectedException {
    FirewalledRequest fwr = new RequestWrapper(request);
    if (!isNormalized(fwr.getServletPath())
                    || !isNormalized(fwr.getPathInfo()))  {
        throw new RequestRejectedException("Un-normalized paths are not
            supported: " + fwr.getServletPath()
            + (fwr.getPathInfo() != null ? fwr.getPathInfo() : ""));
    }
    String requestURI = fwr.getRequestURI();
    if (containsInvalidUrlEncodedSlash(requestURI)) {
        throw new RequestRejectedException("The requestURI cannot contain
                                encoded slash. Got " + requestURI);
    }
    return fwr;
}
```

可以看到，首先构建了 RequestWrapper 对象对原始请求中的功能进行了增强。在 RequestWrapper 构建的过程中，主要做了两件事：

（1）将请求地址中//格式化为/。

（2）将请求中 servletPath 和 pathInfo 中用分号隔开的参数提取出来，只保留路径即可。

举个简单例子，例如你的请求地址是 http://localhost:8080//get，假设 contextPath 为空，那么在原始请求中获取到的 servletPath 是//get，而在 fwr 对象中获取到的 servletPath 则是/get。

需要注意的是，Tomcat 容器本身会自动将//转为/，而 Jetty 和 Undertow 则不会；这三个容器默认都会自动将用分号隔开的参数剔除，只保留请求路径。

获取到 fwr 对象之后，接下来调用 isNormalized 方法判断 servletPath 和 pathInfo 是否规范，判断逻辑和 StrictHttpFirewall 中的逻辑一致，这里不再赘述。

containsInvalidUrlEncodedSlash 用来判断 requestURI 中是否包含编码后的斜杠，即%2f 或者%2F，默认不允许包含此字符。如果开发者有需求，可以通过 setAllowUrlEncodedSlash 方法设置允许%2f 或者%2F 存在。

一般来说，并不建议开发者在项目中使用 DefaultHttpFirewall，因为相比于 StrictHttpFirewall，DefaultHttpFirewall 的安全性要差很多。如果开发者一定要使用，只需要提供一个 DefaultHttpFirewall 的实例即可：

```
@Bean
HttpFirewall httpFirewall() {
    return new DefaultHttpFirewall();
}
```

8.4 小　结

本章主要分析了 Spring Security 中的防火墙机制。其实 HttpFirewall 所做的事情有一部分在 Servlet 容器中已经做了，但是，由于不同 Servlet 容器的差异性带来了一定的安全风险，而 HttpFirewall 则屏蔽了这种风险，将所有的请求地址统一规范，确保进入到 Spring Security 过滤器链中的请求地址都是合法地址。

第 9 章

漏洞保护

对于一名普通的 Java 工程师来说，在处理系统保护时可能仅仅只是考虑到认证与授权，这无形中为系统带来巨大的安全风险。Spring Security 的优势之一就是为各种可能存在的漏洞提供了保护机制，而这些保护机制默认都是开启的，这也是我们选择 Spring Security 的原因之一，即不用开发者考虑太多事情，就可以开发出一套安全的权限管理系统。本章主要从 CSRF 攻击与防御、HTTP 响应头处理以及 HTTP 通信三个方面来介绍 Spring Security 在漏洞保护方面所做出的努力。

本章涉及的主要知识点有：

- CSRF 攻击与防御。
- HTTP 响应头处理。
- HTTP 通信安全。

9.1 CSRF 攻击与防御

9.1.1 CSRF 简介

CSRF（Cross-Site Request Forgery，跨站请求伪造），也可称为一键式攻击（one-click attack），通常缩写为 CSRF 或者 XSRF。

CSRF 攻击是一种挟持用户在当前已登录的浏览器上发送恶意请求的攻击方法。相对于 XSS 利用用户对指定网站的信任，CSRF 则是利用网站对用户网页浏览器的信任。简单来说，CSRF 是攻击者通过一些技术手段欺骗用户的浏览器，去访问一个用户曾经认证过的网站并执行恶意请求，例如发送邮件、发消息、甚至财产操作（如转账和购买商品）。由于客户端（浏览器）已经在该网站上认证过，所以该网站会认为是真正的用户在操作而执行请求（而实际上这个请求并非用户的本意）。

举个简单的例子：

假设 javaboy 现在登录了某银行的网站准备完成一项转账操作，转账的链接如下：

```
https://bank.xxx.com/withdraw?account=javaboy&amount=1000&for=zhangsan
```

可以看到，这个链接是想从 javaboy 这个账户下转账 1000 元到 zhangsan 账户下，假设 javaboy 没有注销登录该银行的网站，就在同一个浏览器新的选项卡中打开了一个危险网站，这个危险网站中有一幅图片，代码如下：

```
<img src="https://bank.xxx.com/withdraw?account=javaboy&amount=1000
        &for=lisi" />
```

一旦用户打开了这个网站，这个图片链接中的请求就会自动发送出去。由于是同一个浏览器并且用户尚未注销登录，所以该请求会自动携带上对应的有效的 Cookie 信息，进而完成一次转账操作。

这就是跨站请求伪造。

9.1.2 CSRF 攻击演示

接下来我们通过一个简单的案例来演示一遍 CSRF 攻击。

首先创建一个名为 csrf-1 的 Spring Boot 项目，并引入 Web 和 Spring Security 依赖，这个项目相当于我们前面所说的银行网站。

项目创建成功后，我们提供一个转账接口：

```java
@RestController
public class HelloController {
    @PostMapping("/withdraw")
    public void withdraw() {
        System.out.println("执行了一次转账操作");
    }
}
```

由于这里主要是向读者展示这个接口被调通，简单起见就不接收参数了。

接下来配置 Spring Security：

```java
@Configuration
public class SecurityConfig extends WebSecurityConfigurerAdapter {
    @Override
    protected void configure(AuthenticationManagerBuilder auth)
                                                    throws Exception {
        auth.inMemoryAuthentication()
                .withUser("javaboy")
                .password("{noop}123")
                .roles("admin");
    }
```

```java
    @Override
    protected void configure(HttpSecurity http) throws Exception {
        http.authorizeRequests()
                .anyRequest().authenticated()
                .and()
                .formLogin()
                .and()
                .csrf().disable();
    }
}
```

这段配置读者都很熟悉了，不再赘述。唯一需要强调的是，由于 Spring Security 默认开启了 CSRF 攻击防御，所以我们要将其禁止，即 HttpSecurity 中配置的 .csrf().disable()。

接下来创建 csrf-2 项目，引入 Web 依赖即可，这个相当于前面提到的危险网站。项目创建成功后，首先修改项目端口号：

```
server.port=8081
```

然后在 resources/static 目录下新建 index.html 文件，内容如下：

```html
<!DOCTYPE html>
<html lang="en">
<head>
    <meta charset="UTF-8">
    <title>Title</title>
</head>
<body>
<form action="http://localhost:8080/withdraw" method="post">
    <input type="hidden" value="javaboy" name="name">
    <input type="hidden" value="10000" name="money">
    <input type="submit" value="点我">
</form>
</body>
</html>
```

配置完成后，分别启动 csrf-1、csrf-2 项目进行测试。

首先在浏览器中打开一个选项卡，输入 http://localhost:8080 访问 csrf-1，并完成登录操作，登录成功后，不要注销登录，继续打开一个新的选项卡访问 csrf-2，当用户输入 http://local.javaboy.org:8081/index.html 并单击页面上的按钮发起请求，csrf-1 项目的控制台就有日志打印出来，这就是一个跨站请求伪造（开发者需要修改本机 hosts 文件，将 local.javaboy.org 解析为 127.0.0.1）。

9.1.3 CSRF 防御

CSRF 攻击的根源在于浏览器默认的身份验证机制（自动携带当前网站的 Cookie 信息），这种机制虽然可以保证请求是来自用户的某个浏览器，但是无法确保该请求是用户授权发送的。攻击者和用户发送的请求一模一样,这意味着我们没有办法去直接拒绝这里的某一个请求。如果能在合法请求中额外携带一个攻击者无法获取的参数,就可以成功区分出两种不同的请求,进而拒绝掉恶意请求。

Spring 中提供了两种机制来防御 CSRF 攻击:

- 令牌同步模式。
- 在 Cookie 上指定 SameSite 属性。

无论是哪种方式,前提都是请求方法幂等,即 HTTP 请求中的 GET、HEAD、OPTIONS、TRACE 方法不应该改变应用的状态。

我们对这两种方案分别进行介绍。

9.1.3.1 令牌同步模式

这是目前主流的 CSRF 攻击防御方案。

具体的操作方式就是在每一个 HTTP 请求中,除了默认自动携带的 Cookie 参数之外,再额外提供一个安全的、随机生成的字符串,我们称之为 CSRF 令牌。这个 CSRF 令牌由服务端生成,生成后在 HttpSession 中保存一份。当前端请求到达后,将请求携带的 CSRF 令牌信息和服务端中保存的令牌进行对比,如果两者不相等,则拒绝掉该 HTTP 请求。

考虑到会有一些外部站点链接到我们的网站,所以我们要求请求是幂等的,这样对于 GET、HEAD、OPTIONS、TRACE 等方法就没有必要使用 CSRF 令牌了,强行使用可能会导致令牌泄漏。

Spring Security 对于令牌同步模式提供了比较好的支持,接下来通过一个案例感受一下。

首先创建一个 Spring Boot 项目,引入 Web、Spring Security 以及 Thymeleaf 依赖。项目创建成功后,方便起见,在 application.properties 中配置登录用户名/密码:

```
spring.security.user.name=javaboy
spring.security.user.password=123
```

然后在 resources/templates 目录下新建一个 index.html 页面,内容如下:

```html
<!DOCTYPE html>
<html lang="en" xmlns:th="http://www.thymeleaf.org">
<head>
    <meta charset="UTF-8">
    <title>Title</title>
</head>
<body>
<form action="/hello" method="post">
    <input type="hidden" th:value="${_csrf.token}"
                         th:name="${_csrf.parameterName}">
    <input type="submit" value="hello">
</form>
</body>
</html>
```

Form 表单中有一个很简单的 POST 请求，需要注意的是，这个请求中包含了一个隐藏域，隐藏域对应的参数 key 和 value 都是服务端默认返回的变量，开发者在此只需要填充变量名即可。

接下来在 Controller 中提供一个/hello 接口，并提供一个页面的映射，代码如下：

```
@Controller
public class HelloController {
    @PostMapping("/hello")
    @ResponseBody
    public String hello() {
        return "hello csrf! ";
    }
    @GetMapping("/index.html")
    public String index() {
        return "index";
    }
}
```

根据前面所讲，请求方法要幂等，所以在 Spring Security 中，默认不会对 GET、HEAD、OPTIONS 以及 TRACE 请求进行 CSRF 令牌校验，这也是/hello 接口是 Post 请求的原因。

至此，我们整个项目就配置完成了。有读者会说，好像没有看到了 CSRF 相关的配置，是的！Spring Security 中默认就开启了 CSRF 攻击防御。

启动项目，先进行登录。登录页面和前面章节中的登录页面是一样的，这个无须多说。读者可以在浏览器中按 F12 键，查看登录请求参数，如图 9-1 所示。

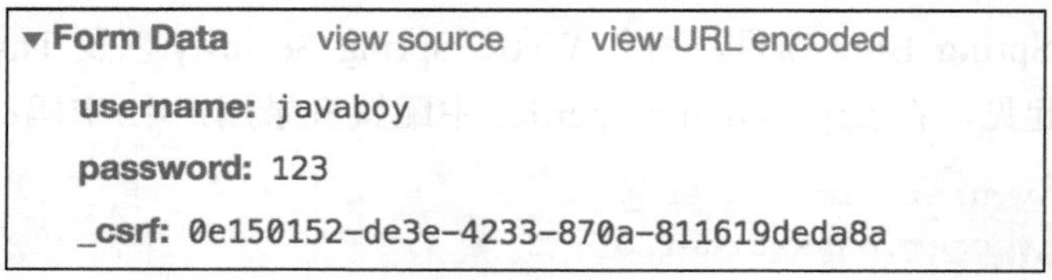

图 9-1　请求参数中多了一个_csrf 参数

这个就是请求额外携带的 CSRF 令牌参数。在前面的章节中，我们在配置 HttpSecurity 时总是加上一句.csrf().disable()，表示关闭 CSRF 攻击防御功能，因此请求参数中就没有 CSRF 令牌。如果我们什么都不配置，默认 CSRF 攻击防御就是开启的，登录参数中就自动存在_csrf 参数。

登录成功后，再去访问 index.html 页面，并单击页面上的按钮，可以看到，/hello 接口请求成功。

如果我们将 index.html 页面中表单的隐藏域注释掉，再去单击提交按钮，此时就会报错，如图 9-2 所示。

上面这个案例是将服务端返回的 CSRF 令牌，放在 request 属性中返回到前端的，开发者通过动态页面模版中的变量渲染可以将其显示出来。那如果是 Ajax 请求呢？很显然上面的方式就行不通了。

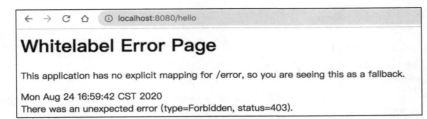

图 9-2　不携带 CSRF 令牌时，请求出错

针对 Ajax 请求，Spring Security 也提供了相应的方案：即将 CSRF 令牌放在响应头 Cookie 中，开发者自行从 Cookie 中提取出 CSRF 令牌信息，然后再作为参数提交到服务端。

我们通过一个简单的例子展示一下其具体的用法。

首先新建一个 Spring Boot 工程，引入 Spring Security 依赖和 Web 依赖，项目创建成功后，添加 SecurityConfig 配置，代码如下：

```java
@Configuration
public class SecurityConfig extends WebSecurityConfigurerAdapter {
    @Override
    protected void configure(AuthenticationManagerBuilder auth)
                                                    throws Exception {
        auth.inMemoryAuthentication()
            .withUser("javaboy")
            .password("{noop}123")
            .roles("admin");
    }
    @Override
    protected void configure(HttpSecurity http) throws Exception {
        http.authorizeRequests()
            .anyRequest().authenticated()
            .and()
            .formLogin()
            .loginProcessingUrl("/login.html")
            .successHandler((req,resp,auth)->{
                resp.getWriter().write("login success");
            })
            .permitAll()
            .and()
            .csrf()
            .csrfTokenRepository(CookieCsrfTokenRepository
                                        .withHttpOnlyFalse());
    }
}
```

需要注意的是，这里将 csrfTokenRepository 配置为 CookieCsrfTokenRepository，并设置 httpOnly 属性为 false，否则前端将无法读取到 Cookie 中的 CSRF 令牌。

接下来创建一个 login.html 登录页面，内容如下：

```
<!DOCTYPE html>
```

```html
<html lang="en">
<head>
    <meta charset="UTF-8">
    <title>Title</title>
    <script
     src="https://cdnjs.cloudflare.com/ajax/libs/jquery/3.5.1/jquery.min.js">
    </script>
    <script    src="https://cdnjs.cloudflare.com/ajax/libs/jquery-cookie/1.4.1/jquery.cookie.min.js"></script>
</head>
<body>
<div>
    <input type="text" id="username">
    <input type="password" id="password">
    <input type="button" value="登录" id="loginBtn">
</div>
<script>
    $("#loginBtn").click(function () {
        let _csrf = $.cookie('XSRF-TOKEN');
        $.post('/login.html', {
            username: $("#username").val(),
            password: $("#password").val(),
            _csrf: _csrf
        }, function (data) {
            alert(data);
        })
    })
</script>
</body>
</html>
```

在这个简易的登录页面中，引入了两个 js 库，jQuery 和 jQuery Cookie，后者用来简化 Cookie 操作。当用户单击了 loginBtn 之后，除了提交正常的用户名和密码之外，我们还将从 Cookie 中获取到的 XSRF-TOKEN 也一并提交到服务端进行验证，注意提交时的参数 key 为 _csrf。

配置完成后，启动项目访问 http://localhost:8080/login.html 页面进行登录测试，登录成功后，页面会弹出 alert，如图 9-3 所示。

图 9-3　登录成功后的提示

读者可能会有疑问，CSRF 令牌放在 Cookie 中会造成 CSRF 攻击吗？

当然不会！CSRF 攻击的根源在于浏览器默认的身份认证机制，即发送请求时会自动携带上网站的 Cookie，但是 Cookie 的内容是什么黑客是不知道的。所以即使非法请求携带了含有 CSRF 令牌的 Cookie 也没用，只有将 CSRF 令牌从 Cookie 中解析出来，并放到请求头或者请求参数中，才有用。

> **注 意**
>
> 使用了 Cookie 来保存 CSRF 令牌，页面上也可以继续通过页面渲染的方式获取 CSRF 令牌。前面两个案例唯一的区别在于第一个案例服务端将 CSRF 令牌保存在 HttpSession 中，第二个案例服务端将 CSRF 令牌放在 Cookie 中，所以对于第二个案例而言，既可以通过页面渲染获取 CSRF 令牌，也可以通过解析 Cookie 获取 CSRF 令牌。

9.1.3.2 SameSite

SameSite 是最近几年才出现的一个解决方案，是 Chrome 51 开始支持的一个属性，用来防止 CSRF 攻击和用户追踪。

这种方式通过在 Cookie 上指定 SameSite 属性，要求浏览器从外部站点发送请求时，不应携带 Cookie 信息，进而防止 CSRF 攻击。添加了 SameSite 属性的响应头类似下面这样：

```
Set-Cookie: JSESSIONID=randomid; Domain=javaboy.org; HttpOnly; SameSite=Lax
```

SameSite 属性值有三种：

- Strict：只有同一站点发送的请求才包含 Cookie 信息，不同站点发送的请求将不会包含 Cookie 信息。
- Lax：同一站点发送的请求或者导航到目标地址的 GET 请求会自动包含 Cookie 信息，否则不包含 Cookie 信息。
- None：Cookie 将在所有上下文中发送，即允许跨域发送。

Strict 是一种非常严格的模式，可能会带来不好的用户体验。举一个简单例子：假设用户登录了 www.javaboy.org 网站，并保持了登录状态，现在用户在 email.qq.com 上收到一封电子邮件，电子邮件中有一个超链接指向 www.javaboy.org，当用户单击这个超链接，理所应当地携带 Cookie 并自动进行 www.javaboy.org 站点的身份认证，然而 Strict 会阻止单击超链接时携带 Cookie，进而造成用户身份认证失败。而 Lax 则稍微友好一些，允许 GET 请求携带 Cookie。

使用 SameSite 还有一个需要考虑的因素就是浏览器的兼容性。虽然大部分现代浏览器都支持 SameSite 属性，但是可能还是存在一些古董级浏览器不支持该属性。所以，如果使用 SameSite 来处理 CSRF 攻击，建议作为一个备选方案，而不是主要方案。

Spring Security 对于 SameSite 并未直接提供支持，但是 Spring Session 提供了，因此，在使用时，需要首先引入 Spring Session 和 Redis 依赖，代码如下：

```xml
<dependency>
    <groupId>org.springframework.boot</groupId>
    <artifactId>spring-boot-starter-data-redis</artifactId>
</dependency>
<dependency>
```

```xml
        <groupId>org.springframework.session</groupId>
        <artifactId>spring-session-data-redis</artifactId>
</dependency>
```

然后在 application.properties 中配置 Redis 连接信息：

```
spring.redis.password=123
spring.redis.host=127.0.0.1
spring.redis.port=6379
```

最后，提供一个 CookieSerializer 实例即可，并配置 SameSite 属性值为 strict：

```java
@Bean
public CookieSerializer httpSessionIdResolver(){
    DefaultCookieSerializer cookieSerializer = new DefaultCookieSerializer();
    cookieSerializer.setSameSite("strict");
    return cookieSerializer;
}
```

配置完成后，启动项目完成登录操作，此时返回的 Cookie 如图 9-4 所示，可以看到已经包含 SameSite 属性了。

```
Expires: 0
Keep-Alive: timeout=60
Location: http://localhost:8080/withdraw
Pragma: no-cache
Set-Cookie: SESSION=ZGMxNTU3NzYtYWM1NC00YzJkLWFiN2ItYzM0ZGM4NDY5OWMx; Path=/; HttpOnly; SameSite=strict
X-Content-Type-Options: nosniff
X-Frame-Options: DENY
X-XSS-Protection: 1; mode=block
```

图 9-4　响应头中包含 SameSite 属性

9.1.3.3　需要注意的问题

会话超时

CSRF 令牌生成后，往往都保存在 HttpSession 中，但是 HttpSession 可能会因为超时而失效，导致前端请求传来的 CSRF 令牌无法得到验证，解决这一问题有如下几种方式：

（1）最佳方案是在表单提交时，通过 js 获取 CSRF 令牌，然后将获取到的 CSRF 令牌跟随表单一起提交。

（2）当会话快要过期时，前端通过 js 提醒用户刷新页面，以给会话"续命"。

（3）将令牌存储在 Cookie 中而不是 HttpSession 中。

登录和注销

为了保护用户的敏感信息，登录请求和注销请求需要注意 CSRF 攻击防护。

文件上传

文件上传请求比较特殊，因此需要额外注意。如果将 CSRF 放在请求体中，就会面临一个"鸡和蛋"的问题。服务端需要先验证 CSRF 令牌以确认请求是否合法，而这也意味需要先读取请求体以获取 CSRF 令牌，这就陷入一个死循环了。

一般来说，将 CSRF 防御与 multipart/form-data 一起使用，我们有两种不同的策略：

- 将 CSRF 令牌放在请求体中。
- 将 CSRF 令牌放在请求 URL 中。

将 CSRF 令牌放在请求体中，意味着任何人都可以向我们的服务器上传临时文件，但是只有 CSRF 令牌验证通过的用户，才能真正提交一个文件，这也是目前推荐的方案，因为上传临时文件对服务器的影响可以忽略不计。如果不希望未经授权的用户上传临时文件，那么可以将 CSRF 令牌放在请求 URL 地址中，但是这种方式可能带来令牌泄漏的风险。

9.1.4　源码分析

接下来我们再来对 Spring Security 中的 CSRF 攻击防御的源码进行分析。

CsrfToken

Spring Security 中提供了 CsrfToken 接口用来描述 CSRF 令牌信息，接口如下：

```java
public interface CsrfToken extends Serializable {
    String getHeaderName();
    String getParameterName();
    String getToken();
}
```

（1）getHeaderName：当 CSRF 令牌被放置在请求头时，获取参数名。
（2）getParameterName：当 CSRF 令牌被当作请求参数传递时，获取参数名。
（3）getToken：获取具体的 CSRF 令牌。

CsrfToken 一共有两个实现类，如图 9-5 所示。

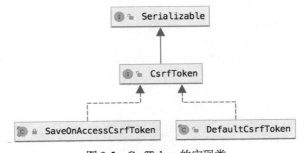

图 9-5　CsrfToken 的实现类

- DefaultCsrfToken 是一个默认的实现类，该类为三个接口提供了对应的属性，属性值通过构造方法传入，再通过各自的 get 方法返回。
- SaveOnAccessCsrfToken 是一个代理类，由于 CsrfToken 只有两个实现类，所以正常来说 SaveOnAccessCsrfToken 代理的就是 DefaultCsrfToken。代理类中主要是对 getToken 方法做了改变，当调用 getToken 方法时，才去执行 CSRF 令牌的保存操作，这样可以避免很多无用的保存操作（后文会有详细解释）。

CsrfTokenRepository

CsrfTokenRepository 是 Spring Security 中提供的 CsrfToken 的保存接口，代码如下：

```java
public interface CsrfTokenRepository {
    CsrfToken generateToken(HttpServletRequest request);
    void saveToken(CsrfToken token, HttpServletRequest request,
                HttpServletResponse response);
    CsrfToken loadToken(HttpServletRequest request);
}
```

（1）generateToken：该方法用来生成一个 CSRF 令牌。
（2）saveToken：该方法用来保存 CSRF 令牌。
（3）loadToken：该方法用来读取一个 CSRF 令牌。

CsrfTokenRepository 一共有三个实现类，如图 9-6 所示。

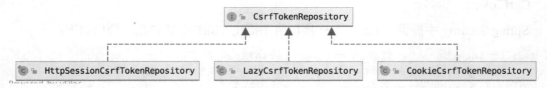

图 9-6　CsrfTokenRepository 的三个实现类

HttpSessionCsrfTokenRepository 是将 CsrfToken 保存在 HttpSession 中，我们来看一下该类的三个核心方法：

```java
public final class HttpSessionCsrfTokenRepository implements
                                            CsrfTokenRepository {
    public void saveToken(CsrfToken token, HttpServletRequest request,
            HttpServletResponse response) {
        if (token == null) {
            HttpSession session = request.getSession(false);
            if (session != null) {
                session.removeAttribute(this.sessionAttributeName);
            }
        }
        else {
            HttpSession session = request.getSession();
            session.setAttribute(this.sessionAttributeName, token);
        }
    }
    public CsrfToken loadToken(HttpServletRequest request) {
        HttpSession session = request.getSession(false);
        if (session == null) {
            return null;
        }
        return (CsrfToken) session.getAttribute(this.sessionAttributeName);
```

```java
    }
    public CsrfToken generateToken(HttpServletRequest request) {
        return new DefaultCsrfToken(this.headerName, this.parameterName,
            createNewToken());
    }
    private String createNewToken() {
        return UUID.randomUUID().toString();
    }
}
```

（1）saveToken：如果传入的 CsrfToken 为 null，就从 HttpSession 中移除 CsrfToken 令牌；否则就将 CsrfToken 令牌保存到 HttpSession 中。

（2）loadToken：该方法返回 HttpSession 中保存的令牌信息。

（3）generateToken：该方法生成一个默认的 DefaultCsrfToken 令牌，headerName 和 parameterName 都是默认的，而具体的令牌则是一个 UUID 字符串。

CookieCsrfTokenRepository 则是将 CsrfToken 保存在 Cookie 中，我们来看一下该类的三个核心方法：

```java
public final class CookieCsrfTokenRepository implements CsrfTokenRepository {
    public CookieCsrfTokenRepository() {
    }
    @Override
    public CsrfToken generateToken(HttpServletRequest request) {
        return new DefaultCsrfToken(this.headerName, this.parameterName,
            createNewToken());
    }
    @Override
    public void saveToken(CsrfToken token, HttpServletRequest request,
            HttpServletResponse response) {
        String tokenValue = token == null ? "" : token.getToken();
        Cookie cookie = new Cookie(this.cookieName, tokenValue);
        cookie.setSecure(request.isSecure());
        if (this.cookiePath != null && !this.cookiePath.isEmpty()) {
            cookie.setPath(this.cookiePath);
        } else {
            cookie.setPath(this.getRequestContext(request));
        }
        if (token == null) {
            cookie.setMaxAge(0);
        }
        else {
            cookie.setMaxAge(-1);
        }
        cookie.setHttpOnly(cookieHttpOnly);
        if (this.cookieDomain != null && !this.cookieDomain.isEmpty()) {
            cookie.setDomain(this.cookieDomain);
        }
```

```java
            response.addCookie(cookie);
        }
        @Override
        public CsrfToken loadToken(HttpServletRequest request) {
            Cookie cookie = WebUtils.getCookie(request, this.cookieName);
            if (cookie == null) {
                return null;
            }
            String token = cookie.getValue();
            if (!StringUtils.hasLength(token)) {
                return null;
            }
            return new DefaultCsrfToken(this.headerName, this.parameterName, token);
        }
        public static CookieCsrfTokenRepository withHttpOnlyFalse() {
            CookieCsrfTokenRepository result = new CookieCsrfTokenRepository();
            result.setCookieHttpOnly(false);
            return result;
        }
    }
```

（1）CookieCsrfTokenRepository 可以通过两种方式获取其实例，第一种方式是直接新建一个实例，这种情况下生成的 Cookie 中 HttpOnly 属性默认为 true，即前端不能通过 js 操作 Cookie；第二种方式是调用静态方法 withHttpOnlyFalse，该方法也会返回一个 CookieCsrfTokenRepository 实例，并且设置 HttpOnly 属性为 false，即允许前端通过 js 操作 Cookie。

（2）generateToken：该方法的逻辑和 HttpSessionCsrfTokenRepository 中的一致，不再赘述。

（3）saveToken：保存 CSRF 令牌，具体方式就是生成 Cookie 并添加到响应头中。

（4）loadToken：从请求头中提取出 Cookie，进而解析出 CSRF 令牌信息。

LazyCsrfTokenRepository 是一个代理类，可以代理 HttpSessionCsrfTokenRepository 或者 CookieCsrfTokenRepository，代理的目的是延迟保存生成的 CsrfToken。我们来看一下该类的核心方法：

```java
public final class LazyCsrfTokenRepository implements CsrfTokenRepository {
    public LazyCsrfTokenRepository(CsrfTokenRepository delegate) {
        this.delegate = delegate;
    }
    public CsrfToken generateToken(HttpServletRequest request) {
        return wrap(request, this.delegate.generateToken(request));
    }
    public void saveToken(CsrfToken token, HttpServletRequest request,
            HttpServletResponse response) {
        if (token == null) {
            this.delegate.saveToken(token, request, response);
        }
    }
```

```java
    public CsrfToken loadToken(HttpServletRequest request) {
        return this.delegate.loadToken(request);
    }
    private CsrfToken wrap(HttpServletRequest request, CsrfToken token) {
        HttpServletResponse response = getResponse(request);
        return new SaveOnAccessCsrfToken(this.delegate, request,
                                        response, token);
    }
}
```

（1）generateToken：在生成 CsrfToken 时，代理类生成的 CsrfToken 类型是 DefaultCsrfToken，这里将 DefaultCsrfToken 装饰为 SaveOnAccessCsrfToken，当调用 SaveOnAccessCsrfToken 中的 getToken 方法时，才会去保存 CsrfToken。

（2）saveToken：只有当 token 为 null 时，才会去执行代理类的 saveToken 方法（相当于只执行移除 CsrfToken 操作）。

CsrfFilter

CsrfFilter 是 Spring Security 过滤器链中的一环，在过滤器中校验客户端传来的 CSRF 令牌是否有效。CsrfFilter 继承自 OncePerRequestFilter，所以对它来说最重要的方法是 doFilterInternal，我们来看一下该方法：

```java
protected void doFilterInternal(HttpServletRequest request,
        HttpServletResponse response, FilterChain filterChain)
            throws ServletException, IOException {
    request.setAttribute(HttpServletResponse.class.getName(), response);
    CsrfToken csrfToken = this.tokenRepository.loadToken(request);
    final boolean missingToken = csrfToken == null;
    if (missingToken) {
        csrfToken = this.tokenRepository.generateToken(request);
        this.tokenRepository.saveToken(csrfToken, request, response);
    }
    request.setAttribute(CsrfToken.class.getName(), csrfToken);
    request.setAttribute(csrfToken.getParameterName(), csrfToken);
    if (!this.requireCsrfProtectionMatcher.matches(request)) {
        filterChain.doFilter(request, response);
        return;
    }
    String actualToken = request.getHeader(csrfToken.getHeaderName());
    if (actualToken == null) {
        actualToken = request.getParameter(csrfToken.getParameterName());
    }
    if (!csrfToken.getToken().equals(actualToken)) {
        if (missingToken) {
            this.accessDeniedHandler.handle(request, response,
                new MissingCsrfTokenException(actualToken));
        }
        else {
            this.accessDeniedHandler.handle(request, response,
```

```
                new InvalidCsrfTokenException(csrfToken, actualToken));
    }
    return;
}
filterChain.doFilter(request, response);
```

（1）首先调用 tokenRepository.loadToken 方法去加载出 CsrfToken 对象，默认使用的 tokenRepository 对象类型是 LazyCsrfTokenRepository。

（2）如果 CsrfToken 对象不存在，则立马生成 CsrfToken 对象并保存起来。需要注意，如果 tokenRepository 类型是 LazyCsrfTokenRepository，则这里并未真正将 CsrfToken 令牌保存起来。

（3）将生成的 CsrfToken 对象设置到 request 属性中，这样我们在前端页面中就可以渲染出生成的令牌信息了。

（4）调用 requireCsrfProtectionMatcher.matches 方法进行请求判断，该方法主要是判断当前请求方法是否是 GET、HEAD、TRACE 以及 OPTIONS。我们前面讲过，如果当前请求方法是这四种之一，则请求直接过，不用进行接下来的 CSRF 令牌校验，这也意味着上一步没有必要进行 CsrfToken 保存操作。此时使用 LazyCsrfTokenRepository 的优势就体现出来了，在第 2 步中生成了 CsrfToken 令牌，但是没有立即保存，而是到后面调用时才保存。

（5）如果请求方法不是 GET、HEAD、TRACE 以及 OPTIONS，则先从请求头中提取出 CSRF 令牌；请求头没有，则从请求参数中提取出 CSRF 令牌，将拿到的 CSRF 令牌和第 1 步中通过 loadToken 加载出来的令牌进行比对，判断请求传来的 CSRF 令牌是否合法。

看完这个过滤器，我们再来把这个流程捋一捋：

请求到达后，会经过 CsrfFilter，在该过滤器中，首先加载出保存的 CsrfToken，可以是从 HttpSession 中加载，也可以是从请求头携带的 Cookie 中加载，默认是从 HttpSession 加载。如果加载出来的 CsrfToken 为 null，则立即生成一个 CsrfToken 并保存起来，由于默认的 tokenRepository 类型是 LazyCsrfTokenRepository，所以这里的保存并不是真正的保存，之所以这么做的原因在于，如果请求方法是 GET、HEAD、TRACE 以及 OPTIONS，就没有必要保存。然后将生成的 CsrfToken 放到请求对象中，以方便前端渲染。接下来判断请求方法是否是需要进行 CSRF 令牌校验的方法，如果不是，则直接执行后面的过滤器，否则就从请求中拿出 CSRF 令牌信息和一开始加载出来的令牌进行比对。

CsrfFilter 过滤器是由 CsrfConfigurer 进行配置的，而 CsrfConfigurer 则是在 WebSecurityConfigurerAdapter#getHttp 方法中添加进 HttpSecurity 中的。CsrfConfigurer 的原理和前面讲的 SessionManagementConfigurer 原理基本一致，这里不再赘述。

CsrfAuthenticationStrategy

CsrfAuthenticationStrategy 实现了 SessionAuthenticationStrategy 接口，默认也是由 CompositeSessionAuthenticationStrategy 代理执行，在用户登录成功后触发执行，具体可以参考 7.2.2.3 小节。

CsrfAuthenticationStrategy 主要用于在登录成功后，删除旧的 CsrfToken 并生成一个新的 CsrfToken，代码如下：

```java
public final class CsrfAuthenticationStrategy implements
                                    SessionAuthenticationStrategy {
    public CsrfAuthenticationStrategy(CsrfTokenRepository csrfTokenRepository) {
        this.csrfTokenRepository = csrfTokenRepository;
    }
    @Override
    public void onAuthentication(Authentication authentication,
            HttpServletRequest request, HttpServletResponse response)
                throws SessionAuthenticationException {
        boolean containsToken = this.csrfTokenRepository
                                            .loadToken(request) != null;
        if (containsToken) {
            this.csrfTokenRepository.saveToken(null, request, response);
            CsrfToken newToken =
                    this.csrfTokenRepository.generateToken(request);
            this.csrfTokenRepository.saveToken(newToken, request, response);
            request.setAttribute(CsrfToken.class.getName(), newToken);
            request.setAttribute(newToken.getParameterName(), newToken);
        }
    }
}
```

可以看到，在 onAuthentication 方法中，首先调用 loadToken 方法去加载令牌，如果加载到了，则先删除已经存在的令牌（saveToken 方法第一个参数为 null），然后生成新的令牌并重新保存起来。

至此，Spring Security 中和 CSRF 攻击防御相关的几个类的源码就分析完了。通过源码的分析，相信大家对于 CSRF 攻击的防御策略有了一个更加深刻的理解。

9.2 HTTP 响应头处理

HTTP 响应头中的许多属性都可以用来提高 Web 安全。本节我们来看一下 Spring Security 中提供显式支持的一些 HTTP 响应头。

Spring Security 默认情况下，显式支持的 HTTP 响应头主要有如下几种：

```
Cache-Control: no-cache, no-store, max-age=0, must-revalidate
Pragma: no-cache
Expires: 0
X-Content-Type-Options: nosniff
Strict-Transport-Security: max-age=31536000 ; includeSubDomains
X-Frame-Options: DENY
X-XSS-Protection: 1; mode=block
```

这里一共有七个响应头，前三个都是与缓存相关的，因此一共可以分为五大类。

这些响应头都是在 HeaderWriterFilter 中添加的，默认情况下，该过滤器就会添加到 Spring Security 过滤器链中，HeaderWriterFilter 是通过 HeadersConfigurer 进行配置的，我们来看一下 HeadersConfigurer 中几个关键的方法：

```java
public class HeadersConfigurer<H extends HttpSecurityBuilder<H>> extends
        AbstractHttpConfigurer<HeadersConfigurer<H>, H> {
    @Override
    public void configure(H http) {
        HeaderWriterFilter headersFilter = createHeaderWriterFilter();
        http.addFilter(headersFilter);
    }
    private HeaderWriterFilter createHeaderWriterFilter() {
        List<HeaderWriter> writers = getHeaderWriters();
        HeaderWriterFilter headersFilter = new HeaderWriterFilter(writers);
        headersFilter = postProcess(headersFilter);
        return headersFilter;
    }
    private List<HeaderWriter> getHeaderWriters() {
        List<HeaderWriter> writers = new ArrayList<>();
        addIfNotNull(writers, contentTypeOptions.writer);
        addIfNotNull(writers, xssProtection.writer);
        addIfNotNull(writers, cacheControl.writer);
        addIfNotNull(writers, hsts.writer);
        addIfNotNull(writers, frameOptions.writer);
        addIfNotNull(writers, hpkp.writer);
        addIfNotNull(writers, contentSecurityPolicy.writer);
        addIfNotNull(writers, referrerPolicy.writer);
        addIfNotNull(writers, featurePolicy.writer);
        writers.addAll(headerWriters);
        return writers;
    }
    private <T> void addIfNotNull(List<T> values, T value) {
        if (value != null) {
            values.add(value);
        }
    }
}
```

可以看到，这里在 configure 方法中创建了 HeaderWriterFilter 过滤器，在过滤器创建时，通过 getHeaderWriters 方法获取到所有需要添加的响应头传入过滤器中。getHeaderWriters 方法执行时，只会添加不为 null 的实例，默认情况下，只有前五个不为 null，其中：

- contentTypeOptions.writer：负责处理 X-Content-Type-Options 响应头。
- xssProtection.writer：负责处理 X-XSS-Protection 响应头。
- cacheControl.writer：负责处理 Cache-Control、Pragma 以及 Expires 响应头。
- hsts.writer：负责处理 Strict-Transport-Security 响应头。

- frameOptions.writer：负责处理 X-Frame-Options 响应头。

了解到这响应头的来源之后，接下来我们来对其逐个进行分析。

9.2.1 缓存控制

和缓存控制相关的响应头一共有三个：

```
Cache-Control: no-cache, no-store, max-age=0, must-revalidate
Pragma: no-cache
Expires: 0
```

可能有的读者对这几个响应头还不太熟悉，这里稍微解释一下。

Cache-Control

Cache-Control 是 HTTP/1.1 中引入的缓存字段，无论是请求头还是响应头都支持该字段。其中 no-store 表示不做任何缓存，每次请求都会从服务端完整地下载内容。no-cache 则表示缓存但是需要重新验证，这种情况下，数据虽然缓存在客户端，但是当需要使用该数据时，还是会向服务端发送请求，服务端则验证请求中所描述的缓存是否过期，如果没有过期，则返回 304，客户端使用缓存；如果已经过期，则返回最新数据。max-age 则表示缓存的有效期，这个有效期并非一个时间戳，而是一个秒数，指从请求发起后多少秒内缓存有效。must-revalidate 表示当缓存在使用一个陈旧的资源时，必须先验证它的状态，已过期的将不被使用。

Pragma

Pragma 是 HTTP/1.0 中定义的响应头，作用类似于 Cache-Control: no-cache，但是并不能代替 Cache-Control，该字段主要用来兼容 HTTP/1.0 的客户端。

Expires

Expires 响应头指定了一个日期，即在指定日期之后，缓存过期。如果日期值为 0 的话，表示缓存已经过期。

从上面的解释可以看到，Spring Security 默认就是不做任何缓存。但是需要注意，这个是针对经过 Spring Security 过滤器的请求，如果请求本身都没经过 Spring Security 过滤器，那么该缓存的还是会缓存的。例如如下代码：

```
@Configuration
public class SecurityConfig extends WebSecurityConfigurerAdapter {
    @Override
    public void configure(WebSecurity web) throws Exception {
        web.ignoring().antMatchers("/hello.html");
    }
}
```

当访问/hello.html 时，请求就不会经过 Spring Security 过滤器，所以该资源还是会缓存的（回顾本书 4.5 节）。

如果请求经过 Spring Security 过滤器，同时开发者又希望开启缓存功能，那么可以关闭 Spring Security 中关于缓存的默认配置，代码如下：

```java
@Configuration
public class SecurityConfig extends WebSecurityConfigurerAdapter {
    @Override
    protected void configure(HttpSecurity http) throws Exception {
        http.authorizeRequests()
                .anyRequest().authenticated()
                .and()
                .headers()
                .cacheControl()
                .disable()
                .and()
                .formLogin()
                .and()
                .csrf().disable();
    }
}
```

调用 .cacheControl().disable() 方法之后，Spring Security 就不会配置 Cache-Control、Pragma 以及 Expires 三个缓存相关的响应头了。

9.2.2　X-Content-Type-Options

要理解 X-Content-Type-Options 响应头，得先了解 MIME 嗅探。

一般来说，浏览器通过响应头 Content-Type 来确定响应报文类型，但是在早期浏览器中，为了提高用户体验，并不会严格根据 Content-Type 的值来解析响应报文，当 Content-Type 的值缺失，或者浏览器认为服务端给出了错误的 Content-Type 值，此时就会对响应报文进行自我解析，即自动判断报文类型然后进行解析，在这个过程中就有可能触发 XSS 攻击。

X-Content-Type-Options 响应头相当于一个提示标志，被服务器用来提示客户端一定要遵循在 Content-Type 中对 MIME 类型的设定，而不能对其进行修改。这就禁用了客户端的 MIME 类型嗅探行为，换言之，就是服务端告诉客户端其对于 MIME 类型的设置没有任何问题。

配置后响应头如下：

```
X-Content-Type-Options: nosniff
```

如果开发者不想禁用 MIME 嗅探，可以通过如下方式从响应头中移除 X-Content-Type-Options。

```java
@Configuration
public class SecurityConfig extends WebSecurityConfigurerAdapter {
    @Override
    protected void configure(HttpSecurity http) throws Exception {
        http.authorizeRequests()
                .anyRequest().authenticated()
```

```
                .and()
                .headers()
                .contentTypeOptions()
                .disable()
                .and()
                .formLogin()
                .and()
                .csrf().disable();
    }
}
```

调用.contentTypeOptions().disable()方法即可移除 X-Content-Type-Options 响应头。

9.2.3 Strict-Transport-Security

Strict-Transport-Security 用来指定当前客户端只能通过 HTTPS 访问服务端，而不能通过 HTTP 访问。

```
Strict-Transport-Security: max-age=31536000 ; includeSubDomains
```

（1）max-age：设置在浏览器收到这个请求后的多少秒的时间内，凡是访问这个域名下的请求都使用 HTTPS 请求。

（2）includeSubDomains：这个参数是可选的，如果被指定，表示第 1 条规则也适用于子域名。

这个响应头并非总是会添加，如果当前请求是 HTTPS 请求，这个请求头才会添加，否则该请求头就不会添加，具体实现逻辑在 HstsHeaderWriter#writeHeaders 方法中：

```java
public void writeHeaders(HttpServletRequest request,
                         HttpServletResponse response) {
    if (this.requestMatcher.matches(request)) {
        if (!response.containsHeader(HSTS_HEADER_NAME)) {
            response.setHeader(HSTS_HEADER_NAME, this.hstsHeaderValue);
        }
    }
}
```

可以看到，向 response 中添加响应头之前，会先调用 requestMatcher.matches 方法对当前请求进行判断，判断当前请求是否是 HTTPS 请求，如果是 HTTPS 请求，则添加该响应头，否则不添加。

为了看到该响应头的效果，我们可以使用 Java 自带的 keytool 工具来生成一个 HTTPS 证书供我们测试使用，具体步骤如下：

（1）确保本地 Java 已经安装好，环境变量也已经配置好，我们首先在命令行执行如下命令生成 HTTPS 证书：

```
keytool -genkey -alias tomcathttps -keyalg RSA -keysize 2048 -keystore javaboy.p12 -validity 365
```

命令含义如下：

- genkey：表示要创建一个新的密钥。
- alias：表示 keystore 的别名。
- keyalg：表示使用的加密算法是 RSA ，一种非对称加密算法。
- keysize：表示密钥的长度。
- keystore：表示生成的密钥存放位置。
- validity：表示密钥的有效时间，单位为天。

具体生成过程如图 9-7 所示。

```
sang-3:~ sang$ keytool -genkey -alias tomcathttps -keyalg RSA -keysize 2048  -keystore javaboy.p12 -validity 365
[Enter keystore password:
[Re-enter new password:
[What is your first and last name?
  [Unknown]:
What is the name of your organizational unit?
  [Unknown]:
What is the name of your organization?
  [Unknown]:
What is the name of your City or Locality?
  [Unknown]:
What is the name of your State or Province?
  [Unknown]:
What is the two-letter country code for this unit?
  [Unknown]:
Is CN=Unknown, OU=Unknown, O=Unknown, L=Unknown, ST=Unknown, C=Unknown correct?
  [no]:  yes
```

图 9-7　HTTPS 证书生成过程

（2）接下来将生成的 javaboy.p12 证书复制到 Spring Boot 项目的 resources 目录下，并在 application.properties 中添加如下配置：

```
server.ssl.key-store=classpath:javaboy.p12
server.ssl.key-alias=tomcathttps
server.ssl.key-store-password=111111
```

- key-store：表示密钥文件位置。
- key-alias：表示密钥别名。
- key-store-password：就是在密钥生成过程中输入的口令。

（3）配置完成后，启动项目。浏览器中输入 https://localhost:8080/login 进行访问，由于这个 HTTPS 证书是我们自己生成的，并不被浏览器认可，所以在访问的时候会有安全提示，大家单击继续访问即可，如图 9-8 所示。

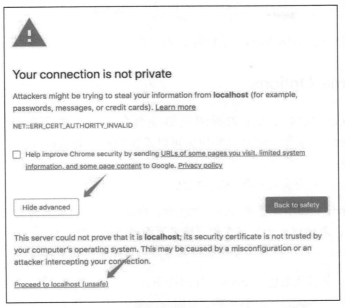

图 9-8　选择继续前往 localhost 即可

请求成功后，查看响应头，发现已经有了 Strict-Transport-Security 字段，如图 9-9 所示。

图 9-9　响应头中的 Strict-Transport-Security 字段

如果需要对 Strict-Transport-Security 的值进行具体配置，例如关闭 includeSubDomains 属性并重新设置 max-age，方式如下：

```
@Configuration
public class SecurityConfig extends WebSecurityConfigurerAdapter {
    @Override
    protected void configure(HttpSecurity http) throws Exception {
        http.authorizeRequests()
                .anyRequest().authenticated()
                .and()
                .formLogin()
                .and()
                .csrf().disable()
                .headers()
                .httpStrictTransportSecurity()
                .includeSubDomains(false)
                .maxAgeInSeconds(3600);
    }
}
```

}
```

当然也可以直接调用.disable()方法移除该响应头。

### 9.2.4 X-Frame-Options

X-Frame-Options 响应头用来告诉浏览器是否允许一个页面在<frame>、<iframe>、<embed>或者<object>中展现，通过该响应头可以确保网站没有被嵌入到其他站点里面，进而避免发生单击劫持。

X-Frame-Options 响应头有三种不同的取值：

- deny：表示该页面不允许在 frame 中展示，即便是在相同域名的页面中嵌套也不允许。
- sameorigin：表示该页面可以在相同域名页面的 frame 中展示。
- allow-from uri：表示该页面可以在指定来源的 frame 中展示。

Spring Security 中默认取值是 deny，代码如下：

```
X-Frame-Options: DENY
```

如果项目需要，开发者也可以对此进行修改，例如将 deny 改为 sameorigin，方式如下：

```java
@Configuration
public class SecurityConfig extends WebSecurityConfigurerAdapter {
    @Override
    protected void configure(HttpSecurity http) throws Exception {
        http.authorizeRequests()
                .anyRequest().authenticated()
                .and()
                .formLogin()
                .and()
                .csrf().disable()
                .headers()
                .frameOptions()
                .sameOrigin();
    }
}
```

当然也可以直接调用.disable()方法移除该响应头。

> **什么是单击劫持？**
>
> 单击劫持是一种视觉上的欺骗手段。攻击者将被劫持的网页放在一个 iframe 标签中，设置该 iframe 标签透明不可见，然后将 iframe 标签覆盖在另一个网页上，最后诱使用户在该网页上进行操作，通过调整 iframe 页面的位置，可以诱使用户恰好单击在 iframe 页面的一些功能性按钮上。
```

9.2.5 X-XSS-Protection

X-XSS-Protection 响应头告诉浏览器，当检测到跨站脚本攻击（XSS）时，浏览器将停止加载页面，该响应头有四种不同的取值：

（1）0 表示禁止 XSS 过滤。

（2）1 表示启用 XSS 过滤（通常浏览器是默认的）。如果检测到跨站脚本攻击，浏览器将清除页面（删除不安全的部分）。

（3）1;mode=block 表示启用 XSS 过滤。如果检测到攻击，浏览器将不会清除页面，而是阻止页面加载。

（4）1; report=<reporting-URI> 表示启用 XSS 过滤。如果检测到跨站脚本攻击，浏览器将清除页面，并使用 CSP report-uri 指令的功能发送违规报告（Chrome 支持）。

Spring Security 中设置的 X-XSS-Protection 响应头如下：

```
X-XSS-Protection: 1; mode=block
```

当然开发者也可以对此进行配置，例如想去除 mode=block 部分，方式如下：

```
@Configuration
public class SecurityConfig extends WebSecurityConfigurerAdapter {
    @Override
    protected void configure(HttpSecurity http) throws Exception {
        http.authorizeRequests()
                .anyRequest().authenticated()
                .and()
                .formLogin()
                .and()
                .csrf().disable()
                .headers()
                .xssProtection()
                .block(false);
    }
}
```

当然也可以直接调用.disable()方法移除该响应头。

> **什么是 XSS 攻击?**
>
> 跨站脚本攻击（Cross-Site Scripting, XSS）是一种安全漏洞，攻击者可以利用这种漏洞在网站上注入恶意的 JavaScript 代码，而浏览器无法区分出这是恶意的 JavaScript 代码还是正常的 JavaScript 代码。当被攻击者登录网站时，就会自动运行这些恶意代码，攻击者可以利用这些恶意代码去窃取 Cookie 信息、监听用户行为以及修改 DOM 结构。

前面介绍这些响应头是 Spring Security 默认会自动配置的响应头。还有其他一些安全相关的响应头，需要我们手动配置，一起来看一下。

9.2.6 Content-Security-Policy

内容安全策略（Content Security Policy，CSP）是一个额外的安全层，用于检测并削弱某些特定类型的攻击，例如跨站脚本（XSS）和数据注入攻击等。

CSP 相当于通过一个白名单明确告诉客户端，哪些外部资源可以加载和执行。举一个简单例子：

```
Content-Security-Policy: default-src 'self';script-src 'self';
object-src 'none';style-src cdn.javaboy.org; img-src *; child-src https:
```

这个响应头含义如下：

- default-src 'self'：默认情况下所有资源只能从当前域中加载。接下来细化的配置会覆盖 default-src，没有细化的选项则使用 default-src。
- script-src 'self'：表示脚本文件只能从当前域名加载。
- object-src 'none'：表示 object 标签不加载任何资源。
- style-src cdn.javaboy.org：表示只加载来自 cdn.javaboy.org 的样式表。
- img-src *：表示可以从任意地址加载图片。
- child-src https：表示必须使用 HTTPS 来加载 frame。

CSP 其他可选值，读者可以参考 https://www.w3.org/TR/CSP2 一文。

Spring Security 为 Content-Security-Policy 提供了配置方法，如果我们需要配置，则方式如下：

```java
@Configuration
public class SecurityConfig extends WebSecurityConfigurerAdapter {
    @Override
    protected void configure(HttpSecurity http) throws Exception {
        http.authorizeRequests()
                .anyRequest().authenticated()
                .and()
                .formLogin()
                .and()
                .csrf().disable()
                .headers()
                .contentSecurityPolicy("default-src 'self'; script-src 'self'; object-src 'none';style-src cdn.javaboy.org; img-src *; child-src https:");
    }
}
```

配置完成后，重启项目，此时默认的登录页面变了，如图 9-10 所示。

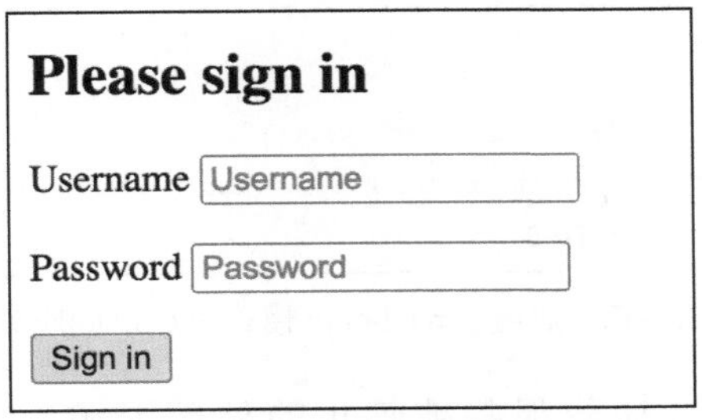

图 9-10　失去了 CSS 样式的登录页面

默认登录页面中加载了外部样式表，现在由于 CSP 限制，外部样式表加载失败，如图 9-11 所示。

Name	Status	Type	Initiator	Size	Time
bootstrap.min.css	(blocked:csp)	stylesheet	login	0 B	0 ms
signin.css	(blocked:csp)	stylesheet	login	0 B	0 ms

图 9-11　外部样式表加载失败

CSP 还有一种报告模式——report-only。在此模式下，CSP 策略不是强制性的，如果出现违规行为，还是会继续加载相应的脚本或者样式表，但是会将违规行为报告给一个指定的 URI 地址。

配置方式如下：

```
@Configuration
public class SecurityConfig extends WebSecurityConfigurerAdapter {
    @Override
    protected void configure(HttpSecurity http) throws Exception {
        http.authorizeRequests()
                .anyRequest().authenticated()
                .and()
                .formLogin()
                .and()
                .csrf().disable()
                .headers()
                .contentSecurityPolicy(contentSecurityPolicyConfig -> {
                  contentSecurityPolicyConfig.policyDirectives("default-src
'self'; script-src 'self'; object-src 'none';style-src cdn.javaboy.org; img-src *;
child-src https:;report-uri http://localhost:8081/report");
                    contentSecurityPolicyConfig.reportOnly();
                });
    }
}
```

这段配置最终生成的响应头如图 9-12 所示。

```
Connection: keep-alive
Content-Length: 1324
Content-Security-Policy-Report-Only: default-src 'self'; script-src 'self'; object-src 'none';sty
le-src cdn.javaboy.org; img-src *; child-src https:;report-uri http://localhost:8081/report
Content-Type: text/html;charset=UTF-8
```

图 9-12　启用了 read-only 模式的 CSP 响应头

此时，浏览器还是会去加载那些被禁止的外部资源，同时会将违规行为发送到 http://localhost:8081/report 地址，开发者收到违规行为报告后可以自行处理。

9.2.7　Referrer-Policy

Referrer-Policy 描述了用户从哪里进入到当前网页。

浏览器默认的取值如下：

```
Referrer Policy: no-referrer-when-downgrade
```

这个表示如果是从 HTTPS 网址链接到 HTTP 网址，就不发送 Referer 字段，其他情况发送。开发者可以通过 Spring Security 中提供的方法对此进行修改，方式如下：

```java
@Configuration
public class SecurityConfig extends WebSecurityConfigurerAdapter {
    @Override
    protected void configure(HttpSecurity http) throws Exception {
        http.authorizeRequests()
                .anyRequest().authenticated()
                .and()
                .formLogin()
                .and()
                .csrf().disable()
                .headers()
                .referrerPolicy()
                .policy(ReferrerPolicyHeaderWriter.ReferrerPolicy.ORIGIN);
    }
}
```

这个配置取值是 origin，表示总是发送源信息（源信息仅包含请求协议和域名，不包含其他路径信息，与之相对的是完整的 URL）。其他的取值还有：

- no-referrer：表示从请求头中移除 Referer 字段。
- same-origin：表示链接到同源地址时，发送文件源信息作为引用地址，否则不发送。
- strict-origin：表示从 HTTPS 链接到 HTTP 时不发送源信息，否则发送。
- origin-when-cross-origin：表示对于同源请求会发送完整的 URL 作为引用地址，但是对于非同源请求，则只发送源信息。
- strict-origin-when-cross-origin：表示对于同源的请求，会发送完整的 URL 作为引用地址；跨域时，如果是从 HTTPS 链接到 HTTP，则不发送 Referer 字段，否则发送文件的源信

- unsafe-url：表示无论是同源请求还是非同源请求，都发送完整的 URL（移除参数信息之后）作为引用地址。

9.2.8 Feature-Policy

Feature-Policy 响应头提供了一种可以在本页面或包含的 iframe 上启用或禁止浏览器特性的机制（移动端开发使用较多）。举一个简单例子，如果想要禁用震动和定位 API，那么可以在响应头中添加如下内容：

```
Feature-Policy: vibrate 'none'; geolocation 'none'
```

Spring Security 中配置如下：

```
@Configuration
public class SecurityConfig extends WebSecurityConfigurerAdapter {
    @Override
    protected void configure(HttpSecurity http) throws Exception {
        http.authorizeRequests()
                .anyRequest().authenticated()
                .and()
                .formLogin()
                .and()
                .csrf().disable()
                .headers()
                .featurePolicy("vibrate 'none'; geolocation 'none'");
    }
}
```

该功能使用较少，这里不做过多介绍。

9.2.9 Clear-Site-Data

Clear-Site-Data 一般用在注销登录响应头中，表示告诉浏览器清除当前网站相关的数据（cookie、cache、storage 等）。可以通过具体的参数指定想要清除的数据，如 cookies、cache、storage 等，也可以通过 "*" 表示清除所有数据。

Spring Security 中配置如下：

```
@Configuration
public class SecurityConfig extends WebSecurityConfigurerAdapter {
    @Override
    protected void configure(HttpSecurity http) throws Exception {
        http.authorizeRequests()
                .anyRequest().authenticated()
                .and()
                .formLogin()
                .and()
```

```
                    .logout()
                    .addLogoutHandler(new HeaderWriterLogoutHandler(
new ClearSiteDataHeaderWriter(ClearSiteDataHeaderWriter.Directive.ALL)))
                    .and()
                    .csrf().disable();
    }
}
```

在注销登录的处理器中，设置了清除浏览器所有和当前网站相关的数据。配置完成后，当浏览器发起注销登录请求时，响应头中就会有 Clear-Site-Data，如图 9-13 所示。

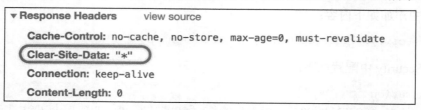

图 9-13　注销登录时的响应头

9.3　HTTP 通信安全

HTTP 通信安全主要从三个方面入手：

（1）使用 HTTPS 代替 HTTP。
（2）Strict-Transport-Security 配置。
（3）代理服务器配置。

其中第 2 点我们前面已经讲过，这里主要和大家分享第 1 点和第 3 点。

9.3.1　使用 HTTPS

作为一个框架，Spring Security 不处理 HTTP 连接问题，因此不直接提供对 HTTPS 的支持。但是，它提供了许多有助于 HTTPS 使用的功能。

接下来我们通过一个简单的案例来演示其具体用法。

首先创建一个 Spring Boot 项目，引入 Spring Security 和 Web，然后参考 9.2.3 小节中的方式创建 HTTPS 证书，并配置到 Spring Boot 项目中。

配置完成后，我们再在 application.properties 中添加如下配置修改项目端口号：

```
spring.security.user.name=javaboy
spring.security.user.password=123
server.port=8443
```

此时我们的项目就支持 HTTPS 访问了，HTTPS 的访问端口是 8443。为了更好地演示 Spring Security 的功能，我们需要项目同时支持 HTTPS 和 HTTP，所以还需要在项目中添加如下配置：

```
@Configuration
public class TomcatConfig {
    @Bean
    TomcatServletWebServerFactory tomcatServletWebServerFactory() {
        TomcatServletWebServerFactory factory =
                            new TomcatServletWebServerFactory();
        factory.addAdditionalTomcatConnectors(createTomcatConnector());
        return factory;
    }
    private Connector createTomcatConnector() {
        Connector connector = new
                Connector("org.apache.coyote.http11.Http11NioProtocol");
        connector.setScheme("http");
        connector.setPort(8080);
        return connector;
    }
}
```

这相当于又添加了一个 Connector，让 Tomcat 同时监听 8080 端口。

配置完成后，启动项目，控制台可以看到如下日志，表示项目现在同时支持 HTTP 和 HTTPS 了：

```
Tomcat started on port(s): 8443 (https) 8080 (http) with context path ''
```

接下来，我们创建两个测试接口，代码如下：

```
@RestController
public class HelloController {
    @GetMapping("/https")
    public String https() {
        return "https";
    }
    @GetMapping("/http")
    public String http() {
        return "http";
    }
}
```

/http 接口表示可以直接使用 HTTP 协议访问，/https 接口表示只可以通过 HTTPS 协议访问。我们来看一下 SecurityConfig 中的配置：

```
@Configuration
public class SecurityConfig extends WebSecurityConfigurerAdapter {
    @Override
    protected void configure(HttpSecurity http) throws Exception {
        http.authorizeRequests()
                .anyRequest().authenticated()
                .and()
                .formLogin()
```

```
            .and()
            .requiresChannel()
            .antMatchers("/https").requiresSecure()
            .antMatchers("/http").requiresInsecure()
            .and()
            .csrf().disable();
    }
}
```

通过 requiresChannel()方法开启配置,requiresSecure()表示该请求是 HTTPS 协议,如果不是,则重定向到 HTTPS 协议请求;requiresInsecure()则要求请求是 HTTP 协议,如果不是,则重定向到 HTTP 协议请求。没有列举出来的请求,则是两种协议都可以。

配置完成后,重启项目。使用如下两个地址都可以访问到登录页面:

- http://localhost:8080/login
- https://localhost:8443/login

现在我们使用 http://localhost:8080/login 地址登录成功后,访问/http 没问题,访问/https 则会自动重定向到 https://localhost:8443/https。之所以重定向到 8443 端口,并非因为我们项目端口是 8443,而是因为 8443 是 HTTPS 的默认监听端口,无论项目端口号是多少,这里都会重定向到 8443 端口。

现在修改 application.properties 中的配置,将 server.port 改为 8444,即将项目中 HTTPS 的监听端口改为 8444,那么如何让这里重定向到 8444 呢?配置如下:

```
@Configuration
public class SecurityConfig extends WebSecurityConfigurerAdapter {
    @Override
    protected void configure(HttpSecurity http) throws Exception {
        http.authorizeRequests()
            .anyRequest().authenticated()
            .and()
            .formLogin()
            .and()
            .portMapper()
            .http(8080).mapsTo(8444)
            .and()
            .requiresChannel()
            .antMatchers("/https").requiresSecure()
            .antMatchers("/http").requiresInsecure()
            .and()
            .csrf().disable();
    }
}
```

通过 portMapper()方法开启端口的映射配置,这里我们配置将 HTTP 端口 8080 转发到 HTTPS 端口 8444。配置完成后,当用户再次访问 http://localhost:8080/https 时,就会自动重定向到 https://localhost:8444/https 地址。

> **提示**
>
> 在测试时有一个问题需要注意。如果一开始使用 HTTP 协议登录，则登录成功访问/http、/https 都没有问题，都会自动进行重定向，但是如果一开始使用了 HTTPS 协议登录，则在登录成功后，从 HTTPS 协议重定向到 HTTP 协议时，会让用户重新登录。出现这一问题的原因在于，如果用户使用 HTTPS 协议登录，则返回的 Cookie 中包含了 Secure 标记（见图 9-14），该属性表示该 Cookie 只可以在安全环境下传输（即 HTTPS 协议中传输），当从 HTTPS 重定向到 HTTP 时，HTTP 请求并不会自动携带该 Cookie，所以就会让用户重新登录。反之，如果一开始登录使用了 HTTP 协议，则返回的 Cookie 中没有 Secure 标记，该 Cookie 在 HTTPS 和 HTTP 环境下都可以传输，因此可以无缝重定向。同时，由于我们的两个测试地址域名都是 localhost，而 Cookie 是不区分端口号的，如果 Cookie 名相同，会自动覆盖，并且读取的是相同的数据。所以，当从 HTTPS 协议重定向到 HTTP 协议时，浏览器上 HTTPS 的 JSESSIONID 还在，但是 HTTP 协议又用不了该 Cookie，就会导致 HTTP 协议一直登录失败，此时只要清除浏览器缓存即可。

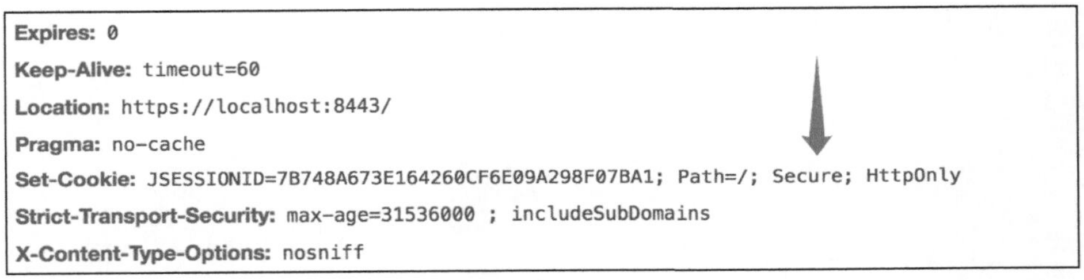

```
Expires: 0
Keep-Alive: timeout=60
Location: https://localhost:8443/
Pragma: no-cache
Set-Cookie: JSESSIONID=7B748A673E164260CF6E09A298F07BA1; Path=/; Secure; HttpOnly
Strict-Transport-Security: max-age=31536000 ; includeSubDomains
X-Content-Type-Options: nosniff
```

图 9-14　HTTPS 环境下，Cookie 中含有 Secure 标记

这就是 Spring Security 中关于 HTTP 请求转发到 HTTPS 的配置，当然，HTTP 自动转发到 HTTPS 也可以在 Servlet 容器层面进行配置，这个不属于本书的范畴，这里不做过多介绍，感兴趣的读者可以自行查找资料学习。

9.3.2　代理服务器配置

在分布式环境下或者集群环境下，项目中可能会引入 Nginx 作为负载均衡服务器，这个时候需要确保自己的代理服务器和 Spring Security 的配置是正确的，以便 Spring Security 能够准确获取请求的真实 IP，避免各种潜在的威胁或应用程序错误。

开发者需要在代理服务器中配置 X-Forwarded-*以便将客户端信息转发到真实的后端，后端收到请求后，不同的 Servlet 容器以及 Spring 框架都有各自的方式从请求头中获取客户端真实的 IP 地址、Host 以及请求协议等，并重新设置在相应的 request.getXXX 方法上，开发者调用这些方法就可以成功获取到客户端的真实信息，并不会感觉到有代理服务器存在。例如，在 Tomcat 中是通过 RemoteIpValve 类来处理，在 Jetty 中是通过 ForwardedRequestCustomizer 来处理，Spring 框架则是通过 ForwardedHeaderFilter 过滤器来处理。实际项目中，选择一种方式即可。

由于这里并不涉及 Spring Security 的相关知识，因此不做过多介绍，读者在实际项目开发中注意此问题即可。

9.4 小　结

　　本章主要介绍了 Spring Security 在处理各种漏洞方面所做的努力。开发者考虑到的安全问题、没考虑到的安全问题，Spring Security 都提供了相应的处理方案，从 CSRF 攻击防御到各种 HTTP 响应头处理，再到 HTTPS 通信，Spring Security 考虑到了各种可能存在的漏洞，并提供了相应的解决方案。这也是我们选择 Spring Security 的原因之一，即不用开发者考虑很多，只要选择了 Spring Security，即使我们只做了一点点配置，我们的系统也足够安全。

第 10 章

HTTP 认证

HTTP 提供了一个用于权限控制和认证的通用方式，这种认证方式通过 HTTP 请求头来提供认证信息，而不是通过表单登录。最为读者所熟知的应该是 HTTP Basic authentication，另外还有一种相对更安全的 HTTP Digest authentication。Spring Security 中对于这两种认证方式都提供了相应的支持。本章介绍这两种认证的具体用法以及各自的优缺点。

本章涉及的主要知识点有：

- HTTP Basic authentication。
- HTTP Digest authentication。

10.1　HTTP Basic authentication

10.1.1　简介

HTTP Basic authentication 中文译作 HTTP 基本认证，在这种认证方式中，将用户的登录用户名/密码经过 Base64 编码之后，放在请求头的 Authorization 字段中，从而完成用户身份的认证。

这是一种在 RFC7235（https://tools.ietf.org/html/rfc7235）规范中定义的认证方式，当客户端发起一个请求之后，服务端可以针对该请求返回一个质询信息，然后客户端再提供用户的凭证信息。具体的质询与应答流程如图 10-1 所示。

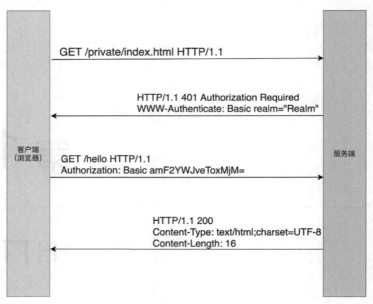

图 10-1　HTTP 基本认证质询应答流程

从图 10-1 中我们可以看到，HTTP 基本认证的流程是下面这样的。

首先客户端（浏览器）发起请求，类似下面这样：

```
GET /hello HTTP/1.1
Host: localhost:8080
```

服务端收到请求后，发现用户还没有认证，于是给出如下响应：

```
HTTP/1.1 401
WWW-Authenticate: Basic realm="Realm"
```

状态码 401 表示用户未认证，WWW-Authenticate 响应头则定义了使用何种验证方式去完成身份认证。最简单、最常见的就是我们使用的 HTTP 基本认证（Basic），除了这种认证方式之外，还有 Bearer（OAuth2.0 认证）、Digest（HTTP 摘要认证）等取值。

客户端收到服务端的响应之后，将用户名/密码使用 Base64 编码之后，放在请求头中，再次发起请求：

```
GET /hello HTTP/1.1
Host: localhost:8080
Authorization: Basic amF2YWJveToxMjM=
```

服务端解析 Authorization 字段，完成用户身份的校验，最后将资源返回给客户端：

```
HTTP/1.1 200
Content-Type: text/html;charset=UTF-8
Content-Length: 16
```

这就是整个 HTTP 基本认证流程。

可以看到，这种认证方式实际上非常简单，基本上所有的浏览器都支持这种认证方式。

但是我们在实际应用中，似乎很少见到这种认证方式，有的读者可能在一些老旧路由器中见过这种认证方式；另外，在一些非公开访问的 Web 应用中，可能也会见到这种认证方式。为什么很少见到这种认证方式的应用场景呢？主要还是安全问题。

HTTP 基本认证没有对传输的凭证信息进行加密，仅仅只是进行了 Base64 编码，这就造成了很大的安全隐患，所以如果用到了 HTTP 基本认证，一般都是结合 HTTPS 一起使用；同时，一旦使用 HTTP 基本认证成功后，除非用户关闭浏览器或者清空浏览器缓存，否则没有办法退出登录。

10.1.2 具体用法

Spring Security 中开启 HTTP 基本认证非常容易，配置如下：

```java
@Configuration
public class SecurityConfig extends WebSecurityConfigurerAdapter {
    @Override
    protected void configure(HttpSecurity http) throws Exception {
        http.authorizeRequests()
                .anyRequest().authenticated()
                .and()
                .httpBasic()
                .and()
                .csrf().disable();
    }
}
```

通过 httpBasic()方法即可开启 HTTP 基本认证。配置完成后，启动项目，此时如果需要访问一个受保护的资源，浏览器就会自动弹出认证框，如图 10-2 所示。输入用户名/密码完成认证之后，就可以访问受保护的资源了。

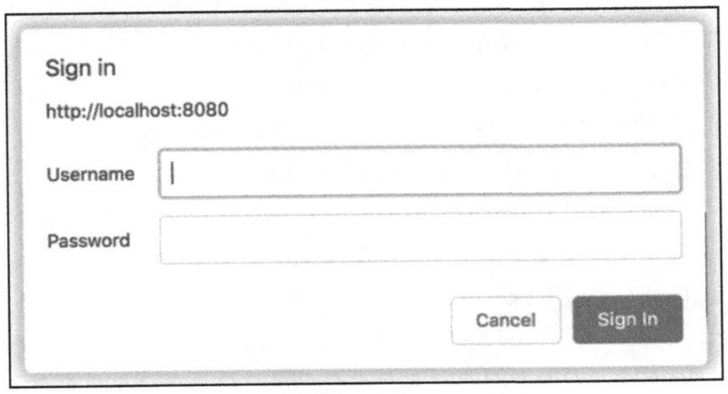

图 10-2　HTTP 基本认证登录框

10.1.3 源码分析

代码实现很简单，接下来我们来看一下 Spring Security 中是如何实现 HTTP 基本认证的。实现整体上分为两部分：

（1）对未认证的请求发出质询。

（2）解析携带了认证信息的请求。

我们分别从这两部分来分析。

10.1.3.1 质询

httpBasic() 方法开启了 HTTP 基本认证的配置，具体的配置通过 HttpBasicConfigurer 类来完成。在 HttpBasicConfigurer 配置类的 init 方法中调用了 registerDefaultEntryPoint 方法，该方法完成了失败请求失败处理类 AuthenticationEntryPoint 的配置，代码如下：

```java
private void registerDefaultEntryPoint(B http,
                                    RequestMatcher preferredMatcher) {
    ExceptionHandlingConfigurer<B> exceptionHandling = http
            .getConfigurer(ExceptionHandlingConfigurer.class);
    if (exceptionHandling == null) {
        return;
    }
    exceptionHandling.defaultAuthenticationEntryPointFor(
            postProcess(this.authenticationEntryPoint), preferredMatcher);
}
```

可以看到，这里调用了 exceptionHandling 对象的方法进行配置，该对象的最终目的是配置异常过滤器 ExceptionTranslationFilter，关于该过滤器中的执行逻辑在本书第 12 章将会详细介绍，这里先不赘述。

这里配置到 exceptionHandling 中的 authenticationEntryPoint 是一个代理对象，该代理对象是在 HttpBasicConfigurer 构造方法中创建的，具体代理的就是 BasicAuthenticationEntryPoint。简而言之，如果一个请求没有携带认证信息，最终将被 BasicAuthenticationEntryPoint 的实例处理，我们来看一下该类的实现（部分）：

```java
public class BasicAuthenticationEntryPoint implements
                            AuthenticationEntryPoint,InitializingBean {
    public void commence(HttpServletRequest request,
                    HttpServletResponse response,
            AuthenticationException authException) throws IOException {
        response.addHeader("WWW-Authenticate",
                                "Basic realm=\"" + realmName + "\"");
        response.sendError(HttpStatus.UNAUTHORIZED.value(),
                        HttpStatus.UNAUTHORIZED.getReasonPhrase());
    }
}
```

可以看到，这个类的处理逻辑还是非常简单的，响应头中添加 WWW-Authenticate 字段，然后发送错误响应，响应码为 401。

这就是发出质询的代码。总结一下，就是一个未经认证的请求，在经过 Spring Security 过滤器链时会抛出异常，该异常会在 ExceptionTranslationFilter 过滤器中调用 BasicAuthentication

EntryPoint#commence 方法进行处理。

10.1.3.2 请求解析

HttpBasicConfigurer 类的 configure 方法中，向 Spring Security 过滤器链添加一个过滤器 BasicAuthenticationFilter，该过滤器专门用来处理 HTTP 基本认证相关的事情，我们来看一下它核心的 doFilterInternal 方法：

```java
protected void doFilterInternal(HttpServletRequest request,
        HttpServletResponse response, FilterChain chain)
            throws IOException, ServletException {
    try {
        UsernamePasswordAuthenticationToken authRequest =
                            authenticationConverter.convert(request);
        if (authRequest == null) {
            chain.doFilter(request, response);
            return;
        }
        String username = authRequest.getName();
        if (authenticationIsRequired(username)) {
            Authentication authResult =
                    this.authenticationManager.authenticate(authRequest);
            SecurityContextHolder.getContext().setAuthentication(authResult);
            this.rememberMeServices.loginSuccess(request, response, authResult);
            onSuccessfulAuthentication(request, response, authResult);
        }
    }
    catch (AuthenticationException failed) {
        SecurityContextHolder.clearContext();
        this.rememberMeServices.loginFail(request, response);
        onUnsuccessfulAuthentication(request, response, failed);
        if (this.ignoreFailure) {
            chain.doFilter(request, response);
        }
        else {
            this.authenticationEntryPoint.commence(request, response, failed);
        }
        return;
    }
    chain.doFilter(request, response);
}
```

该方法执行流程如下：

（1）首先调用 authenticationConverter.convert 方法，对请求头中的 Authorization 字段进行解析，经过 Base64 解码后的用户名/密码是一个用":"隔开的字符串，例如用户使用 javaboy/123 进行登录，那么这里解码后的结果就是 javaboy:123，然后根据拿到的用户名/密码，构造一个 UsernamePasswordAuthenticationToken 对象出来。

（2）如果 authRequest 变量为 null，说明请求头中没有包含认证信息，那么直接执行接下来的过滤器即可，该方法也到此为止。在执行接下来的过滤器时，最终就会通过 ExceptionTranslationFilter 过滤器进入到 BasicAuthenticationEntryPoint#commence 方法中；如果 authRequest 变量不为 null，说明请求是携带了认证信息的，那么就对请求携带的认证信息进行校验。

（3）从 authRequest 对象中提取出用户名，然后调用 authenticationIsRequired 方法判断是否有必要进行认证，如果没有必要，则直接执行剩下的过滤器即可；如果有必要进行认证，则进行用户认证。authenticationIsRequired 方法的具体逻辑就是，从 SecurityContextHolder 中取出当前登录对象，判断使用是否已经登录过了，同时判断是否就是当前用户。

（4）如果有必要进行认证，则调用 authenticationManager.authenticate 方法完成用户认证，同时将用户信息存入 SecurityContextHolder；如果配置了 rememberMeServices，也进行相应的处理，最后还有一个登录成功的回调方法 onSuccessfulAuthentication，不过该方法并未做任何实现。

（5）如果认证过程抛出异常，则进行相应处理即可，这里逻辑比较简单。

（6）最后继续执行接下来的过滤器。在后续过滤器的执行过程中，由于 SecurityContextHolder 中已经保存了登录用户信息了，相当于用户已经完成登录了，因此就和普通的请求一致，不会被"半路拦截"。

这就是整个 HTTP 基本认证的实现逻辑，可以看到实现还是比较容易的。

10.2　HTTP Digest authentication

10.2.1　简介

HTTP 基本认证虽然简单易用，但是在安全方面问题突出，于是又推出了 HTTP 摘要认证。HTTP 摘要认证最早在 RFC2069 中被定义，随后被 RFC2617（https://tools.ietf.org/html/rfc2617）所取代，在 RFC2617 中引入了一系列增强安全性的参数，以防止各种可能存在的网络攻击。

相比于 HTTP 基本认证，HTTP 摘要认证的安全性有了很大提高，但是依然存在问题，例如不支持 bCrypt、PBKDF2、SCrypt 等加密方式。

图 10-3 所示描述了 HTTP 摘要认证的具体流程，从图中可以看出，这个认证流程和 HTTP 基本认证流程一致，不同的是每次传递的参数有所差异。

第 10 章 HTTP 认证 | 261

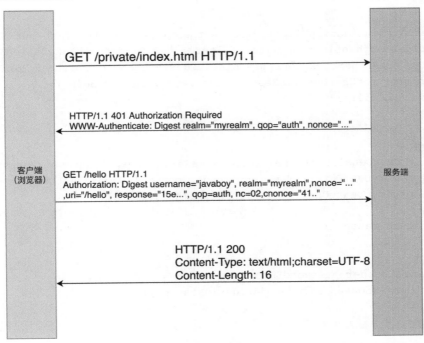

图 10-3 HTTP 摘要认证流程图

10.2.2 具体用法

Spring Security 中为 HTTP 摘要认证提供了相应的 AuthenticationEntryPoint 和 Filter，但是没有自动化配置，需要我们手动配置，配置方式如下：

```
@Configuration
public class SecurityConfig extends WebSecurityConfigurerAdapter {
    @Override
    protected void configure(HttpSecurity http) throws Exception {
        http.authorizeRequests()
                .anyRequest().authenticated()
                .and()
                .csrf().disable()
                .exceptionHandling()
                .authenticationEntryPoint(digestAuthenticationEntryPoint())
                .and()
                .addFilter(digestAuthenticationFilter());
    }
    DigestAuthenticationEntryPoint digestAuthenticationEntryPoint() {
        DigestAuthenticationEntryPoint entryPoint =
                                    new DigestAuthenticationEntryPoint();
        entryPoint.setNonceValiditySeconds(3600);
        entryPoint.setRealmName("myrealm");
        entryPoint.setKey("javaboy");
        return entryPoint;
```

```
    }
    DigestAuthenticationFilter digestAuthenticationFilter() throws Exception {
        DigestAuthenticationFilter filter = new DigestAuthenticationFilter();
        filter.setAuthenticationEntryPoint(digestAuthenticationEntryPoint());
        filter.setUserDetailsService(userDetailsServiceBean());
        return filter;
    }
    @Override
    @Bean
    public UserDetailsService userDetailsServiceBean() throws Exception {
        InMemoryUserDetailsManager manager = new InMemoryUserDetailsManager();
        manager.createUser(User.withUsername("javaboy")
                             .password("123").roles("admin").build());
        return manager;
    }
    @Bean
    PasswordEncoder passwordEncoder() {
        return NoOpPasswordEncoder.getInstance();
    }
}
```

（1）首先由开发者提供一个 DigestAuthenticationEntryPoint 实例（相当于 HTTP 基本认证中的 BasicAuthenticationEntryPoint），当用户发起一个没有认证的请求时，由该实例进行处理。配置该实例时，我们需要提供一个随机数的有效期，RealmName 以及一个 Key。

（2）创建一个 DigestAuthenticationFilter 实例，并添加到 Spring Security 过滤器链中，DigestAuthenticationFilter 的作用类似于 HTTP 基本认证中 BasicAuthenticationFilter 过滤器的作用。

（3）配置一个 UserDetailsService 实例。

（4）配置一个 PasswordEncoder 实例。

需要注意的是，由于客户端是对明文密码进行 Hash 运算，所以服务端也需要保存用户的明文密码，因此这里提供的 PasswordEncoder 实例是 NoOpPasswordEncoder 的实例。

在 DigestAuthenticationFilter 过滤器中有一个 passwordAlreadyEncoded 属性，表示用户密码是否已经编码，该属性默认为 false，表示密码未进行编码。开发者可以对密码进行编码，只需要先将该属性设置为 true，然后将 username + ":" + realm + ":" + password 使用 MD5 算法计算其消息摘要，将计算结果作为用户密码即可。举个简单例子，例如用户名是 javaboy，realm 是 myrealm，用户密码是 123，则计算器消息摘要代码如下：

```
String rawPassword = "javaboy:myrealm:123";
MessageDigest digest = MessageDigest.getInstance("MD5");
String s = new String(Hex.encode(digest.digest(rawPassword.getBytes())));
System.out.println(s);
```

计算结果如下：

```
e7ecfd3f08e6960f154e1ff29079fbd3
```

然后修改配置类:

```
DigestAuthenticationFilter digestAuthenticationFilter() throws Exception {
    DigestAuthenticationFilter filter = new DigestAuthenticationFilter();
    filter.setAuthenticationEntryPoint(digestAuthenticationEntryPoint());
    filter.setUserDetailsService(userDetailsServiceBean());
    filter.setPasswordAlreadyEncoded(true);
    return filter;
}
@Override
@Bean
public UserDetailsService userDetailsServiceBean() throws Exception {
    InMemoryUserDetailsManager manager = new InMemoryUserDetailsManager();
    manager.createUser(User.withUsername("javaboy")
            .password("e7ecfd3f08e6960f154e1ff29079fbd3")
            .roles("admin").build());
    return manager;
}
```

调用 DigestAuthenticationFilter 的 setPasswordAlreadyEncoded 方法，将 passwordAlreadyEncoded 属性设置为 true，然后设置用户密码为编码后的密码即可。

注意，这样配置完成后，PasswordEncoder 的实例依然是 NoOpPasswordEncoder，具体原因将在下一小节的源码分析中介绍。

配置完成后，启动项目，访问页面时，浏览器就会弹出输入框要求输入用户名/密码信息，具体流程和 HTTP 基本认证一致，这里不再赘述。

10.2.3 源码分析

接下来对 HTTP 摘要认证的源码进行简单分析，我们从质询、客户端处理以及请求解析三个方面入手。需要说明的是，Spring Security 源码中关于 HTTP 摘要认证并未严格遵守 RFC2617，下面的分析以 Spring Security 源码为准。

10.2.3.1 质询

HTTP 摘要认证的质询是由 DigestAuthenticationEntryPoint#commence 方法负责处理的，源码如下：

```
public void commence(HttpServletRequest request,
                HttpServletResponse response,
                AuthenticationException authException) throws IOException {
    HttpServletResponse httpResponse = response;
    long expiryTime = System.currentTimeMillis() + (nonceValiditySeconds * 1000);
    String signatureValue = DigestAuthUtils.md5Hex(expiryTime + ":" + key);
    String nonceValue = expiryTime + ":" + signatureValue;
    String nonceValueBase64 =
            new String(Base64.getEncoder().encode(nonceValue.getBytes()));
```

```
        String authenticateHeader = "Digest realm=\"" + realmName + "\", "
            + "qop=\"auth\", nonce=\"" + nonceValueBase64 + "\"";
        if (authException instanceof NonceExpiredException) {
            authenticateHeader = authenticateHeader + ", stale=\"true\"";
        }
        httpResponse.addHeader("WWW-Authenticate", authenticateHeader);
        httpResponse.sendError(HttpStatus.UNAUTHORIZED.value(),
            HttpStatus.UNAUTHORIZED.getReasonPhrase());
    }
```

和 HTTP 基本认证一样，这里的响应码也是 401，响应头中也包含 WWW-Authenticate 字段，不同的是 WWW-Authenticate 字段的值有所区别：

- Digest：表示这里使用 HTTP 摘要认证。
- Realm：服务端返回的标识访问资源的安全域。
- qop：服务端返回的保护级别，客户端据此选择合适的摘要算法，如果值为 auth，则表示只进行身份认证；如果取值为 auth-int，则除了身份认证之外，还要校验内容完整性。
- nonce：服务端生成的一个随机字符串，在客户端生成摘要信息时会用到该随机字符串。
- stale：一个标记，当随机字符串 nonce 过期时，会包含该标记。stale=true 表示客户端不必再次弹出输入框，只需要带上已有的认证信息，重新发起认证请求即可。

随机字符串 nonce 的生成过程是，先对过期时间和 key 组成的字符串 expiryTime + ":" + key 计算出消息摘要 signatureValue，然后再对 expiryTime + ":" + signatureValue 进行 Base64 编码，进而获取 nonce。

经过上面的分析，我们可以得出，响应头内容如下：

```
HTTP/1.1 401
WWW-Authenticate: Digest realm="myrealm", qop="auth",
nonce="MTU5OTIyNDE4NDg1NDowZGIzOWU0NGM2MTA5ZDVmZDkyNWYzMzRmNmYxZjg1ZA=="
```

10.2.3.2　客户端处理

当客户端（浏览器）收到质询请求后，弹出输入框，用户输入用户名/密码，然后客户端会对用户名/密码进行 Hash 运算。根据响应头中 qop 值的不同，运算过程会略有差异。

如果服务端响应头中不包含 qop 参数，则运算过程如下：

（1）对 username + ":" + realm + ":" + password 计算其消息摘要得到 digest1。
（2）对 HttpMethod+ ":" + uri 计算其消息摘要得到 digest2。
（3）对 digest1 + ":" + nonce + ":" + digest2 计算其消息摘要得到 response。

如果服务端响应头中 qop="auth"，则前两步计算步骤一致，第 3 步不同，第 3 步计算方式如下：

对 digest1 + ":" + nonce + ":" + nc + ":" + cnonce + ":" + qop + ":" + digest2 计算其消息摘要，得到 response。

这里的几个参数和大家解释下：

- nonce 和 qop：就是服务端返回的数据。
- nc：表示请求次数，该参数在防止重放攻击时有用。
- cnonce：表示客户端生成的随机数。

这里计算出来的 response 就是客户端提交给服务端的重要认证信息，服务端主要据此判断用户身份是否合法。

最终客户端提交的请求头如下：

```
GET /hello HTTP/1.1
Host: localhost:8080
Authorization: Digest username="javaboy", realm="myrealm",
nonce="MTU5OTIyNDE4NDg1NDowZGIzOWU0NGM2MTA5ZDVmZDkyNWYzMzRmNmYxZjg1ZA=="
,uri="/hello", response="f14cbc00cc461092c3f6d392d234f5b1", qop=auth
, nc=00000002, cnonce="2867d826762e8b56"
```

用户名放在请求头中，用户密码则经过各种 MD5 运算之后，现在包含在 response 中，生成 response 时所需要的 cnonce、nonce、nc、qop、realm 以及 uri 也都包含在请求头中一并发送给服务端，服务端拿到这些参数之后，再根据用户名去数据库中查询到用户密码，然后进行 MD5 运算，将运算结果和 response 进行比对，就能知道请求是否合法。

> **什么是重放攻击？**
>
> 重放攻击（Replay attack）也称为回放攻击，这是一种通过重复或者延迟有效数据的网络攻击形式，是一种低级别的"中间人攻击"。举个简单例子，当用户和服务端进行数据交互时，为了向服务端证明身份，传递了一个经过 MD5 运算的字符串，该字符串被黑客窃取到。黑客就可以通过该字符串冒充受害者。

10.2.3.3 请求解析

请求解析主要是在 DigestAuthenticationFilter 过滤器中完成的，我们来看一下其 doFilter 方法：

```java
public void doFilter(ServletRequest req, ServletResponse res,
            FilterChain chain)throws IOException, ServletException {
    HttpServletRequest request = (HttpServletRequest) req;
    HttpServletResponse response = (HttpServletResponse) res;
    String header = request.getHeader("Authorization");
    if (header == null || !header.startsWith("Digest ")) {
        chain.doFilter(request, response);
        return;
    }
    DigestData digestAuth = new DigestData(header);
    try {
        digestAuth.validateAndDecode(this.authenticationEntryPoint.getKey(),
                this.authenticationEntryPoint.getRealmName());
    }
    catch (BadCredentialsException e) {
```

```java
        fail(request, response, e);
        return;
    }
    boolean cacheWasUsed = true;
    UserDetails user =
            this.userCache.getUserFromCache(digestAuth.getUsername());
    String serverDigestMd5;
    try {
        if (user == null) {
            cacheWasUsed = false;
            user = this.userDetailsService
                    .loadUserByUsername(digestAuth.getUsername());
            if (user == null) {
                throw new AuthenticationServiceException("");
            }
            this.userCache.putUserInCache(user);
        }
        serverDigestMd5 = digestAuth
          .calculateServerDigest(user.getPassword(),request.getMethod());
        if (!serverDigestMd5.equals(digestAuth.getResponse())&& cacheWasUsed) {
            user = this.userDetailsService
                    .loadUserByUsername(digestAuth.getUsername());
            this.userCache.putUserInCache(user);
            serverDigestMd5 = digestAuth
              .calculateServerDigest(user.getPassword(),request.getMethod());
        }
    }
    catch (UsernameNotFoundException notFound) {
        fail(request, response,new BadCredentialsException(""));
        return;
    }
    if (!serverDigestMd5.equals(digestAuth.getResponse())) {
        fail(request, response,new BadCredentialsException(""));
        return;
    }
    if (digestAuth.isNonceExpired()) {
        fail(request, response,new NonceExpiredException(""));
        return;
    }
    Authentication authentication =
                        createSuccessfulAuthentication(request, user);
    SecurityContext context = SecurityContextHolder.createEmptyContext();
    context.setAuthentication(authentication);
    SecurityContextHolder.setContext(context);
    chain.doFilter(request, response);
}
```

这个 doFilter 方法比较冗长，我们逐步进行分析一下：

（1）首先从请求头中获取 Authorization 字段，如果该字段不存在，或者该字段的值不是以 Digest 开头，则直接执行剩下的过滤器。在执行剩下的过滤器时，最终会进入到质询环节。

（2）根据获取到的 Authorization 字段信息，构造出一个 DigestData 对象。这个过程就是将请求头中的 username、realm、nonce、uri、response、qop、nc、cnonce 等字段解析出来，设置给 DigestData 对象中对应的属性，方便后续处理。

（3）接下来调用 validateAndDecode 方法，对刚刚解析出来的数据进行初步的验证。这里的验证代码比较简单，此处就不一一列出来了，主要介绍一下方法的执行逻辑：①首先判断 username、realm、nonce、uri 以及 response 是否为 null，如果存在为 null 的数据，则直接抛出异常；②判断 qop 的值是否为 auth，如果为 auth，则 nc 和 cnonce 都不能为 null，否则抛出异常；③检验请求传来的 realm 和 authenticationEntryPoint 中的 realm 是否相等，如果不相等，则直接抛出异常；④尝试对 nonce 进行 Base64 解码，如果解码失败，则抛出异常；⑤对 nonce 进行 Base64 解码，将解码的结果拆分成一个名为 nonceTokens 的数组，如果数组的长度不为 2，则抛出异常；⑥取出 nonceTokens 数组中的第 0 项，就是 nonce 的过期时间，将其赋值给 nonceExpiryTime 属性；⑦根据 nonceExpiryTime 以及 authenticationEntryPoint 中的 key，进行 MD5 运算，并将运算结果和 nonceTokens 数组中的第 1 项进行比较，如果不相等，则抛出异常。至此，就完成了对请求参数的初步校验。

（4）根据请求传来的用户名去加载用户对象，先去缓存中加载，缓存中没有，则调用 userDetailsService 实例去加载（这也是为什么我们在配置 DigestAuthenticationFilter 过滤器时，需要指定 userDetailsService 实例的原因）。

（5）接下来调用 calculateServerDigest 方法去计算服务端的摘要信息，该方法内部又调用了 DigestAuthUtils.generateDigest 方法。计算过程比较简单，这里主要说下计算流程：①首先是 a1Md5 的计算，如果设置了 passwordAlreadyEncoded，则直接将用户密码赋值给 a1Md5，否则根据 username、realm 以及 password 计算出 a1Md5；②a2Md5 计算方式是固定的，通过 httpMethod 以及请求 uri 计算出 a2Md5；③如果 qop 为 null，则 digest = a1Md5 + ":" + nonce + ":" + a2Md5；如果 qop 的值为 auth，则 digest = a1Md5 + ":" + nonce + ":" + nc + ":" + cnonce + ":" + qop + ":" + a2Md5；④对 digest 进行 MD5 运算，并将运算结果返回（注意，这里的流程和上一小节所讲的客户端处理是对应的）。

（6）如果服务端基于缓存用户计算出来的摘要信息不等于请求传来的 response 字段的值，则重新从 userDetailsService 中加载用户信息，并重新完成第 5 步的运算。

（7）如果服务端计算出来的摘要信息不等于请求传来的 response 字段的值，则抛出异常。

（8）如果随机字符串 nonce 过期，则抛出异常。

（9）如果前面的步骤都顺利，没有抛出异常，则认证成功。将登录成功的用户信息存入 SecurityContext 中，同时继续执行接下来的过滤器。存入 SecurityContext 中的 Authentication 实例里边用户密码，就是从 userDetailsService 中查询出来的用户密码，在以后的过滤器中，如果还需要进行密码校验，由于 SecurityContext 中的 Authentication 实例中的用户密码和 userDetailsService 对象提供的用户密码一模一样，所以在 10.2.2 小节中配置的 PasswordEncoder

实例只能是 NoOpPasswordEncoder，否则就会校验失败。

这就是整个 HTTP 摘要认证的工作流程。

和 HTTP 基本认证相比，这里最大的亮点是不明文传输用户密码，由客户端对密码进行 MD5 运算，并将运算所需的参数以及运算结果发送到服务端，服务端再去校验数据是否正确，这样可以避免密码泄漏。

这里，大家也能发现 HTTP 摘要认证存在的问题，例如，密码最多只能进行 MD5 运算后存储，或者就只能存储明文密码，无论哪种方式，都存在一定安全隐患。同时，由于使用的复杂性，HTTP 摘要认证在实际项目中使用并不多。

10.3 小　结

本章主要介绍了 HTTP 认证，包括 HTTP 基本认证和 HTTP 摘要认证。HTTP 基本认证简单易用但是安全问题突出；HTTP 摘要认证一定程度上解决了安全问题，但是又带来了密码存储的安全性问题。两种方式各有优缺点。由于安全问题，这种认证方式在实际项目中使用并不多，如果一定要使用，一般也要结合 HTTPS 一起用，可以最大程度地保证数据安全。

第 11 章

跨域问题

跨域问题是实际应用开发中一个非常常见的需求，在 Spring 框架中对于跨域问题的处理方案有好几种，引入了 Spring Security 之后，跨域问题的处理方案又增加了。本章将和读者分享如何处理跨域问题，以及不同方案之间的区别与联系。

本章涉及的主要知识点有：

- Spring 处理方案。
- Spring Security 处理方案。

11.1 什么是 CORS

CORS（Cross-Origin Resource Sharing）是由 W3C 制定的一种跨域资源共享技术标准，其目的就是为了解决前端的跨域请求。在 JavaEE 开发中，最常见的前端跨域请求解决方案是 JSONP，但是 JSONP 只支持 GET 请求，这是一个很大的缺陷，而 CORS 则支持多种 HTTP 请求方法，也是目前主流的跨域解决方案。

CORS 中新增了一组 HTTP 请求头字段，通过这些字段，服务器告诉浏览器，哪些网站通过浏览器有权限访问哪些资源。同时规定，对那些可能修改服务器数据的 HTTP 请求方法（如 GET 以外的 HTTP 请求等），浏览器必须首先使用 OPTIONS 方法发起一个预检请求（preflight request），预检请求的目的是查看服务端是否支持即将发起的跨域请求，如果服务端允许，才能发起实际的 HTTP 请求。在预检请求的返回中，服务器端也可以通知客户端，是否需要携带身份凭证（如 Cookies、HTTP 认证信息等）。

以 GET 请求为例，如果需要发起一个跨域请求，则请求头如下：

```
Host: localhost:8080
Origin: http://localhost:8081
Referer: http://localhost:8081/index.html
……
```

如果服务端支持该跨域请求，那么返回的响应头中将包含如下字段：

```
Access-Control-Allow-Origin: http://localhost:8081
```

Access-Control-Allow-Origin 字段用来告诉浏览器可以访问该资源的域，当浏览器收到这样的响应头信息之后，提取出 Access-Control-Allow-Origin 字段中的值，发现该值包含当前页面所在的域，就知道这个跨域是被允许的，因此就不再对前端的跨域请求进行限制。

这属于简单请求，即不需要进行预检请求的跨域。

对于一些非简单请求，会首先发送一个预检请求。预检请求类似下面这样：

```
OPTIONS /put HTTP/1.1
Host: localhost:8080
Connection: keep-alive
Accept: */*
Access-Control-Request-Method: PUT
Origin: http://localhost:8081
Referer: http://localhost:8081/index.html
……
```

请求方法是 OPTIONS，请求头 Origin 字段告诉服务端当前页面所在的域，请求头 Access-Control-Request-Method 告诉服务端即将发起的跨域请求所使用的方法。服务端对此进行判断，如果允许即将发起的跨域请求，则会给出如下响应：

```
HTTP/1.1 200
Access-Control-Allow-Origin: http://localhost:8081
Access-Control-Allow-Methods: PUT
Access-Control-Max-Age: 3600
……
```

Access-Control-Allow-Methods 字段表示允许的跨域方法；Access-Control-Max-Age 字段表示预检请求的有效期，单位为秒，在有效期内如果发起该跨域请求，则不用再次发起预检请求。预检请求结束后，接下来就会发起一个真正的跨域请求，跨域请求和前面的 GET 请求步骤类似，此处不再赘述。

这是我们关于 CORS 的一个简单介绍。

11.2 Spring 处理方案

首先回顾一下 Spring 中关于跨域的处理方案，这有助于我们理解 Spring Security 中的跨域。Spring 中对于跨域的处理一共有三种不同的方案，我们分别来看。

11.2.1 @CrossOrigin

Spring 中第一种处理跨域的方式是通过@CrossOrigin 注解来标记支持跨域，该注解可以添加在方法上，也可以添加在 Controller 上。当添加在 Controller 上时，表示 Controller 中的所有接口都支持跨域，具体配置如下：

```
@RestController
public class HelloController {
    @CrossOrigin(origins = "http://localhost:8081")
    @PostMapping("/post")
    public String post() {
        return "hello post";
    }
}
```

@CrossOrigin 注解各属性含义如下：

- allowCredentials：浏览器是否应当发送凭证信息，如 Cookie。
- allowedHeaders：请求被允许的请求头字段，*表示所有字段。
- exposedHeaders：哪些响应头可以作为响应的一部分暴露出来。注意，这里只可以一一列举，通配符*在这里是无效的。
- maxAge：预检请求的有效期，有效期内不必再次发送预检请求，默认是 1800 秒。
- methods：允许的请求方法，*表示允许所有方法。
- origins：允许的域，*表示允许所有域。

该注解的实现原理属于 Spring 范畴，这里不做过多介绍，本书仅和大家简单梳理一下 @CrossOrigin 注解执行过程，这有助于我们理解后面的内容（读者可按照下面的叙述自行 Debug 以加深理解）。

（1）@CrossOrigin 注解在 AbstractHandlerMethodMapping 的内部类 MappingRegistry 的 register 方法中完成解析的，@CrossOrigin 注解中的内容会被解析成一个配置对象 CorsConfiguration。

（2）将@CrossOrigin 所标记的请求方法对象 HandlerMethod 和 CorsConfiguration 一一对应存入一个名为 corsLookup 的 Map 集合中。

（3）当请求到达 DispatcherServlet#doDispatch 方法之后，调用 AbstractHandlerMapping #getHandler 方法获取执行链 HandlerExecutionChain 时，会从 corsLookup 集合中获取到 CorsConfiguration 对象。

（4）根据获取到的 CorsConfiguration 对象构建一个 CorsInterceptor 拦截器。

（5）在 CorsInterceptor 拦截器中触发对 DefaultCorsProcessor#processRequest 的调用，跨域请求的校验工作将在该方法中完成。

11.2.2　addCorsMappings

@CrossOrigin 注解需要添加在不同的 Controller 上。所以还有一种全局的配置方法，就是通过重写 WebMvcConfigurerComposite#addCorsMappings 方法来实现，具体配置如下：

```
@Configuration
public class WebMvcConfig implements WebMvcConfigurer {
   @Override
   public void addCorsMappings(CorsRegistry registry) {
       registry.addMapping("/**")
               .allowedMethods("*")
               .allowedOrigins("*")
               .allowedHeaders("*")
               .allowCredentials(false)
               .exposedHeaders("")
               .maxAge(3600);
   }
}
```

addMapping 表示要处理的请求地址，接下来的方法含义和@CrossOrigin 注解中属性的含义都一一对应，这里不再赘述。

这种配置方式最终的处理方式和@CrossOrigin 注解相同，都是在 CorsInterceptor 拦截器中触发对 DefaultCorsProcessor#processRequest 的调用，并最终在该方法中完成对跨域请求的校验工作。不过在源码执行过程中略有差异。

（1）registry.addMapping("/**")方法配置了一个 CorsRegistration 对象，该对象中包含了一个路径拦截规则，拦截规则的值就是 addMapping 方法的参数，同时 CorsRegistration 中还包含了一个 CorsConfiguration 配置对象，该对象用来保存这里跨域相关的配置。

（2）在 WebMvcConfigurationSupport#requestMappingHandlerMapping 方法中触发了 addCorsMappings 方法的执行，将获取到的 CorsRegistration 对象重新组装成一个 UrlBasedCors ConfigurationSource 对象，该对象中定义了一个 corsConfigurations 变量（Map<String, CorsConfiguration>），该变量保存了拦截规则和 CorsConfiguration 对象的映射关系。

（3）将新建的 UrlBasedCorsConfigurationSource 对象赋值给 AbstractHandlerMapping#cors ConfigurationSource 属性。

（4）当请求到达时的处理方法和@CrossOrigin 注解处理流程的第 3 步一样，都是在 AbstractHandlerMapping#getHandler 方法中进行处理，不同的是，这里是从 corsConfiguration Source 中获取 CorsConfiguration 配置对象，而@CrossOrigin 注解则从 corsLookup 集合中获取到 CorsConfiguration 配置对象。如果两处都可以获取到 CorsConfiguration 对象，则对获取到的对象属性值进行合并。

（5）根据获取到的 CorsConfiguration 对象构建一个 CorsInterceptor 拦截器。

（6）在 CorsInterceptor 拦截器中触发对 DefaultCorsProcessor#processRequest 的调用，跨

域请求的校验工作将在该方法中完成。

这两种跨域配置方式殊途同归，最终目的都是配置了一个 CorsConfiguration 对象，并根据该对象创建 CorsInterceptor 拦截器，然后在 CorsInterceptor 拦截器中触发 DefaultCorsProcessor#processRequest 方法的执行，完成跨域的校验。另外还需要注意的是，这里的跨域校验是由 DispatcherServlet 中的方法触发的，而 DispatcherServlet 的执行是在 Filter 之后，这一点需要牢记，我们后面将会用到。

11.2.3 CorsFilter

CorsFilter 是 Spring Web 中提供的一个处理跨域的过滤器，开发者也可以通过该过滤器处理跨域：

```java
@Configuration
public class WebMvcConfig {
    @Bean
    FilterRegistrationBean<CorsFilter> corsFilter() {
        FilterRegistrationBean<CorsFilter> registrationBean =
                                        new FilterRegistrationBean<>();
        CorsConfiguration corsConfiguration = new CorsConfiguration();
        corsConfiguration.setAllowedHeaders(Arrays.asList("*"));
        corsConfiguration.setAllowedMethods(Arrays.asList("*"));
        corsConfiguration.
                setAllowedOrigins(Arrays.asList("http://localhost:8081"));
        corsConfiguration.setMaxAge(3600L);
        UrlBasedCorsConfigurationSource source =
                                new UrlBasedCorsConfigurationSource();
        source.registerCorsConfiguration("/**", corsConfiguration);
        registrationBean.setFilter(new CorsFilter(source));
        registrationBean.setOrder(-1);
        return registrationBean;
    }
}
```

CorsFilter 过滤器的配置也很容易：

- 由于是在 Spring Boot 项目中，这里通过 FilterRegistrationBean 来配置一个过滤器，这种配置方式既可以设置拦截规则，又可以为配置的过滤器设置优先级。
- 在这里依然离不开 CorsConfiguration 对象，不同的是我们自己手动创建该对象，并逐个设置跨域的各项处理规则。
- 我们还需要创建一个 UrlBasedCorsConfigurationSource 对象，将过滤器的拦截规则和 CorsConfiguration 对象之间的映射关系由 UrlBasedCorsConfigurationSource 中的 corsConfigurations 变量保存起来。
- 最后创建一个 CorsFilter，并为其配置一个优先级。

在 CorsFilter 过滤器的 doFilterInternal 方法中，触发对 DefaultCorsProcessor#processRequest 的调用，进而完成跨域请求的校验。

和前面两种方式不同的是，CorsFilter 是在过滤器中处理跨域的，而前面两种方案则是在 DispatcherServlet 中触发跨域处理，从处理时间上来说，CorsFilter 对于跨域的处理时机要早于前面两种。

这就是 Spring 中为我们提供的三种不同的跨域解决方案，三种方式都能解决问题，选择其中任意一种即可。需要说明的是：

- @CrossOrigin 注解 + 重写 addCorsMappings 方法同时配置，这两种方式中关于跨域的配置会自动合并，跨域在 CorsInterceptor 中只处理了一次。
- @CrossOrigin 注解 + CorsFilter 同时配置，或者重写 addCorsMappings 方法 + CorsFitler 同时配置，都会导致跨域在 CorsInterceptor 和 CorsFilter 中各处理了一次，降低程序运行效率，这种组合不可取。

11.3　Spring Security 处理方案

当我们为项目添加了 Spring Security 依赖之后，发现上面三种跨域方式有的失效了，有的则可以继续使用，这是怎么回事？

通过@CrossOrigin 注解或者重写 addCorsMappings 方法配置跨域，统统失效了；通过 CorsFilter 配置的跨域，有没有失效则要看过滤器的优先级，如果过滤器优先级高于 Spring Security 过滤器，即先于 Spring Security 过滤器执行，则 CorsFilter 所配置的跨域处理依然有效；如果过滤器优先级低于 Spring Security 过滤器，则 CorsFilter 所配置的跨域处理就会失效。

为了理清楚这个问题，我们先简略了解一下 Filter、DispatcherServlet 以及 Interceptor 执行顺序。图 11-1 所示描述了请求从浏览器到达 Controller 的过程，Filter、Servlet 以及 Interceptor 执行顺序一目了然。

理清楚了执行顺序，我们再来看跨域请求过程。

由于非简单请求都要首先发送一个预检请求（preflight request），而预检请求并不会携带认证信息，所以预检请求就有被 Spring Security 拦截的可能。如果通过@CrossOrigin 注解或者重写 addCorsMappings 方法配置跨域，最终都是在 CorsInterceptor 中对跨域请求进行校验的。要进入 CorsInterceptor 拦截器，肯定要先过 Spring Security 过滤器链，而在经过 Spring Security 过滤器链时，由于预检请求没

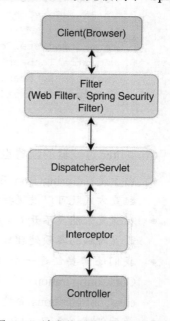

图 11-1　请求从浏览器达到 Controller

有携带认证信息，就会被拦截下来。

如果使用了 CorsFilter 配置跨域，只要过滤器的优先级高于 Spring Security 过滤器，即在 Spring Security 过滤器之前执行了跨域请求校验，那么就不会有问题。如果 CorsFilter 的优先级低于 Spring Security 过滤器，则预检请求一样需要先经过 Spring Security 过滤器，由于没有携带认证信息，在经过 Spring Security 过滤器时就会被拦截下来。

搞清楚了问题所在，接下来我们来看问题如何解决。

11.3.1 特殊处理 OPTIONS 请求

在引入 Spring Security 之后，如果还想继续通过@CrossOrigin 注解或者重写 addCorsMappings 方法配置跨域，那么可以通过给 OPTIONS 请求单独放行，来解决预检请求被拦截的问题，具体配置如下：

```
@Configuration
public class SecurityConfig extends WebSecurityConfigurerAdapter {
    @Override
    protected void configure(HttpSecurity http) throws Exception {
        http.authorizeRequests()
                .antMatchers(HttpMethod.OPTIONS).permitAll()
                .anyRequest().authenticated()
                .and()
                .httpBasic()
                .and()
                .csrf().disable();
    }
}
```

在 configure(HttpSecurity)方法中直接指定所有的 OPTIONS 请求直接通过。

这种方案既不安全，也不优雅，所以并不推荐在实际开发中使用，读者仅作了解即可。

11.3.2 继续使用 CorsFilter

第二种方案则是使用 CorsFilter 来处理跨域，只需要将 CorsFilter 的优先级设置高于 Spring Security 过滤器优先级，配置如下：

```
@Bean
FilterRegistrationBean<CorsFilter> corsFilter() {
    FilterRegistrationBean<CorsFilter> registrationBean =
                                        new FilterRegistrationBean<>();
    CorsConfiguration corsConfiguration = new CorsConfiguration();
    corsConfiguration.setAllowedHeaders(Arrays.asList("*"));
    corsConfiguration.setAllowedMethods(Arrays.asList("*"));
    corsConfiguration
        .setAllowedOrigins(Arrays.asList("http://local.javaboy.org:8081"));
```

```
        corsConfiguration.setMaxAge(3600L);
        UrlBasedCorsConfigurationSource source =
                            new UrlBasedCorsConfigurationSource();
        source.registerCorsConfiguration("/**", corsConfiguration);
        registrationBean.setFilter(new CorsFilter(source));
        registrationBean.setOrder(Ordered.HIGHEST_PRECEDENCE);
        return registrationBean;
    }
```

过滤器的优先级，数字越小，优先级越高，这里我们配置了 CorsFilter 的优先级为最高。

当然这里也可以不设置最高优先级，我们只需要了解到 Spring Security 中 FilterChainProxy 过滤器的优先级，只要 CorsFilter 的优先级高于 FilterChainProxy 即可。

Spring Security 中关于 FilterChainProxy 优先级的配置在 SecurityFilterAutoConfiguration 类中，部分源码如下：

```
@Bean
@ConditionalOnBean(name = DEFAULT_FILTER_NAME)
public DelegatingFilterProxyRegistrationBean
            securityFilterChainRegistration(
                            SecurityProperties securityProperties) {
    DelegatingFilterProxyRegistrationBean registration =
        new DelegatingFilterProxyRegistrationBean(DEFAULT_FILTER_NAME);
    registration.setOrder(securityProperties.getFilter().getOrder());
    registration.setDispatcherTypes(getDispatcherTypes(securityProperties));
    return registration;
}
```

可以看到，过滤器的优先级是从 SecurityProperties 对象中读取的，该对象中默认的过滤器优先级是-100，即开发者配置的 CorsFilter 过滤器优先级只需要小于-100 即可（开发者也可以在 application.properties 文件中，通过 spring.security.filter.order 配置去修改 FilterChainProxy 过滤器的默认优先级）。

11.3.3 专业解决方案

Spring Security 中也提供了更加专业的方式来解决预检请求所面临的问题。我们来看一下具体配置：

```
@Configuration
public class SecurityConfig extends WebSecurityConfigurerAdapter {
    @Override
    protected void configure(HttpSecurity http) throws Exception {
        http.authorizeRequests()
                .anyRequest().authenticated()
                .and()
                .httpBasic()
                .and()
```

```
            .cors()
                .configurationSource(corsConfigurationSource())
                .and()
            .csrf().disable();
    }
    CorsConfigurationSource corsConfigurationSource() {
        CorsConfiguration corsConfiguration = new CorsConfiguration();
        corsConfiguration.setAllowedHeaders(Arrays.asList("*"));
        corsConfiguration.setAllowedMethods(Arrays.asList("*"));
        corsConfiguration
            .setAllowedOrigins(Arrays.asList("http://local.javaboy.org:8081"));
        corsConfiguration.setMaxAge(3600L);
        UrlBasedCorsConfigurationSource source =
                            new UrlBasedCorsConfigurationSource();
        source.registerCorsConfiguration("/**", corsConfiguration);
        return source;
    }
}
```

首先需要提供一个 CorsConfigurationSource 实例,将跨域的各项配置都填充进去,然后在 configure(HttpSecurity) 方法中,通过 cors() 开启跨域配置,并将一开始配置好的 CorsConfigurationSource 实例设置进去。这样我们就完成了 Spring Security 中的跨域配置。

那么,这段配置的原理是什么呢?cors() 方法开启了对 CorsConfigurer 的配置,对 CorsConfigurer 而言最重要的就是 configure 方法(回顾 4.1.5 小节),我们一起来看一下:

```
public void configure(H http) {
    ApplicationContext context = http.getSharedObject(ApplicationContext.class);
    CorsFilter corsFilter = getCorsFilter(context);
    if (corsFilter == null) {
        throw new IllegalStateException("");
    }
    http.addFilter(corsFilter);
}
```

可以看到,configure 方法中就是获取了一个 CorsFilter 并添加到 Spring Security 过滤器链中。我们先来看 CorsFilter 是如何获取到的:

```
private CorsFilter getCorsFilter(ApplicationContext context) {
    if (this.configurationSource != null) {
        return new CorsFilter(this.configurationSource);
    }
    boolean containsCorsFilter = context
            .containsBeanDefinition(CORS_FILTER_BEAN_NAME);
    if (containsCorsFilter) {
        return context.getBean(CORS_FILTER_BEAN_NAME, CorsFilter.class);
    }
    boolean containsCorsSource = context
            .containsBean(CORS_CONFIGURATION_SOURCE_BEAN_NAME);
```

```
    if (containsCorsSource) {
        CorsConfigurationSource configurationSource = context.getBean(
            CORS_CONFIGURATION_SOURCE_BEAN_NAME,
            CorsConfigurationSource.class);
        return new CorsFilter(configurationSource);
    }
    boolean mvcPresent = ClassUtils.isPresent(HANDLER_MAPPING_INTROSPECTOR,
        context.getClassLoader());
    if (mvcPresent) {
        return MvcCorsFilter.getMvcCorsFilter(context);
    }
    return null;
}
```

可以看到，这里一共有四种不同的方式获取 CorsFilter：

（1）如果 configurationSource 不为 null，则直接根据 configurationSource 创建一个 CorsFilter 对象并返回。我们前面的配置最终就是通过这种方式获取到 CorsFilter 实例的。

（2）Spring 容器中是否包含一个名为 corsFilter 的实例，如果包含，则从 Spring 容器中获取该实例并返回，这意味着我们也可以直接向 Spring 容器中注入一个 corsFilter。

（3）Spring 容器中是否包含一个名为 corsConfigurationSource 的实例，如果包含，则根据该实例创建一个 CorsFilter 并返回。这意味着在前面的配置中，我们可以将自己创建的 CorsConfigurationSource 实例直接注入到 Spring 容器中（添加@Bean 注解即可），然后在 configure(HttpSecurity) 方法中通过 cors() 方法开启跨域配置即可，不用再手动指定 CorsConfigurationSource 实例。

（4）HandlerMappingIntrospector 是 Spring Web 中提供的一个类，该类实现了 CorsConfigurationSource 接口，所以也可以据此创建一个 CorsFilter。

这是四种获取 CorsFilter 实例的方式。

拿到 CorsFilter 之后，调用 http.addFilter 方法将其添加到 Spring Security 过滤器链中，在过滤器链构建之前，会先对所有的过滤器进行排序（回顾 4.1.4 小节关于 HttpSecurity 的讲解），排序的依据在 FilterComparator 中已经定义好了：

```
FilterComparator() {
    Step order = new Step(INITIAL_ORDER, ORDER_STEP);
    put(ChannelProcessingFilter.class, order.next());
    put(ConcurrentSessionFilter.class, order.next());
    put(WebAsyncManagerIntegrationFilter.class, order.next());
    put(SecurityContextPersistenceFilter.class, order.next());
    put(HeaderWriterFilter.class, order.next());
    put(CorsFilter.class, order.next());
    put(CsrfFilter.class, order.next());
    put(LogoutFilter.class, order.next());
    ......
```

可以看到，CorsFilter 的位置在 HeaderWriterFilter 之后，在 CsrfFilter 之前，这个时候还没到认证过滤器。

至此，Spring Security 中关于跨域问题的处理就清晰了，Spring Security 根据开发者提供的 CorsConfigurationSource 对象构建出一个 CorsFilter，并将该 CorsFilter 置于认证过滤器之前。

Spring Security 中关于跨域的这三种处理方式，在实际项目中推荐使用第三种。

11.4 小 结

本章主要介绍了 Spring Security 对于跨域的处理方案，一个核心的思路就是将处理跨域过滤器 CorsFilter 放在 Spring Security 认证过滤器之前即可。通过本章的学习，大家对于 Spring 中的跨域会有一个更加深刻的认识。

第 12 章

异常处理

异常也算是一个开发中不可避免的问题，Spring Security 中关于异常的处理主要是两个方面：认证异常处理、权限异常处理。除此之外的异常则抛出，交给 Spring 去处理。本章主要介绍一下 Spring Security 中关于异常的配置。

本章涉及的主要知识点有：

- Spring Security 异常体系。
- ExceptionTranslationFilter 原理分析。
- 自定义异常配置。

12.1 Spring Security 异常体系

Spring Security 中的异常主要分为两大类：

- AuthenticationException：认证异常。
- AccessDeniedException：权限异常。

其中认证异常涉及的异常类型比较多，表 12-1 展示了 Spring Security 中的所有认证异常。

表 12-1 Spring Security 认证异常类

异常类型	备注
AuthenticationException	认证异常的父类，抽象类
BadCredentialsException	登录凭证（密码）异常
InsufficientAuthenticationException	登录凭证不够充分而抛出的异常

异常类型	备注
SessionAuthenticationException	会话并发管理时抛出的异常，例如会话总数超出最大限制数
UsernameNotFoundException	用户名不存在异常
PreAuthenticatedCredentialsNotFoundException	身份预认证失败异常
ProviderNotFoundException	未配置 AuthenticationProvider 异常
AuthenticationServiceException	由于系统问题而无法处理认证请求异常。
InternalAuthenticationServiceException	由于系统问题而无法处理认证请求异常。和 AuthenticationServiceException 不同之处在于，如果外部系统出错，则不会抛出该异常
AuthenticationCredentialsNotFoundException	SecurityContext 中不存在认证主体时抛出的异常
NonceExpiredException	HTTP 摘要认证时随机数过期异常
RememberMeAuthenticationException	RememberMe 认证异常
CookieTheftException	RememberMe 认证时 Cookie 被盗窃异常
InvalidCookieException	RememberMe 认证时无效的 Cookie 异常
AccountStatusException	账户状态异常
LockedException	账户被锁定异常
DisabledException	账户被禁用异常
CredentialsExpiredException	登录凭证（密码）过期异常
AccountExpiredException	账户过期异常

相比于认证异常，权限异常类就要少很多了，表 12-2 展示了 Spring Security 中的权限异常。

表 12-2 Spring Security 权限异常类

异常类型	备注
AccessDeniedException	权限异常的父类
AuthorizationServiceException	由于系统问题而无法处理权限时抛出异常
CsrfException	Csrf 令牌异常
MissingCsrfTokenException	Csrf 令牌缺失异常
InvalidCsrfTokenException	Csrf 令牌无效异常

在实际项目中，如果 Spring Security 提供的这些异常类无法满足需要，开发者也可以根据实际需要自定义异常类。

12.2 ExceptionTranslationFilter 原理分析

Spring Security 中的异常处理主要是在 ExceptionTranslationFilter 过滤器中完成的，该过滤器主要处理 AuthenticationException 和 AccessDeniedException 类型的异常，其他异常则会继续

抛出，交给上一层容器去处理。

接下来我们来分析 ExceptionTranslationFilter 的工作原理。

在 WebSecurityConfigurerAdapter#getHttp 方法中进行 HttpSecurity 初始化的时候，就调用了 exceptionHandling()方法去配置 ExceptionTranslationFilter 过滤器：

```
protected final HttpSecurity getHttp() throws Exception {
    ...
    ...
    if (!disableDefaults) {
        http
            .csrf().and()
            .addFilter(new WebAsyncManagerIntegrationFilter())
            .exceptionHandling().and()
            .headers().and()
            .sessionManagement().and()
            .securityContext().and()
            .requestCache().and()
            .anonymous().and()
            .servletApi().and()
            .apply(new DefaultLoginPageConfigurer<>()).and()
            .logout();
        ...
        ...
    }
    configure(http);
    return http;
}
```

exceptionHandling()方法就是调用 ExceptionHandlingConfigurer 去配置 ExceptionTranslationFilter。对 ExceptionHandlingConfigurer 配置类而言，最重要的当然就是它里边的 configure 方法了，我们来看一下该方法：

```
@Override
public void configure(H http) {
    AuthenticationEntryPoint entryPoint = getAuthenticationEntryPoint(http);
    ExceptionTranslationFilter exceptionTranslationFilter =
        new ExceptionTranslationFilter(entryPoint, getRequestCache(http));
    AccessDeniedHandler deniedHandler = getAccessDeniedHandler(http);
    exceptionTranslationFilter.setAccessDeniedHandler(deniedHandler);
    exceptionTranslationFilter = postProcess(exceptionTranslationFilter);
    http.addFilter(exceptionTranslationFilter);
}
```

可以看到，这里首先获取到一个 entryPoint 对象，这个就是认证失败时的处理器，然后创建 ExceptionTranslationFilter 过滤器并传入 entryPoint。接下来还会获取到一个 deniedHandler 对象设置给 ExceptionTranslationFilter 过滤器，这个 deniedHandler 就是权限异常处理器。最后调用 postProcess 方法将 ExceptionTranslationFilter 过滤器注册到 Spring 容器中，然后调用

addFilter 方法再将其添加到 Spring Security 过滤器链中。

AuthenticationEntryPoint

AuthenticationEntryPoint 实例是通过 getAuthenticationEntryPoint 方法获取到的，我们来具体看一下：

```
AuthenticationEntryPoint getAuthenticationEntryPoint(H http) {
    AuthenticationEntryPoint entryPoint = this.authenticationEntryPoint;
    if (entryPoint == null) {
        entryPoint = createDefaultEntryPoint(http);
    }
    return entryPoint;
}
private AuthenticationEntryPoint createDefaultEntryPoint(H http) {
    if (this.defaultEntryPointMappings.isEmpty()) {
        return new Http403ForbiddenEntryPoint();
    }
    if (this.defaultEntryPointMappings.size() == 1) {
        return this.defaultEntryPointMappings.values().iterator().next();
    }
    DelegatingAuthenticationEntryPoint entryPoint =
        new DelegatingAuthenticationEntryPoint(this.defaultEntryPointMappings);
    entryPoint.setDefaultEntryPoint(
            this.defaultEntryPointMappings.values().iterator().next());
    return entryPoint;
}
```

默认情况下，系统的 authenticationEntryPoint 属性值为 null，所以最终还是通过 createDefaultEntryPoint 方法来获取 AuthenticationEntryPoint 实例。在 createDefaultEntryPoint 方法中有一个 defaultEntryPointMappings 变量，它是一个 LinkedHashMap<RequestMatcher, AuthenticationEntryPoint>类型。

可以看到，这个 LinkedHashMap 的 key 是一个 RequestMatcher，即一个请求匹配器，而 value 则是一个 AuthenticationEntryPoint 认证失败处理器，即一个请求匹配器对应一个认证失败处理器。换句话说，针对不同的请求，可以给出不同的认证失败处理器。如果 defaultEntryPointMappings 变量为空，则返回一个 Http403ForbiddenEntryPoint 类型的处理器；如果 defaultEntryPointMappings 变量中只有一项，则将这一项取出来返回即可；如果 defaultEntryPointMappings 变量中有多项，则使用 DelegatingAuthenticationEntryPoint 代理类，在代理类中，会遍历 defaultEntryPointMappings 变量中的每一项，查看当前请求是否满足其 RequestMatcher，如果满足，则使用对应的认证失败处理器来处理。

当我们新建一个 Spring Security 项目，不做任何配置时，在 WebSecurityConfigurerAdapter#configure(HttpSecurity)方法中默认会配置表单登录和 HTTP 基本认证，表单登录和 HTTP 基本认证在配置的过程中，会分别向 defaultEntryPointMappings 变量中添加认证失败处理器：

- 表单登录在 AbstractAuthenticationFilterConfigurer#registerAuthenticationEntryPoint 方法中

向 defaultEntryPointMappings 变量添加的处理器，对应的 AuthenticationEntryPoint 实例就是 LoginUrlAuthenticationEntryPoint，默认情况下访问需要认证才能访问的页面时，会自动跳转到登录页面，就是通过 LoginUrlAuthenticationEntryPoint 实现的。

- HTTP 基本认证在 HttpBasicConfigurer#registerDefaultEntryPoint 方法中向 defaultEntryPointMappings 变量添加处理器，对应的 AuthenticationEntryPoint 实例则是 BasicAuthenticationEntryPoint（具体参考 10.1.3.1 小节）。

所以默认情况下，defaultEntryPointMappings 变量中将存在两个认证失败处理器。

AccessDeniedHandler

我们再来看 AccessDeniedHandler 实例的获取，AccessDeniedHandler 实例是通过 getAccessDeniedHandler 方法获取到的：

```
AccessDeniedHandler getAccessDeniedHandler(H http) {
    AccessDeniedHandler deniedHandler = this.accessDeniedHandler;
    if (deniedHandler == null) {
        deniedHandler = createDefaultDeniedHandler(http);
    }
    return deniedHandler;
}
private AccessDeniedHandler createDefaultDeniedHandler(H http) {
    if (this.defaultDeniedHandlerMappings.isEmpty()) {
        return new AccessDeniedHandlerImpl();
    }
    if (this.defaultDeniedHandlerMappings.size() == 1) {
        return this.defaultDeniedHandlerMappings.values().iterator().next();
    }
    return new RequestMatcherDelegatingAccessDeniedHandler(
            this.defaultDeniedHandlerMappings,
            new AccessDeniedHandlerImpl());
}
```

可以看到，AccessDeniedHandler 实例的获取流程和 AuthenticationEntryPoint 的获取流程基本上一模一样，这里也有一个类似的 defaultDeniedHandlerMappings 变量，也可以为不同的路径配置不同的鉴权失败处理器；如果存在多个鉴权失败处理器，则可以通过代理类统一处理。

不同的是，默认情况下这里的 defaultDeniedHandlerMappings 变量是空的，所以最终获取到的实例是 AccessDeniedHandlerImpl。在 AccessDeniedHandlerImpl#handle 方法中处理鉴权失败的情况，如果存在错误页面，就跳转到到错误页面，并设置响应码为 403；如果没有错误页面，则直接给出错误响应即可。

AuthenticationEntryPoint 和 AccessDeniedHandler 都有了之后，接下来就是 ExceptionTranslationFilter 中的处理逻辑了。

ExceptionTranslationFilter

默认情况下，ExceptionTranslationFilter 过滤器在整个 Spring Security 过滤器链中排名倒数

第二，倒数第一是 FilterSecurityInterceptor。在 FilterSecurityInterceptor 中将会对用户的身份进行校验，如果用户身份不合法，就会抛出异常，抛出来的异常，刚好就在 ExceptionTranslationFilter 过滤器中进行处理了。我们一起来看一下 ExceptionTranslationFilter 中的 doFilter 方法：

```java
public void doFilter(ServletRequest req, ServletResponse res,
            FilterChain chain) throws IOException, ServletException {
    HttpServletRequest request = (HttpServletRequest) req;
    HttpServletResponse response = (HttpServletResponse) res;
    try {
        chain.doFilter(request, response);
    }
    catch (IOException ex) {
        throw ex;
    }
    catch (Exception ex) {
        Throwable[] causeChain = throwableAnalyzer.determineCauseChain(ex);
        RuntimeException ase = (AuthenticationException) throwableAnalyzer
          .getFirstThrowableOfType(AuthenticationException.class, causeChain);
        if (ase == null) {
            ase = (AccessDeniedException) throwableAnalyzer
              .getFirstThrowableOfType(AccessDeniedException.class, causeChain);
        }
        if (ase != null) {
            if (response.isCommitted()) {
                throw new ServletException("", ex);
            }
            handleSpringSecurityException(request, response, chain, ase);
        }
        else {
            if (ex instanceof ServletException) {
                throw (ServletException) ex;
            }
            else if (ex instanceof RuntimeException) {
                throw (RuntimeException) ex;
            }
            throw new RuntimeException(ex);
        }
    }
}
```

可以看到，在该过滤器中，直接执行了 chain.doFilter 方法，让当前请求继续执行剩下的过滤器（即 FilterSecurityInterceptor），然后用一个 try{…}catch(){…}代码块将 chain.doFilter 包裹起来，如果后面有异常抛出，就直接在这里捕获到了。

throwableAnalyzer 对象是一个异常分析器，由于异常在抛出的过程中可能被"层层转包"，我们需要还原最初的异常，通过 throwableAnalyzer.determineCauseChain 方法可以获得整个异常链。有的读者可能对此不太理解，这里举一个简单例子，例如如下一段代码：

```java
NullPointerException aaa = new NullPointerException("aaa");
ServletException bbb = new ServletException(aaa);
IOException ccc = new IOException(bbb);
ThrowableAnalyzer throwableAnalyzer = new ThrowableAnalyzer();
Throwable[] causeChain = throwableAnalyzer.determineCauseChain(ccc);
for (int i = 0; i < causeChain.length; i++) {
    System.out.println("causeChain[i].getClass() = "
                                    + causeChain[i].getClass());
}
```

打印信息如下:

```
causeChain[i].getClass() = class java.io.IOException
causeChain[i].getClass() = class javax.servlet.ServletException
causeChain[i].getClass() = class java.lang.NullPointerException
```

throwableAnalyzer.determineCauseChain 方法的功能就很清楚了：把"层层转包"的异常再解析出来形成一个数组。

所以在 catch 块中捕获到异常之后，首先获取异常链，然后调用 getFirstThrowableOfType 方法查看异常链中是否有认证失败类型的异常 AuthenticationException，如果不存在，再去查看是否有鉴权失败类型的异常 AccessDeniedException。注意这个查找顺序，先找认证异常，再找鉴权异常。如果存在这两种类型的异常，则调用 handleSpringSecurityException 方法进行异常处理，否则将异常抛出交给上层容器去处理。

我们来看 handleSpringSecurityException 方法：

```java
private void handleSpringSecurityException(HttpServletRequest request,
                HttpServletResponse response, FilterChain chain,
        RuntimeException exception)   throws IOException, ServletException {
    if (exception instanceof AuthenticationException) {
        sendStartAuthentication(request, response, chain,
            (AuthenticationException) exception);
    }
    else if (exception instanceof AccessDeniedException) {
        Authentication authentication =
                SecurityContextHolder.getContext().getAuthentication();
        if (authenticationTrustResolver.isAnonymous(authentication) ||
            authenticationTrustResolver.isRememberMe(authentication)) {
            sendStartAuthentication(
                request,
                response,
                chain,
                new InsufficientAuthenticationException(""));
        }
        else {
            accessDeniedHandler.handle(request, response,
                (AccessDeniedException) exception);
        }
```

```
        }
    }
```

在 handleSpringSecurityException 方法中，首先判断异常类型是不是 AuthenticationException，如果是，则进入 sendStartAuthentication 方法中处理认证失败；如果异常类型是 AccessDeniedException，那么先从 SecurityContextHolder 中取出当前认证主体；如果当前认证主体是一个匿名用户，或者当前认证是通过 RememberMe 完成的，那么也认为是认证异常，需要重新创建一个 InsufficientAuthenticationException 类型的异常对象，然后进入 sendStartAuthentication 方法进行处理，否则就认为是鉴权异常，调用 accessDeniedHandler.handle 方法进行处理。

最后我们再来看看 sendStartAuthentication 方法：

```
protected void sendStartAuthentication(HttpServletRequest request,
        HttpServletResponse response, FilterChain chain,
        AuthenticationException reason) throws ServletException, IOException {
    SecurityContextHolder.getContext().setAuthentication(null);
    requestCache.saveRequest(request, response);
    authenticationEntryPoint.commence(request, response, reason);
}
```

这里做了三件事：

（1）清除 SecurityContextHolder 中保存的认证主体。
（2）保存当前请求。
（3）调用 authenticationEntryPoint.commence 方法完成认证失败处理。

至此，我们前面所说的 AuthenticationEntryPoint 和 AccessDeniedHandler 在这里就派上用场了。

12.3　自定义异常配置

Spring Security 中默认提供的异常处理器不一定满足我们的需求，如果开发者需要自定义，也是可以的，定义方式如下：

```
@Configuration
public class SecurityConfig extends WebSecurityConfigurerAdapter {
    @Override
    protected void configure(HttpSecurity http) throws Exception {
        http.authorizeRequests()
                .antMatchers("/admin").hasRole("admin")
                .anyRequest().authenticated()
                .and()
                .exceptionHandling()
```

```
                .authenticationEntryPoint((req,resp,e)->{
                    resp.setStatus(HttpStatus.UNAUTHORIZED.value());
                    resp.getWriter().write("please login");
                })
                .accessDeniedHandler((req,resp,e)->{
                    resp.setStatus(HttpStatus.FORBIDDEN.value());
                    resp.getWriter().write("forbidden");
                })
                .and()
                .formLogin()
                .and()
                .csrf().disable();
    }
}
```

首先我们设置了访问/admin 接口必须具备 admin 角色，其他接口只需要认证就可以访问。然后我们对 exceptionHandling 分别配置了 authenticationEntryPoint 和 accessDeniedHandler。回顾上一小节的源码分析，这里配置完成后，defaultEntryPointMappings 和 defaultDeniedHandlerMappings 中的处理器就会失效。

接下来我们启动项目，如果用户未经登录就访问/hello 接口，则结果如图 12-1 所示。

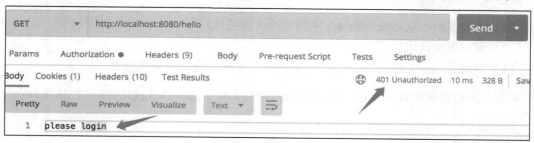

图 12-1　认证失败响应

当用户登录成功后，但是不具备 admin 角色，此时如果访问/admin 接口，则结果如图 12-2 所示。

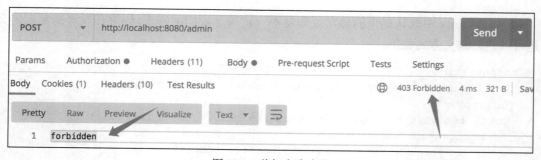

图 12-2　鉴权失败响应

当然，开发者也可以为不同的接口配置不同的异常处理器，配置方式如下：

```java
@Configuration
public class SecurityConfig extends WebSecurityConfigurerAdapter {
    @Override
    protected void configure(HttpSecurity http) throws Exception {
        AntPathRequestMatcher matcher1 = new AntPathRequestMatcher("/qq/**");
        AntPathRequestMatcher matcher2 = new AntPathRequestMatcher("/wx/**");
        http.authorizeRequests()
                .antMatchers("/wx/**").hasRole("wx")
                .antMatchers("/qq/**").hasRole("qq")
                .anyRequest().authenticated()
                .and()
                .exceptionHandling()
                .defaultAuthenticationEntryPointFor((req, resp, e) -> {
                    resp.setContentType("text/html;charset=utf-8");
                    resp.setStatus(HttpStatus.UNAUTHORIZED.value());
                    resp.getWriter().write("请登录,QQ用户");
                }, matcher1)
                .defaultAuthenticationEntryPointFor((req, resp, e) -> {
                    resp.setContentType("text/html;charset=utf-8");
                    resp.setStatus(HttpStatus.UNAUTHORIZED.value());
                    resp.getWriter().write("请登录,WX用户");
                }, matcher2)
                .defaultAccessDeniedHandlerFor((req, resp, e) -> {
                    resp.setContentType("text/html;charset=utf-8");
                    resp.setStatus(HttpStatus.FORBIDDEN.value());
                    resp.getWriter().write("权限不足,QQ用户");
                }, matcher1)
                .defaultAccessDeniedHandlerFor((req, resp, e) -> {
                    resp.setContentType("text/html;charset=utf-8");
                    resp.setStatus(HttpStatus.FORBIDDEN.value());
                    resp.getWriter().write("权限不足,WX用户");
                }, matcher2)
                .and()
                .formLogin()
                .and()
                .csrf().disable();
    }
}
```

一开始我们定义了两个路径匹配器 matcher1 和 matcher2,然后配置/wx/**格式的路径需要有 wx 角色才能访问,/qq/**格式的路径则需要有 qq 角色才可以访问。接下来分别调用 defaultAuthenticationEntryPointFor 方法和 defaultAccessDeniedHandlerFor 方法向 defaultEntryPointMappings 和 defaultDeniedHandlerMappings 两个变量中添加异常处理器即可。

配置完成后,启动项目进行测试,不同接口将会给出不同的异常响应,这里不再赘述。

12.4 小　结

本章主要介绍了 Spring Security 中关于异常的处理方式。总结一下就是在 ExceptionTranslationFilter 过滤器中分别对 AuthenticationException 和 AccessDeniedException 类型的异常进行处理，如果异常不是这两种类型的，则将异常抛出交给上层容器处理。AuthenticationException 和 AccessDeniedException 两种不同类型的异常，分别对应了 AuthenticationEntryPoint 和 AccessDeniedHandler 两种不同的异常处理器。如果系统提供的异常处理器不能满足需求，开发者也可以自定义异常处理器，并且可以为不同的请求指定不同的异常处理器。

第 13 章

权限管理

认证和授权是 Spring Security 中的两大核心功能，所谓授权也就是我们日常所说的权限管理。Spring Security 中对这两者做了很好的解耦，无论使用哪种认证方式，都不影响权限管理功能的使用，这也是 Spring Security 受欢迎的原因之一。同时 Spring Security 中对于 RBAC、ACL 等不同权限模型也都有很好的支持，本章我们将学习 Spring Security 中的权限管理。

本章涉及的主要知识点有：

- 什么是权限管理。
- Spring Security 权限管理策略。
- 权限管理核心概念。
- 基于 URL 地址的权限管理。
- 基于方法的权限管理。

13.1 什么是权限管理

和其他安全管理框架类似，Spring Security 提供的功能主要包含两方面：认证和授权。在前面的章节中，我们对于认证已经做了很多介绍，Spring Security 支持多种不同的认证方式，但是无论开发者使用哪种认证方式，都不会影响授权功能的使用，Spring Security 很好地实现了认证和授权两大功能的解耦，这也是它受欢迎的原因之一。

认证就是确认用户身份，也就是我们常说的登录。授权则是根据系统提前设置好的规则，给用户分配可以访问某一资源的权限，用户根据自己所具有的权限，去执行相应的操作。在我们所见到的大部分系统中，无论是操作系统还是企业级应用系统，都能看到权限管理功能。一个优秀的认证+授权系统可以为我们的应用系统提供强有力的安全保障功能。

13.2 Spring Security 权限管理策略

从技术上来说，Spring Security 中提供的权限管理功能主要有两种类型：
- 基于过滤器的权限管理（FilterSecurityInterceptor）。
- 基于 AOP 的权限管理（MethodSecurityInterceptor）。

基于过滤器的权限管理主要用来拦截 HTTP 请求，拦截下来之后，根据 HTTP 请求地址进行权限校验。

基于 AOP 的权限管理则主要用来处理方法级别的权限问题。当需要调用某一个方法时，通过 AOP 将操作拦截下来，然后判断用户是否具备相关的权限，如果具备，则允许方法调用，否则禁止方法调用。

本章接下来的介绍都是基于这两种权限管理方式展开的。

13.3 核心概念

为了方便大家理解后面的内容，我们先对 Spring Security 中一些权限相关的核心概念做个简单介绍。

13.3.1 角色与权限

根据前面 1.3.1 小节的介绍，大家已经知道在 Spring Security 中，当用户登录成功后，当前登录用户信息将保存在 Authentication 对象中，Authentication 对象中有一个 getAuthorities 方法，用来返回当前对象所具备的权限信息，也就是已经授予当前登录用户的权限。getAuthorities 方法返回值是 Collection<? extends GrantedAuthority>，即集合中存放的是 GrantedAuthority 的子类，当需要进行权限判断的时候，就会调用该方法获取用户的权限，进而做出判断。

无论用户的认证方式是用户名/密码形式、RememberMe 形式，还是其他如 CAS、OAuth2 等认证方式，最终用户的权限信息都可以通过 getAuthorities 方法获取。这就是我们前面所讲的，无论用户采用何种认证方式，都不影响授权。

那么对于 Authentication#getAuthorities 方法的返回值，我们应该理解为用户的角色还是用户的权限呢？

从设计层面来讲，角色和权限是两个完全不同的东西：权限就是一些具体的操作，例如针对员工数据的读权限（READ_EMPLOYEE）和针对员工数据的写权限（WRITE_EMPLOYEE）；角色则是某些权限的集合，例如管理员角色 ROLE_ADMIN、普通用户角色 ROLE_USER。

从代码层面来讲，角色和权限并没有太大的不同，特别是在 Spring Security 中，角色和权限的处理的方式基本上是一样的，唯一的区别在于 Spring Security 在多个地方会自动给角色添加一个 ROLE_前缀，而权限则不会自动添加任何前缀。这里读者需要特别注意，我们在后面的讲解中，遇到自动添加 ROLE_前缀的地方也都会和大家重点说明。

至于 Authentication#getAuthorities 方法的返回值，则要分情况来对待：

（1）如果权限系统设计比较简单，就是用户<=>权限<=>资源三者之间的关系，那么 getAuthorities 方法的含义就很明确，就是返回用户的权限。

（2）如果权限系统设计比较复杂，同时存在角色和权限的概念，如用户<=>角色<=>权限<=>资源（用户关联角色、角色关联权限、权限关联资源），此时我们可以将 getAuthorities 方法的返回值当作权限来理解。由于 Spring Security 并未提供相关的角色类，因此这个时候需要我们自定义角色类。

对于第一种情况，大家都好理解，这里不再赘述。对于第二种情况，我们简单介绍一下。

如果系统同时存在角色和权限，我们可以使用 GrantedAuthority 的实现类 SimpleGrantedAuthority 来表示一个权限，在 SimpleGrantedAuthority 类中，我们可以将权限描述为一个字符串，如 READ_EMPLOYEE、WRITE_EMPLOYEE。据此，我们定义角色类如下：

```java
public class Role implements GrantedAuthority {
    private String name;
    private List<SimpleGrantedAuthority> allowedOperations = new ArrayList<>();
    @Override
    public String getAuthority() {
        return name;
    }
    //省略 getter/setter
}
```

角色继承自 GrantedAuthority，一个角色对应多个权限。然后在定义用户类的时候，将角色转为权限即可：

```java
public class User implements UserDetails {
    private List<Role> roles = new ArrayList<>();
    @Override
    public Collection<? extends GrantedAuthority> getAuthorities() {
        List<SimpleGrantedAuthority> authorities = new ArrayList<>();
        for (Role role : roles) {
            authorities.addAll(role.getAllowedOperations());
        }
        return authorities.stream().distinct().collect(Collectors.toList());
    }
}
```

```
    //省略 getter/setter
}
```

整体上来说，设计层面上，角色和权限是两个东西；代码层面上，角色和权限其实差别不大，注意区分即可。

13.3.2 角色继承

角色继承就是指角色存在一个上下级的关系，例如 ROLE_ADMIN 继承自 ROLE_USER，那么 ROLE_ADMIN 就自动具备 ROLE_USER 的所有权限。

Spring Security 中通过 RoleHierarchy 类对角色继承提供支持，我们来看一下它的源码：

```
public interface RoleHierarchy {
    Collection<? extends GrantedAuthority> getReachableGrantedAuthorities(
            Collection<? extends GrantedAuthority> authorities);
}
```

RoleHierarchy 中只有一个 getReachableGrantedAuthorities 方法，该方法返回用户真正"可触达"的权限。举个简单例子，假设用户定义了 ROLE_ADMIN 继承自 ROLE_USER，ROLE_USER 继承自 ROLE_GUEST，现在当前用户角色是 ROLE_ADMIN，但是它实际可访问的资源也包含 ROLE_USER 和 ROLE_GUEST 能访问的资源。getReachableGrantedAuthorities 方法就是根据当前用户所具有的角色，从角色层级映射中解析出用户真正"可触达"的权限。

RoleHierarchy 只有一个实现类 RoleHierarchyImpl，开发者一般通过 RoleHierarchyImpl 类来定义角色的层级关系，如下面代码表示 ROLE_C 继承自 ROLE_D，ROLE_B 继承自 ROLE_C，ROLE_A 继承自 ROLE_B：

```
@Bean
RoleHierarchy roleHierarchy() {
    RoleHierarchyImpl hierarchy = new RoleHierarchyImpl();
    hierarchy.setHierarchy("ROLE_A > ROLE_B > ROLE_C > ROLE_D");
    return hierarchy;
}
```

这样的角色层级，在 RoleHierarchyImpl 类中首先通过 buildRolesReachableInOneStepMap 方法解析成下面这样的 Map 集合（key→value）：

```
ROLE_A→ROLE_B
ROLE_B→ROLE_C
ROLE_C→ROLE_D
```

然后再通过 buildRolesReachableInOneOrMoreStepsMap 方法对上面的集合再次解析，最终解析结果如下：

```
ROLE_A→[ROLE_B,ROLE_C,ROLE_D]
ROLE_B→[ROLE_C,ROLE_D]
ROLE_C→ROLE_D
```

最后通过 getReachableGrantedAuthorities 方法从该 Map 集合中获取用户真正"可触达"的权限。

13.3.3 两种处理器

前面我们讲了，Spring Security 中提供的权限管理功能主要有两种类型：基于过滤器的权限管理（FilterSecurityInterceptor）和基于 AOP 的权限管理（MethodSecurityInterceptor）。

无论是哪种，都涉及一个前置处理器和后置处理器。图 13-1 和图 13-2 所示分别表示基于过滤器的权限管理和基于 AOP 的权限管理中的前置处理器和后置处理器。

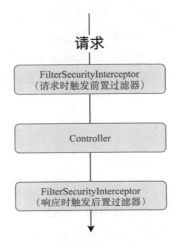

图 13-1 基于过滤器的权限管理中的前置处理器和后置处理器

图 13-2 基于方法的权限管理中的前置处理器和后置处理器

在基于过滤器的权限管理中，请求首先到达过滤器 FilterSecurityInterceptor，在其执行过程中，首先会由前置处理器去判断发起当前请求的用户是否具备相应的权限，如果具备，则请求继续向下走，到达目标方法并执行完毕。在响应时，又会经过 FilterSecurityInterceptor 过滤器，此时由后置处理器再去完成其他收尾工作。在基于过滤器的权限管理中，后置处理器一般是不工作的。这也很好理解，因为基于过滤器的权限管理，实际上就是拦截请求 URL 地址，这种权限管理方式粒度较粗，而且过滤器中拿到的是响应的 HttpServletResponse 对象，对其所返回的数据做二次处理并不方便。

在基于方法的权限管理中，目标方法的调用会被 MethodSecurityInterceptor 拦截下来，实现原理当然就是大家所熟知的 AOP 机制。当目标方法的调用被 MethodSecurityInterceptor 拦截下之后，在其 invoke 方法中首先会由前置处理器去判断当前用户是否具备调用目标方法所需要的权限，如果具备，则继续执行目标方法。当目标方法执行完毕并给出返回结果后，在 MethodSecurityInterceptor#invoke 方法中，由后置处理器再去对目标方法的返回结果进行过滤或者鉴权，然后在 invoke 方法中将处理后的结果返回。

可以看到，无论是基于过滤器的权限管理还是基于方法的权限管理，前置处理器都是重中之重，两者都会用到。而后置处理器则只是在基于方法的权限管理中会用到。

13.3.4 前置处理器

要理解前置处理器，需要先了解投票器。

投票器

投票器是 Spring Security 权限管理功能中的一个组件，顾名思义，投票器的作用就是针对是否允许某一个操作进行投票。当请求的 URL 地址被拦截下来之后，或者当调用的方法被 AOP 拦截下来之后，都会调用投票器对当前操作进行投票，以便决定是否允许当前操作。

在 Spring Security 中，投票器由 AccessDecisionVoter 来定义，我们来看一下这个接口：

```java
public interface AccessDecisionVoter<S> {
    int ACCESS_GRANTED = 1;
    int ACCESS_ABSTAIN = 0;
    int ACCESS_DENIED = -1;
    boolean supports(ConfigAttribute attribute);
    boolean supports(Class<?> clazz);
    int vote(Authentication authentication, S object,
        Collection<ConfigAttribute> attributes);
}
```

- 这个接口首先定义了三个常量：ACCESS_GRANTED 表示投票通过、ACCESS_ABSTAIN 表示弃权、ACCESS_DENIED 表示拒绝。
- supports(ConfigAttribute)方法：用来判断是否支持处理 ConfigAttribute 对象。
- supports(Class)方法：用来判断是否支持处理受保护的安全对象。
- vote 方法：则是具体的投票方法，根据用户所具有的权限以及当前请求需要的权限进行投票。vote 方法有三个参数：第一个参数 authentication 中可以提取出来当前用户所具备的权限。第二个参数 object 表示受保护的安全对象，如果受保护的是 URL 地址，则 object

就是一个 FilterInvocation 对象；如果受保护的是一个方法，则 object 就是一个 MethodInvocation 对象。第三个参数 attributes 表示访问受保护对象所需要的权限。vote 方法的返回值就是前面所定义的三个常量之一。

Spring Security 中为 AccessDecisionVoter 提供了诸多不同的实现类，图 13-3 所示是 AccessDecisionVoter 的所有继承类。

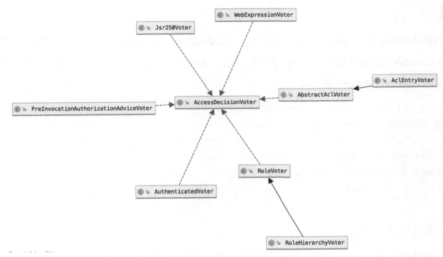

图 13-3　AccessDecisionVoter 的实现类

- RoleVoter：RoleVoter 是根据登录主体的角色进行投票，即判断当前用户是否具备受保护对象所需要的角色。需要注意的是，默认情况下角色需以"ROLE_"开始，否则 supports 方法直接返回 false，不进行后续的投票操作。
- RoleHierarchyVoter：RoleHierarchyVoter 继承自 RoleVoter，投票逻辑和 RoleVoter 一致，不同的是 RoleHierarchyVoter 支持角色的继承，它通过 RoleHierarchyImpl 对象对用户所具有的角色进行解析，获取用户真正"可触达"的角色；而 RoleVoter 则直接调用 authentication.getAuthorities()方法获取用户的角色。
- WebExpressionVoter：基于 URL 地址进行权限控制时的投票器（支持 SpEL）。
- Jsr250Voter：处理 JSR-250 权限注解的投票器，如@PermitAll、@DenyAll 等。
- AuthenticatedVoter：AuthenticatedVoter 用于判断当前用户的认证形式，它有三种取值：IS_AUTHENTICATED_FULLY、IS_AUTHENTICATED_REMEMBERED 以及 IS_AUTHENTICATED_ANONYMOUSLY。其中：IS_AUTHENTICATED_FULLY 要求当前用户既不是匿名用户也不是通过 RememberMe 进行认证；IS_AUTHENTICATED_REMEMBERED 则在前者的基础上，允许用户通过 RememberMe 进行认证；IS_AUTHENTICATED_ANONYMOUSLY 则允许当前用户通过 RememberMe 认证，也允许当前用户是匿名用户。
- AbstractAclVoter：基于 ACL 进行权限控制时的投票器。这是一个抽象类，没有绑定到具体的 ACL 系统（关于 ACL，本书第 14 章会做详细介绍）。
- AclEntryVoter：AclEntryVoter 继承自 AbstractAclVoter，基于 Spring Security 提供的 ACL

权限系统的投票器。
- PreInvocationAuthorizationAdviceVoter：处理@PreFilter和@PreAuthorize注解的投票器。

这就是Spring Security中提供的所有投票器，在具体使用中，可以单独使用一个，也可以多个一起使用。如果上面这些投票器都无法满足需求，开发者也可以自定义投票器。需要注意的是，投票结果并非最终结果（通过或拒绝），最终结果还要看决策器（AccessDecisionManager）。

决策器

决策器由AccessDecisionManager负责，AccessDecisionManager会同时管理多个投票器，由AccessDecisionManager调用投票器进行投票，然后根据投票结果做出相应的决策，所以我们将AccessDecisionManager也称作是一个决策管理器。我们来看一下它的源码：

```java
public interface AccessDecisionManager {
    void decide(Authentication authentication, Object object,
        Collection<ConfigAttribute> configAttributes)
        throws AccessDeniedException,InsufficientAuthenticationException;
    boolean supports(ConfigAttribute attribute);
    boolean supports(Class<?> clazz);
}
```

这里一共有三个方法：
- decide方法：是核心的决策方法，在这个方法中判断是否允许当前URL或者方法的调用，如果不允许，则会抛出AccessDeniedException异常。
- supports(ConfigAttribute)方法：用来判断是否支持处理ConfigAttribute对象。
- supports(Class)方法：用来判断是否支持当前安全对象。

我们来看一幅官方提供的类图，如图13-4所示。

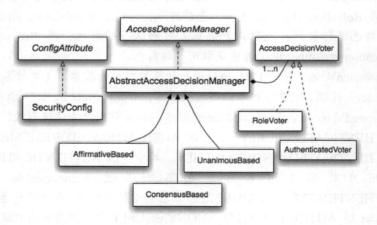

图13-4 基于投票器的AccessDecisionManager

从图中可以看出，AccessDecisionManager有一个实现类AbstractAccessDecisionManager，一个AbstractAccessDecisionManager对应多个投票器。多个投票器针对同一个请求可能会给出

不同的结果，那么听谁的呢？这就要看决策器了。

- AffirmativeBased：一票通过机制，即只要有一个投票器通过就可以访问（默认即此）。
- UnanimousBased：一票否决机制，即只要有一个投票器反对就不可以访问。
- ConsensusBased：少数服从多数机制。如果是平局并且至少有一张赞同票，则根据 allowIfEqualGrantedDeniedDecisions 参数的取值来决定，如果该参数取值为 true，则可以访问，否则不可以访问。

这是 Spring Security 中提供的三个决策器，如果这三个决策器无法满足需求，开发者也可以自定义类继承自 AbstractAccessDecisionManager 实现自己的决策器。

这就是前置处理器中的大致逻辑，无论是基于 URL 地址的权限管理，还是基于方法的权限管理，都是在前置处理器中通过 AccessDecisionManager 调用 AccessDecisionVoter 进行投票，进而做出相应的决策。

13.3.5　后置处理器

后置处理器一般只在基于方法的权限控制中会用到，当目标方法执行完毕后，通过后置处理器可以对目标方法的返回值进行权限校验或者过滤。

后置处理器由 AfterInvocationManager 负责，我们来看一下它的源码：

```
public interface AfterInvocationManager {
    Object decide(Authentication authentication, Object object,
            Collection<ConfigAttribute> attributes, Object returnedObject)
            throws AccessDeniedException;
    boolean supports(ConfigAttribute attribute);
    boolean supports(Class<?> clazz);
}
```

AfterInvocationManager 和 AccessDecisionManager 的源码高度相似，主要的区别在于 decide 方法的参数和返回值。当后置处理器执行时，被权限保护的方法已经执行完毕，后置处理器主要是对执行的结果进行过滤，所以 decide 方法中有一个 returnedObject 参数，这就是目标方法的执行结果，decide 方法的返回值就是对 returnedObject 对象进行过滤/鉴权后的结果。

我们来看官方提供的一幅 AfterInvocationManager 类图，如图 13-5 所示。

有没有感觉眼熟？这就是 AuthenticationManager、ProviderManager 以及 AuthenticationProvider 的翻版。

AfterInvocationManager 只有一个实现类 AfterInvocationProviderManager，一个 AfterInvocationProviderManager 关联多个 AfterInvocationProvider。在 AfterInvocationManager 的 decide 以及 supports 方法执行时，都是遍历 AfterInvocationProvider 并执行它里边对应的方法。

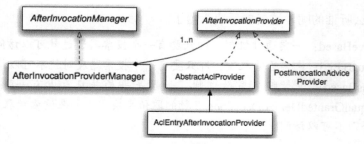

图 13-5　AfterInvocationManager 类图

AfterInvocationProvider 有多个不同的实现类，常见到的是 PostInvocationAdviceProvider，该类主要用来处理 @PostAuthorize 和 @PostFilter 注解配置的过滤器。另外还有 AbstractAclProvider 主要用来处理 ACL 中的验证逻辑，它有两个子类，AclEntryAfterInvocationProvider 用来进行权限校验，AclEntryAfterInvocationCollectionFilteringProvider 则用来做集合/数组过滤（官方未给出该类的类图，读者可以查看源码）。

这就是 Spring Security 中提供的后置处理器。

13.3.6　权限元数据

这一小节主要介绍 ConfigAttribute 和 SecurityMetadataSource。

ConfigAttribute

在 13.3.4 小节中介绍投票器时，在具体的投票方法 vote 中，受保护对象所需要的权限保存在一个 Collection<ConfigAttribute>集合中，集合中的对象是 ConfigAttribute，而不是我们所熟知的 GrantedAuthority，那么这里就需要和大家介绍一下 ConfigAttribute。

ConfigAttribute 用来存储与安全系统相关的配置属性，也就是系统关于权限的配置，通过 ConfigAttribute 来存储，我们来看一下 ConfigAttribute 接口，代码如下：

```
public interface ConfigAttribute extends Serializable {
    String getAttribute();
}
```

该接口只有一个 getAttribute 方法返回具体的权限字符串，而 GrantedAuthority 中则是通过 getAuthority 方法返回用户所具有的权限，两者返回值都是字符串。所以虽然是 ConfigAttribute 和 GrantedAuthority 两个不同的对象，但是最终是可以比较的。

图 13-6 所示表示 ConfigAttribute 的所有继承类。

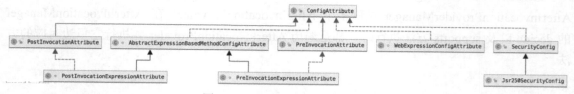

图 13-6　ConfigAttribute 的继承类

接口就不做过多介绍了，主要来说一下最终的五个比较关键的实现类（接下来涉及的一些注解，会在本章后面做详细介绍）：

- WebExpressionConfigAttribute：如果用户是基于 URL 地址来控制权限并且支持 SpEL，那么默认配置的权限控制表达式最终会被封装为 WebExpressionConfigAttribute 对象。
- SecurityConfig：如果用户使用了 @Secured 注解来控制权限，那么配置的权限就会被封装为 SecurityConfig 对象。
- Jsr250SecurityConfig：如果用户使用了 JSR-250 相关的注解来控制权限（如@PermitAll、@DenyAll），那么配置的权限就会被封装为 Jsr250SecurityConfig 对象。
- PreInvocationExpressionAttribute：如果用户使用了 @PreAuthorize、@PreFilter 注解来控制权限，那么相关的配置就会被封装为 PreInvocationExpressionAttribute 对象。
- PostInvocationExpressionAttribute：如果用户使用了 @PostAuthorize、@PostFilter 注解来控制权限，那么相关的配置就会被封装为 PostInvocationExpressionAttribute 对象。

可以看到，针对不同的配置方式，配置数据会以不同的 ConfigAttribute 对象存储。

SecurityMetadataSource

当投票器在投票时，需要两方面的权限：其一是当前用户具备哪些权限；其二是当前访问的 URL 或者方法需要哪些权限才能访问。投票器所做的事情就是对这两种权限进行比较。

根据 13.3.1 小节的介绍，用户具备的权限保存在 authentication 中，那么当前访问的 URL 或者方法所需要的权限如何获取呢？这就和 SecurityMetadataSource 有关了。

从字面上来理解，SecurityMetadataSource 就是安全元数据源，SecurityMetadataSource 所做的事情，就是提供受保护对象所需要的权限。例如，用户访问了一个 URL 地址，该 URL 地址需要哪些权限才能访问？这个就由 SecurityMetadataSource 来提供。

SecurityMetadataSource 本身只是一个接口，我们来看一下它的源码：

```java
public interface SecurityMetadataSource extends AopInfrastructureBean {
    Collection<ConfigAttribute> getAttributes(Object object)
            throws IllegalArgumentException;
    Collection<ConfigAttribute> getAllConfigAttributes();
    boolean supports(Class<?> clazz);
}
```

这里只有三个方法：

- getAttributes：根据传入的安全对象参数返回其所需要的权限。如果受保护的对象是一个 URL 地址，那么传入的参数 object 就是一个 FilterInvocation 对象；如果受保护的是一个方法，那么传入的参数 object 就是一个 MethodInvocation 对象。
- getAllConfigAttributes：getAllConfigAttributes 方法返回所有的角色/权限，以便验证是否支持。不过这个方法并不是必需的，也可以直接返回 null。
- supports：返回当前的 SecurityMetadataSource 是否支持受保护的对象如 FilterInvocation 或者 MethodInvocation。

图 13-7 表示 SecurityMetadataSource 的继承关系。

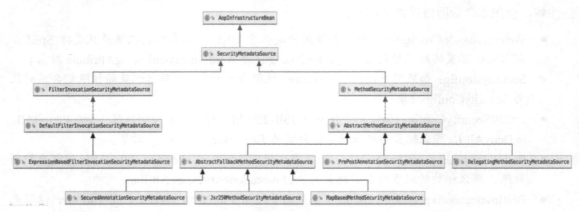

图 13-7　SecurityMetadataSource 的实现类

由图中可以看到，直接继承自 SecurityMetadataSource 的接口主要有两个：FilterInvocationSecurityMetadataSource 和 MethodSecurityMetadataSource。

- FilterInvocationSecurityMetadataSource：这是一个空接口，更像是一个标记。如果被保护的对象是一个 URL 地址，那么将由 FilterInvocationSecurityMetadataSource 的实现类提供访问该 URL 地址所需要的权限。
- MethodSecurityMetadataSource：这也是一个接口，如果受保护的对象是一个方法，那么将通过 MethodSecurityMetadataSource 的实现类来获取受保护对象所需要的权限。

FilterInvocationSecurityMetadataSource 有一个子类 DefaultFilterInvocationSecurityMetadataSource，该类中定义了一个如下格式的 Map 集合：

```
private final Map<RequestMatcher, Collection<ConfigAttribute>> requestMap;
```

可以看到，在这个 Map 集合中，key 是一个请求匹配器，value 则是一个权限集合，也就是说 requestMap 中保存了请求 URL 和其所需权限之间的映射关系。在 Spring Security 中，如果直接在 configure(HttpSecurity) 方法中配置 URL 请求地址拦截，像下面这样：

```
http.authorizeRequests()
    .antMatchers("/admin/**").hasRole("admin")
    .antMatchers("/user/**").access("hasRole('user')")
    .anyRequest().access("isAuthenticated()")
```

这段配置表示访问/admin/**格式的 URL 地址需要 admin 角色，访问/user/**格式的 URL 地址需要 user 角色，其余地址认证后即可访问。这段请求和权限之间的映射关系，会经过 DefaultFilterInvocationSecurityMetadataSource 的子类 ExpressionBasedFilterInvocationSecurityMetadataSource 进行处理，并最终将映射关系保存到 requestMap 变量中，以备后续使用。

在实际开发中，URL 地址以及访问它所需要的权限可能保存在数据库中，此时我们可以自定义类实现 FilterInvocationSecurityMetadataSource 接口，然后重写里边的 getAttributes 方法，在 getAttributes 方法中，根据当前请求的 URL 地址去数据库中查询其所需要的权限，然后将查询结果封装为相应的 ConfigAttribute 集合返回即可。

如果是基于方法的权限管理,那么对应的 MethodSecurityMetadataSource 实现类就比较多了:

- PrePostAnnotationSecurityMetadataSource:@PreAuthorize、@PreFilter、@PostAuthorize、@PostFilter 四个注解所标记的权限规则,将由 PrePostAnnotationSecurityMetadataSource 负责提供。
- SecuredAnnotationSecurityMetadataSource:@Secured 注解所标记的权限规则,将由 SecuredAnnotationSecurityMetadataSource 负责提供。
- MapBasedMethodSecurityMetadataSource:基于 XML 文件配置的方法权限拦截规则(基于 sec:protect 节点),将由 MapBasedMethodSecurityMetadataSource 负责提供。
- Jsr250MethodSecurityMetadataSource:JSR-250 相关的注解(如@PermitAll、@DenyAll)所标记的权限规则,将由 Jsr250MethodSecurityMetadataSource 负责提供。

这就是 SecurityMetadataSource 的作用。总之,不同的权限拦截方式都对应了一个 SecurityMetadataSource 实现类,请求的 URL 或者方法需要什么权限,调用 SecurityMetadataSource#getAttributes 方法就可以获取到。

13.3.7 权限表达式

Spring Security 3.0 引入了 SpEL 表达式进行权限配置,我们可以在请求的 URL 或者访问的方法上,通过 SpEL 来配置需要的权限。

内置的权限表达式如表 13-1 所示。

表 13-1 内置权限表达式

表达式	备注
hasRole(String role)	当前用户是否具备指定角色
hasAnyRole(String…roles)	当前用户是否具备指定角色中的任意一个
hasAuthority(String authority)	当前用户是否具备指定的权限
hasAnyAuthority(String…authorities)	当前用户是否具备指定权限中的任意一个
principal	代表当前登录主体 Principal
authentication	这个是从 SecurityContext 中获取到的 Authentication 对象
permitAll	允许所有的请求/调用
denyAll	拒绝所有的请求/调用
isAnonymous()	当前用户是否是一个匿名用户
isRememberMe()	当前用户是否是通过 RememberMe 自动登录
isAuthenticated()	当前用户是否已经认证成功
isFullyAuthenticated()	当前用户是否既不是匿名用户又不是通过 RememberMe 自动登录的
hasPermission(Object target, Object permission)	当前用户是否具备指定目标的指定权限

（续表）

表达式	备注
hasPermission(Object targetId, String targetType, Object permission)	当前用户是否具备指定目标的指定权限
hasIpAddress(String ipAddress)	当前请求 IP 地址是否为指定 IP

这是 Spring Security 内置的表达式，一般来说就足够使用了。如果这些内置的表达式无法满足项目需求，开发者也可以自定义表达式，后面会介绍自定义表达式的方式。

Spring Security 中通过 SecurityExpressionOperations 接口定义了基本的权限表达式，代码如下：

```
public interface SecurityExpressionOperations {
    Authentication getAuthentication();
    boolean hasAuthority(String authority);
    boolean hasAnyAuthority(String... authorities);
    boolean hasRole(String role);
    boolean hasAnyRole(String... roles);
    boolean permitAll();
    boolean denyAll();
    boolean isAnonymous();
    boolean isAuthenticated();
    boolean isRememberMe();
    boolean isFullyAuthenticated();
    boolean hasPermission(Object target, Object permission);
    boolean hasPermission(Object targetId, String targetType, Object permission);
}
```

返回值为 boolean 类型的就是权限表达式，如果返回 true，则表示权限校验通过，否则表示权限校验失败。

SecurityExpressionOperations 的实现类如图 13-8 所示。

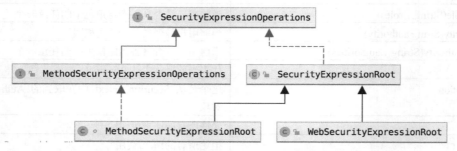

图 13-8　SecurityExpressionOperations 的实现类

SecurityExpressionRoot

SecurityExpressionRoot 对 SecurityExpressionOperations 接口做了基本的实现，并在此基础上增加了 principal。

接口的实现原理都很简单，这里就不展示源码了，我们说一下实现思路。

hasAuthority、hasAnyAuthority、hasRole 以及 hasAnyRole 四个方法主要是将传入的参数和 authentication 对象中保存的用户权限进行比对，如果用户具备相应的权限就返回 true，否则返回 false。permitAll 方法总是返回 true，而 denyAll 方法总是返回 false。isAnonymous、isAuthenticated、isRememberMe 以及 isFullyAuthenticated 四个方法则是根据对 authentication 对象的分析，然后返回 true 或者 false。最后的 hasPermission 则需要调用 PermissionEvaluator 中对应的方法进行计算，然后返回 true 或者 false。

SecurityExpressionRoot 中定义的表达式既可以在基于 URL 地址的权限管理中使用，也可以在基于方法的权限管理中使用。

WebSecurityExpressionRoot

WebSecurityExpressionRoot 继承自 SecurityExpressionRoot，并增加了 hasIpAddress 方法，用来判断请求的 IP 地址是否满足要求。

在 Spring Security 中，如果我们的权限管理是基于 URL 地址的，那么使用的就是 WebSecurityExpressionRoot，换句话说，这时可以使用 hasIpAddress 表达式。

MethodSecurityExpressionOperations

MethodSecurityExpressionOperations 定义了基于方法的权限管理时一些必须实现的接口，主要是参数对象的 get/set、返回对象的 get/set 以及返回受保护的对象。

MethodSecurityExpressionRoot

MethodSecurityExpressionRoot 实现了 MethodSecurityExpressionOperations 接口，并对其定义的方法进行了实现。MethodSecurityExpressionRoot 虽然也继承自 SecurityExpressionRoot，但是并未扩展新的表达式，换句话说，SecurityExpressionRoot 中定义的权限表达式在方法上也可以使用，但是 hasIpAddress 不可以在方法上使用。

通过上面的介绍，相信大家对 Spring Security 中权限相关概念有一个基本认知了，接下来我们就来看看基于 URL 地址的权限管理和基于方法的权限管理。

13.4 基于 URL 地址的权限管理

基于 URL 地址的权限管理主要是通过过滤器 FilterSecurityInterceptor 来实现的。如果开发者配置了基于 URL 地址的权限管理，那么 FilterSecurityInterceptor 就会被自动添加到 Spring Security 过滤器链中，在过滤器链中拦截下请求，然后分析当前用户是否具备请求所需要的权限，如果不具备，则抛出异常。

FilterSecurityInterceptor 将请求拦截下来之后，会交给 AccessDecisionManager 进行处理，AccessDecisionManager 则会调用投票器进行投票，然后对投票结果进行决策，最终决定请求是否通过。

13.4.1 基本用法

首先创建一个 Spring Boot 项目，引入 Web 依赖和 Spring Security 依赖，然后项目中添加如下配置：

```
@Configuration
public class SecurityConfig extends WebSecurityConfigurerAdapter {
    @Override
    protected void configure(AuthenticationManagerBuilder auth)
                                                    throws Exception {
        auth.inMemoryAuthentication()
                .withUser("javaboy")
                .password("{noop}123")
                .roles("ADMIN")
                .and()
                .withUser("江南一点雨")
                .password("{noop}123")
                .roles("USER")
                .and()
                .withUser("itboyhub")
                .password("{noop}123")
                .authorities("READ_INFO");
    }
}
```

在这段配置中，我们定义了三个用户：

- javaboy：具有 ADMIN 角色。
- 江南一点雨：具有 USER 角色。
- itboyhub：具有 READ_INFO 权限。

对于复杂的权限管理系统，用户和角色关联，角色和权限关联，权限和资源关联；对于简单的权限管理系统，用户和权限关联，权限和资源关联。无论是哪种，用户都不会和角色以及权限同时直接关联。反映到代码上就是 roles 方法和 authorities 方法不能同时调用，如果同时调用，后者会覆盖前者。我们来看一下它的源码（org.springframework.security.core.userdetails.User）：

```
public UserBuilder roles(String... roles) {
    List<GrantedAuthority> authorities = new ArrayList<>(roles.length);
    for (String role : roles) {
        authorities.add(new SimpleGrantedAuthority("ROLE_" + role));
    }
    return authorities(authorities);
}
public UserBuilder authorities(GrantedAuthority... authorities) {
    return authorities(Arrays.asList(authorities));
```

```
}
public UserBuilder
        authorities(Collection<? extends GrantedAuthority> authorities) {
    this.authorities = new ArrayList<>(authorities);
    return this;
}
public UserBuilder authorities(String... authorities) {
    return authorities(AuthorityUtils.createAuthorityList(authorities));
}
```

可以看到，无论是给用户设置角色还是设置权限，最终都会来到 authorities(Collection<? extends GrantedAuthority>)方法，在该方法中直接给用户的 authorities 属性重新赋值，所以如果同时调用了 roles 方法和 authorities 方法，那么后者就会覆盖前者。同时大家要注意，Spring Security 会自动给用户角色添加 ROLE_前缀。

接下来我们配置权限拦截规则，重写 configure(HttpSecurity)方法即可：

```
@Override
protected void configure(HttpSecurity http) throws Exception {
    http.authorizeRequests()
            .antMatchers("/admin/**").hasRole("ADMIN")
            .antMatchers("/user/**").access("hasAnyRole('USER','ADMIN')")
            .antMatchers("/getinfo").hasAuthority("READ_INFO")
            .anyRequest().access("isAuthenticated()")
            .and()
            .formLogin()
            .and()
            .csrf().disable();
}
```

这段请求拦截规则的含义如下：

- 用户必须具备 ADMIN 角色才可以访问/admin/**格式的地址。
- 用户必须具备 USER 和 ADMIN 任意一个角色，才可以访问/user/**格式的地址。
- 用户必须具备 READ_INFO 权限，才可以访问/getinfo 接口。
- 剩余的请求只要是认证后的用户就可以访问。

这段配置其实很好理解，但是有一些需要注意的地方：

（1）大部分的表达式都有对应的方法可以直接调用，例如我们上面调用的 hasRole 方法对应的就是 hasRole 表达式。开发者为了方便可以直接调用 hasRole 方法，但是 hasRole 方法最终还是会被转为表达式，当表达式执行结果为 true，这个请求就可以通过，否则请求不通过。

（2）Spring Security 会为 hasRole 表达式自动添加上 ROLE_前缀，例如上面的 hasRole("ADMIN")方法转为表达式之后，就是 hasRole('ROLE_ADMIN')，所以用户的角色也必须有 ROLE_前缀，而我们上面案例中的用户是基于内存创建的，会自动给用户角色加上 ROLE_前缀；hasAuthority 方法并不会添加任何前缀，而在用户定义时设置的用户权限也不会添加任何前缀。一言以蔽之，基于内存定义的用户，会自动给角色添加 ROLE_前缀，而 hasRole

也会自动添加 ROLE_前缀；基于内存定义的用户，不会给权限添加任何前缀，而 hasAuthority 也不会添加任何前缀。如果大家的用户信息是从数据库中读取的，则需要注意 ROLE_前缀的问题。

（3）可以通过 access 方法来使用权限表达式，access 方法的参数就是权限表达式。

（4）代码的顺序很关键，当请求到达后，按照从上往下的顺序依次进行匹配。

配置完成后，我们再提供四个测试接口：

```
@RestController
public class HelloController {
    @GetMapping("/hello")
    public String hello() {
        return "hello";
    }
    @GetMapping("/admin/hello")
    public String admin() {
        return "hello admin";
    }
    @GetMapping("/user/hello")
    public String user() {
        return "hello user";
    }
    @GetMapping("/getinfo")
    public String getInfo() {
        return "getinfo";
    }
}
```

最后启动项目，进行测试。如果使用 javaboy/123 进行登录，则前三个接口都可以访问；如果使用江南一点雨/123 进行登录，则只能访问/hello 和/user/hello 两个接口；如果使用 itboyhub/123 进行登录，则可以访问/getinfo 接口。

13.4.2 角色继承

如果需要配置角色继承，则只需要提供一个 RoleHierarchy 实例即可：

```
@Configuration
public class SecurityConfig extends WebSecurityConfigurerAdapter {
    @Bean
    RoleHierarchy roleHierarchy() {
        RoleHierarchyImpl hierarchy = new RoleHierarchyImpl();
        hierarchy.setHierarchy("ROLE_ADMIN > ROLE_USER");
        return hierarchy;
    }
    @Override
    protected void configure(AuthenticationManagerBuilder auth)
```

```java
                                                  throws Exception {
        //省略
    }
    @Override
    protected void configure(HttpSecurity http) throws Exception {
        http.authorizeRequests()
                .antMatchers("/admin/**").hasRole("ADMIN")
                .antMatchers("/user/**").access("hasRole('USER')")
                .antMatchers("/getinfo").hasAuthority("READ_INFO")
                .anyRequest().access("isAuthenticated()")
                .and()
                .formLogin()
                .and()
                .csrf().disable();
    }
}
```

/user/**需要 USER 角色才能访问，但是由于 ROLE_ADMIN 继承自 ROLE_USER，所以自动具备 ROLE_USER 的权限，因此如果用户具有 ROLE_ADMIN 角色也可以访问/user/**格式的地址。

13.4.3 自定义表达式

如果内置的表达式无法满足需求，开发者也可以自定义表达式。

假设现在有两个接口：

```java
@GetMapping("/hello/{userId}")
public String hello(@PathVariable Integer userId) {
    return "hello " + userId;
}
@GetMapping("/hi")
public String hello2User(String username) {
    return "hello " + username;
}
```

第一个接口，参数 userId 必须是偶数方可请求成功；第二个接口，参数 username 必须是 javaboy 方可请求成功，同时两个接口都必须认证后才能访问（这里主要是展示用法，所以大家不必纠结于业务）。如果我们想通过自定义表达式实现这一功能，只需要按照如下方式定义：

```java
@Component
public class PermissionExpression {
    public boolean checkId(Authentication authentication, Integer userId) {
        if (authentication.isAuthenticated()) {
            return userId % 2 == 0;
        }
        return false;
```

```
    }
    public boolean check(HttpServletRequest req) {
        return "javaboy".equals(req.getParameter("username"));
    }
}
```

自定义 PermissionExpression 类并注册到 Spring 容器中，然后在里边定义相应的方法。

- checkId 方法：用来检查参数 userId，同时传入了 authentication 对象，通过 authentication 对象可以判断出当前用户是否已经登录。如果方法返回 true，则表示校验通过，否则表示校验未通过。
- check 方法：用来检验请求 request，只要拿到了 request，就能拿到所有请求相关的参数，也就可以做任何校验。

最后在 SecurityConfig 中添加如下路径匹配规则：

```
@Override
protected void configure(HttpSecurity http) throws Exception {
    http.authorizeRequests()
            //省略其他
            .antMatchers("/hello/{userId}")
            .access("@permissionExpression.checkId(authentication,#userId)")
            .antMatchers("/hi")
            .access("isAuthenticated() and @permissionExpression.check(request)")
            //省略其他
}
```

在 access 方法中，我们可以通过@符号引用一个 Bean 并调用其中的方法。在 checkId 方法调用时，#userId 就表示前面的 userId 参数；在 check 方法中，我们用了两个表达式，需要同时满足 isAuthenticated()和 check()方法都为 true，该请求才会通过。

13.4.4 原理剖析

有的读者可能觉得权限管理系统很复杂，其实复杂的是系统设计，单纯从技术上来说，还是比较容易的。接下来我们就来简单梳理一下 Spring Security 中基于 URL 地址进行权限管理的一个大致原理。

AbstractSecurityInterceptor

首先处于"上帝视角"的类是 AbstractSecurityInterceptor，该类统筹着关于权限处理的一切。该类中的方法很多，这里只需要关注其中的三个方法：beforeInvocation、afterInvocation 和 finallyInvocation。

在这三个方法中，beforeInvocation 中会调用前置处理器完成权限校验，afterInvocation 中调用后置处理器完成权限校验，finallyInvocation 则主要做一些校验后的清理工作。

我们先来看 beforeInvocation：

```java
protected InterceptorStatusToken beforeInvocation(Object object) {
    if (!getSecureObjectClass().isAssignableFrom(object.getClass())) {
        throw new IllegalArgumentException("");
    }
    Collection<ConfigAttribute> attributes =
            this.obtainSecurityMetadataSource().getAttributes(object);
    if (attributes == null || attributes.isEmpty()) {
        if (rejectPublicInvocations) {
            throw new IllegalArgumentException("");
        }
        publishEvent(new PublicInvocationEvent(object));
        return null;
    }
    if (SecurityContextHolder.getContext().getAuthentication() == null) {
        credentialsNotFound("",  object, attributes);
    }
    Authentication authenticated = authenticateIfRequired();
    try {
        this.accessDecisionManager.decide(authenticated, object, attributes);
    }
    catch (AccessDeniedException accessDeniedException) {
        publishEvent(new AuthorizationFailureEvent(object, attributes,
                            authenticated, accessDeniedException));
        throw accessDeniedException;
    }
    if (publishAuthorizationSuccess) {
        publishEvent(new AuthorizedEvent(object, attributes, authenticated));
    }
    Authentication runAs =
            this.runAsManager.buildRunAs(authenticated, object,attributes);
    if (runAs == null) {
        return new InterceptorStatusToken(
            SecurityContextHolder.getContext(), false,attributes, object);
    } else {
        SecurityContext origCtx = SecurityContextHolder.getContext();
        SecurityContextHolder
                .setContext(SecurityContextHolder.createEmptyContext());
        SecurityContextHolder.getContext().setAuthentication(runAs);
        return new InterceptorStatusToken(origCtx, true, attributes, object);
    }
}
```

方法比较长，我们大概梳理一下：

（1）首先调用 obtainSecurityMetadataSource 方法获取 SecurityMetadataSource 对象，然后调用其 getAttributes 方法获取受保护对象所需要的权限。如果获取到的值为空，此时：如果 rejectPublicInvocations 变量为 true，表示受保护的对象拒绝公开调用，则直接抛出异常；如果 rejectPublicInvocations 变量为 false，表示受保护对象允许公开访问，此时直接返回 null 即可。

（2）接下来到 SecurityContextHolder 中查看当前用户的认证信息是否存在。

（3）调用 authenticateIfRequired 方法检查当前用户是否已经登录。

（4）调用 accessDecisionManager.decide 方法进行决策，该方法中会调用投票器进行投票，如果该方法执行抛出异常，则说明权限不足。

（5）接下来调用 runAsManager.buildRunAs 方法去临时替换用户身份，不过默认情况下，runAsManager 的实例是 NullRunAsManager，即不做任何替换，所以返回的 runAs 对象为 null。如果 runAs 为 null，则直接创建一个 InterceptorStatusToken 对象返回即可；否则将 SecurityContextHolder 中保存的用户信息修改为替换的用户对象，然后返回一个 InterceptorStatusToken 对象。InterceptorStatusToken 对象中保存了当前用户的 SecurityContext 对象，假如进行了临时用户替换，在替换完成后，最终还是要恢复成当前用户身份的，恢复的依据就是 InterceptorStatusToken 中保存的原始 SecurityContext 对象。

这就是 beforeInvocation 的大致工作流程，其实一个核心功能就是调用 accessDecisionManager.decide 方法进行权限验证。

我们再来看 finallyInvocation 方法：

```
protected void finallyInvocation(InterceptorStatusToken token) {
    if (token != null && token.isContextHolderRefreshRequired()) {
        SecurityContextHolder.setContext(token.getSecurityContext());
    }
}
```

如果临时替换了用户身份，那么最终要将用户身份恢复，finallyInvocation 方法所做的事情就是恢复用户身份。这里的参数 token 就是 beforeInvocation 方法的返回值，用户原始的身份信息都保存在 token 中，从 token 中取出用户身份信息，并设置到 SecurityContextHolder 中去即可。

最后我们再来看看 afterInvocation 方法：

```
protected Object afterInvocation(InterceptorStatusToken token,
                                 Object returnedObject) {
    if (token == null) {
        return returnedObject;
    }
    finallyInvocation(token);
    if (afterInvocationManager != null) {
        try {
            returnedObject = afterInvocationManager
                .decide(token.getSecurityContext().getAuthentication(),
                    token.getSecureObject(),
                    token.getAttributes(), returnedObject);
        }catch (AccessDeniedException accessDeniedException) {
            AuthorizationFailureEvent event = new AuthorizationFailureEvent(
                token.getSecureObject(), token.getAttributes(), token
                    .getSecurityContext().getAuthentication(),
```

```
            accessDeniedException);
        publishEvent(event);
        throw accessDeniedException;
    }
  }
  return returnedObject;
}
```

afterInvocation 方法接收两个参数，第一个参数 token 就是 beforeInvocation 方法的返回值，第二个参数 returnObject 则是受保护对象的返回值。afterInvocation 方法的核心工作就是调用 afterInvocationManager.decide 方法对 returnObject 进行过滤，然后将过滤后的结果返回。

这就是 AbstractSecurityInterceptor 类中三大方法的作用。

FilterSecurityInterceptor

在 13.4.1 小节的案例中，我们使用了基于 URL 地址的权限管理，此时最终使用的是 AbstractSecurityInterceptor 的子类 FilterSecurityInterceptor，这是一个过滤器。当我们在 configure(HttpSecurity)方法中调用 http.authorizeRequests() 开启 URL 路径拦截规则配置时，就会通过 AbstractInterceptUrlConfigurer#configure 方法将 FilterSecurityInterceptor 添加到 Spring Security 过滤器链中。

对过滤器而言，最重要的当然就是 doFilter 方法了，我们来看看 FilterSecurityInterceptor#doFilter 方法：

```
public void doFilter(ServletRequest request, ServletResponse response,
        FilterChain chain) throws IOException, ServletException {
    FilterInvocation fi = new FilterInvocation(request, response, chain);
    invoke(fi);
}
public void invoke(FilterInvocation fi) throws IOException, ServletException {
    if ((fi.getRequest() != null)
            && (fi.getRequest().getAttribute(FILTER_APPLIED) != null)
            && observeOncePerRequest) {
        fi.getChain().doFilter(fi.getRequest(), fi.getResponse());
    } else {
        if (fi.getRequest() != null && observeOncePerRequest) {
            fi.getRequest().setAttribute(FILTER_APPLIED, Boolean.TRUE);
        }
        InterceptorStatusToken token = super.beforeInvocation(fi);
        try {
            fi.getChain().doFilter(fi.getRequest(), fi.getResponse());
        }
        finally {
            super.finallyInvocation(token);
        }
        super.afterInvocation(token, null);
    }
}
```

在 doFilter 方法中,首先构建了受保护对象 FilterInvocation,然后调用 invoke 方法。

在 invoke 方法中,如果当前过滤器已经执行过了,则继续执行剩下的过滤器,否则就调用父类的 beforeInvocation 方法进行权限校验,校验通过后继续执行剩余的过滤器,然后在 finally 代码块中调用父类的 finallyInvocation 方法,最后调用父类的 afterInvocation 方法。可以看到,前置处理器和后置处理器都是在 invoke 方法中触发的。

AbstractInterceptUrlConfigurer

AbstractInterceptUrlConfigurer 主要负责创建 FilterSecurityInterceptor 对象,AbstractInterceptUrlConfigurer 有两个不同的子类,两个子类创建出来的 FilterSecurityInterceptor 对象略有差异:

- ExpressionUrlAuthorizationConfigurer
- UrlAuthorizationConfigurer

通过 ExpressionUrlAuthorizationConfigurer 构建出来的 FilterSecurityInterceptor,使用的投票器是 WebExpressionVoter,使用的权限元数据对象是 ExpressionBasedFilterInvocationSecurityMetadataSource,所以它支持权限表达式。

通过 UrlAuthorizationConfigurer 构建出来的 FilterSecurityInterceptor,使用的投票器是 RoleVoter 和 AuthenticatedVoter,使用的权限元数据对象是 DefaultFilterInvocationSecurityMetadataSource,所以它不支持权限表达式。

这是两者最主要的区别。

当我们在 configure(HttpSecurity)方法中开启权限配置时,一般是通过如下方式:

```
@Override
protected void configure(HttpSecurity http) throws Exception {
    http.authorizeRequests()
            .antMatchers("/admin/**").hasRole("ADMIN")
            .antMatchers("/user/**").access("hasAnyRole('USER','ADMIN')")
            …
}
```

http.authorizeRequests()方法实际上就是通过 ExpressionUrlAuthorizationConfigurer 来配置基于 URL 地址的权限管理,所以在配置时可以使用权限表达式。使用 ExpressionUrlAuthorizationConfigurer 进行配置,有一个硬性要求,就是至少配置一对 URL 地址和权限之间的映射关系。如果写成下面这种,就会出错:

```
@Override
protected void configure(HttpSecurity http) throws Exception {
    http.authorizeRequests()
            .and()
            .formLogin()
            .and()
            .csrf().disable();
}
```

这个配置中不存在 URL 地址和权限之间的映射关系,所以当项目启动时,会抛出如下异常：

```
Caused by: java.lang.IllegalStateException: At least one mapping is required
(i.e. authorizeRequests().anyRequest().authenticated())
```

如果使用 UrlAuthorizationConfigurer 去配置 FilterSecurityInterceptor，则不存在此要求，即代码中可以一条映射关系都不用配置，只需要 URL 路径和权限之间的映射关系完整即可，这在动态权限配置中非常有用。

不过在 Spring Security 中，使用 UrlAuthorizationConfigurer 去配置 FilterSecurityInterceptor 并不像使用 ExpressionUrlAuthorizationConfigurer 去配置那么容易，没有现成的方法，需要我们手动创建，代码如下：

```
@Override
protected void configure(HttpSecurity http) throws Exception {
    ApplicationContext applicationContext =
                        http.getSharedObject(ApplicationContext.class);
    http.apply(new UrlAuthorizationConfigurer<>(applicationContext))
            .getRegistry()
            .mvcMatchers("/admin/**").access("ROLE_ADMIN")
            .mvcMatchers("/user/**").access("ROLE_USER");
    http
        .formLogin()
        .and()
        .csrf().disable();
}
```

开发者自己创建一个 UrlAuthorizationConfigurer 对象出来，并调用其 getRegistry()方法去开启 URL 路径和权限之间映射关系的配置。由于 UrlAuthorizationConfigurer 中使用的投票器是 RoleVoter 和 AuthenticatedVoter，所以这里的角色需要自带 ROLE_前缀（因为 RoleVoter 的 supports 方法中会判断角色是否带有 ROLE_前缀）。

使用 UrlAuthorizationConfigurer 去配置 FilterSecurityInterceptor 时，需要确保映射关系完整，如果像下面这样，就会出错：

```
@Override
protected void configure(HttpSecurity http) throws Exception {
    ApplicationContext applicationContext =
                        http.getSharedObject(ApplicationContext.class);
    http.apply(new UrlAuthorizationConfigurer<>(applicationContext))
            .getRegistry()
            .mvcMatchers("/admin/**").access("ROLE_ADMIN")
            .antMatchers("/user/**");
    http
        .formLogin()
        .and()
        .csrf().disable();
}
```

没有配置/user/**所需要的权限，此时启动项目就会报出如下错误：

```
Caused by: java.lang.IllegalStateException: An incomplete mapping was found
for [Ant [pattern='/user/**']]. Try completing it with something like
requestUrls().<something>.hasRole('USER')
```

另外需要注意的是，无论是 ExpressionUrlAuthorizationConfigurer 还是 UrlAuthorizationConfigurer，对于 FilterSecurityInterceptor 的配置来说都在其父类 AbstractInterceptUrlConfigurer #configure 方法中，该方法中并未配置后置处理器 afterInvocationManager，所以在基于 URL 地址的权限管理中，主要是前置处理器在工作。

这就是 ExpressionUrlAuthorizationConfigurer 和 UrlAuthorizationConfigurer 两个配置类的区别。

13.4.5 动态管理权限规则

在前面的案例中，我们配置的 URL 拦截规则和请求 URL 所需要的权限都是通过代码来配置的，这样就比较死板，如果想要调整访问某一个 URL 所需要的权限，就需要修改代码。

动态管理权限规则就是我们将 URL 拦截规则和访问 URL 所需要的权限都保存在数据库中，这样，在不改变源代码的情况下，只需要修改数据库中的数据，就可以对权限进行调整。

13.4.5.1 数据库设计

简单起见，我们这里就不引入权限表了，直接使用角色表，用户和角色关联，角色和资源关联，设计出来的表结构如图 13-9 所示。

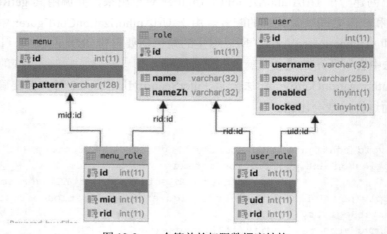

图 13-9　一个简单的权限数据库结构

menu 表是相当于我们的资源表，它里边保存了访问规则，如图 13-10 所示。

id	pattern
1	/admin/**
2	/user/**
3	/guest/**

图 13-10　访问规则

role 是角色表，里边定义了系统中的角色，如图 13-11 所示。

id	name	nameZh
1	ROLE_ADMIN	系统管理员
2	ROLE_USER	普通用户
3	ROLE_GUEST	游客

图 13-11　用户角色表

user 是用户表，如图 13-12 所示。

id	username	password	enabled	locked
1	admin	{noop}123	1	0
2	user	{noop}123	1	0
3	javaboy	{noop}123	1	0

图 13-12　用户表

user_role 是用户角色关联表，用户具有哪些角色，可以通过该表体现出来，如图 13-13 所示。

id	uid	rid
1	1	1
2	2	2
3	3	3

图 13-13　用户角色关联表

menu_role 是资源角色关联表，访问某一个资源，需要哪些角色，可以通过该表体现出来，如图 13-14 所示。

id	mid	rid
1	1	1
2	2	1
3	2	2
4	3	1
5	3	2
6	3	3

图 13-14　资源角色关联表

至此，一个简易的权限数据库就设计好了（在本书提供的案例中，有 SQL 脚本）。

13.4.5.2 实战

项目创建

创建 Spring Boot 项目,由于涉及数据库操作,这里选用目前大家使用较多的 MyBatis 框架,所以除了引入 Web、Spring Security 依赖之外,还需要引入 MyBatis 以及 MySQL 依赖。

最终的 pom.xml 文件内容如下:

```xml
<dependencies>
    <dependency>
        <groupId>org.springframework.boot</groupId>
        <artifactId>spring-boot-starter-security</artifactId>
    </dependency>
    <dependency>
        <groupId>org.springframework.boot</groupId>
        <artifactId>spring-boot-starter-web</artifactId>
    </dependency>
    <dependency>
        <groupId>org.mybatis.spring.boot</groupId>
        <artifactId>mybatis-spring-boot-starter</artifactId>
        <version>2.1.3</version>
    </dependency>
    <dependency>
        <groupId>mysql</groupId>
        <artifactId>mysql-connector-java</artifactId>
        <scope>runtime</scope>
    </dependency>
</dependencies>
```

项目创建完成后,接下来在 application.properties 中配置数据库连接信息:

```
spring.datasource.username=root
spring.datasource.password=123
spring.datasource.url=jdbc:mysql:///security13?useUnicode=true&characterEncoding=UTF-8&serverTimezone=Asia/Shanghai
```

配置完成后,我们的准备工作就算完成了。

创建实体类

根据前面设计的数据库,我们需要创建三个实体类。

首先来创建角色类 Role:

```java
public class Role {
    private Integer id;
    private String name;
    private String nameZh;
    //省略 getter/setter
}
```

然后创建菜单类 Menu:

```java
public class Menu {
    private Integer id;
    private String pattern;
    private List<Role> roles;
    //省略 getter/setter
}
```

菜单类中包含一个 roles 属性,表示访问该项资源所需要的角色。

最后我们创建 User 类:

```java
public class User implements UserDetails {
    private Integer id;
    private String password;
    private String username;
    private boolean enabled;
    private boolean locked;
    private List<Role> roles;
    @Override
    public Collection<? extends GrantedAuthority> getAuthorities() {
        return roles.stream()
                    .map(r -> new SimpleGrantedAuthority(r.getName()))
                    .collect(Collectors.toList());
    }
    @Override
    public String getPassword() {
        return password;
    }
    @Override
    public String getUsername() {
        return username;
    }
    @Override
    public boolean isAccountNonExpired() {
        return true;
    }
    @Override
    public boolean isAccountNonLocked() {
        return !locked;
    }
    @Override
    public boolean isCredentialsNonExpired() {
        return true;
    }
    @Override
    public boolean isEnabled() {
        return enabled;
    }
    //省略其他 getter/setter
}
```

由于数据库中有 enabled 和 locked 字段，所以 isEnabled()和 isAccountNonLocked()两个方法如实返回，其他几个账户状态方法默认返回 true 即可。在 getAuthorities()方法中，我们对 roles 属性进行遍历，组装出新的集合对象返回即可。

创建 Service

接下来我们创建 UserService 和 MenuService，并提供相应的查询方法。

先来看 UserService：

```java
@Service
public class UserService implements UserDetailsService {
    @Autowired
    UserMapper userMapper;
    @Override
    public UserDetails loadUserByUsername(String username)
                                throws UsernameNotFoundException {
        User user = userMapper.loadUserByUsername(username);
        if (user == null) {
            throw new UsernameNotFoundException("用户不存在");
        }
        user.setRoles(userMapper.getUserRoleByUid(user.getId()));
        return user;
    }
}
```

这段代码应该不用多说了，不熟悉的读者可以参考本书 2.4 节。

对应的 UserMapper 如下：

```java
@Mapper
public interface UserMapper {
    List<Role> getUserRoleByUid(Integer uid);
    User loadUserByUsername(String username);
}
```

UserMapper.xml：

```xml
<!DOCTYPE mapper
        PUBLIC "-//mybatis.org//DTD Mapper 3.0//EN"
        "http://mybatis.org/dtd/mybatis-3-mapper.dtd">
<mapper namespace="org.javaboy.base_on_url_dy.mapper.UserMapper">
    <select id="loadUserByUsername"
                    resultType="org.javaboy.base_on_url_dy.model.User">
        select * from user where username=#{username};
    </select>
    <select id="getUserRoleByUid"
                    resultType="org.javaboy.base_on_url_dy.model.Role">
        select r.* from role r,user_role ur where ur.uid=#{uid} and ur.rid=r.id
    </select>
</mapper>
```

再来看 MenuService，该类只需要提供一个方法，就是查询出所有的 Menu 数据，代码如下：

```java
@Service
public class MenuService {
    @Autowired
    MenuMapper menuMapper;
    public List<Menu> getAllMenu() {
        return menuMapper.getAllMenu();
    }
}
```

MenuMapper：

```java
@Mapper
public interface MenuMapper {
    List<Menu> getAllMenu();
}
```

MenuMapper.xml：

```xml
<!DOCTYPE mapper
        PUBLIC "-//mybatis.org//DTD Mapper 3.0//EN"
        "http://mybatis.org/dtd/mybatis-3-mapper.dtd">
<mapper namespace="org.javaboy.base_on_url_dy.mapper.MenuMapper">
    <resultMap id="MenuResultMap" type="org.javaboy.base_on_url_dy.model.Menu">
        <id property="id" column="id"/>
        <result property="pattern" column="pattern"></result>
        <collection property="roles"
                    ofType="org.javaboy.base_on_url_dy.model.Role">
            <id column="rid" property="id"/>
            <result column="rname" property="name"/>
            <result column="rnameZh" property="nameZh"/>
        </collection>
    </resultMap>
    <select id="getAllMenu" resultMap="MenuResultMap">
        select m.*,r.id as rid,r.name as rname,r.nameZh as rnameZh from menu m left join menu_role mr on m.`id`=mr.`mid` left join role r on r.`id`=mr.`rid`
    </select>
</mapper>
```

需要注意，由于每一个 Menu 对象都包含了一个 Role 集合，所以这个查询是一对多，这里通过 resultMap 来进行查询结果映射。

至此，所有基础工作都完成了，接下来配置 Spring Security。

配置 Spring Security

回顾 13.3.6 小节的内容，SecurityMetadataSource 接口负责提供受保护对象所需要的权限。在本案例中，受保护对象所需要的权限保存在数据库中，所以我们可以通过自定义类继承自

FilterInvocationSecurityMetadataSource，并重写 getAttributes 方法来提供受保护对象所需要的权限，代码如下：

```java
@Component
public class CustomSecurityMetadataSource
                    implements FilterInvocationSecurityMetadataSource {
    @Autowired
    MenuService menuService;
    AntPathMatcher antPathMatcher = new AntPathMatcher();
    @Override
    public Collection<ConfigAttribute> getAttributes(Object object)
                                        throws IllegalArgumentException {
        String requestURI =
                ((FilterInvocation) object).getRequest().getRequestURI();
        List<Menu> allMenu = menuService.getAllMenu();
        for (Menu menu : allMenu) {
            if (antPathMatcher.match(menu.getPattern(), requestURI)) {
                String[] roles = menu.getRoles().stream()
                        .map(r -> r.getName()).toArray(String[]::new);
                return SecurityConfig.createList(roles);
            }
        }
        return null;
    }
    @Override
    public Collection<ConfigAttribute> getAllConfigAttributes() {
        return null;
    }
    @Override
    public boolean supports(Class<?> clazz) {
        return FilterInvocation.class.isAssignableFrom(clazz);
    }
}
```

自定义 CustomSecurityMetadataSource 类并实现 FilterInvocationSecurityMetadataSource 接口，然后重写它里边的三个方法：

- getAttributes：该方法的参数是受保护对象，在基于 URL 地址的权限控制中，受保护对象就是 FilterInvocation，该方法的返回值则是访问受保护对象所需要的权限。在该方法里边，我们首先从受保护对象 FilterInvocation 中提取出当前请求的 URL 地址，例如 /admin/hello，然后通过 menuService 对象查询出所有的菜单数据（每条数据中都包含访问该条记录所需要的权限），遍历查询出来的菜单数据，如果当前请求的 URL 地址和菜单中某一条记录的 pattern 属性匹配上了（例如/admin/hello 匹配上/admin/**），那么我们就可以获取当前请求所需要的权限。从 menu 对象中获取 roles 属性，并将其转为一个数组，然后通过 SecurityConfig.createList 方法创建一个 Collection<ConfigAttribute>对象并返回。如果当前请求的 URL 地址和数据库中 menu 表的所有项都匹配不上，那么最终返回 null。如果返回 null，那么受保护对象到底能不能访问呢？这就要看 AbstractSecurityInterceptor 对象中的 rejectPublicInvocations 属性了，该属性默认为 false，表示当 getAttributes 方法返回 null 时，允许访问受保护对象（回顾 13.4.4 小节中关于 AbstractSecurityInterceptor#beforeInvocation 的讲解）。
- getAllConfigAttributes：该方法可以用来返回所有的权限属性，以便在项目启动阶段做校验，如果不需要校验，则直接返回 null 即可。
- supports：该方法表示当前对象支持处理的受保护对象是 FilterInvocation。

CustomSecurityMetadataSource 类配置完成后，接下来我们要用它来代替默认的 SecurityMetadataSource 对象，具体配置如下：

```java
@Configuration
public class SecurityConfig extends WebSecurityConfigurerAdapter {
    @Autowired
    CustomSecurityMetadataSource customSecurityMetadataSource;
    @Autowired
    UserService userService;
    @Override
    protected void configure(AuthenticationManagerBuilder auth)
                                                        throws Exception {
        auth.userDetailsService(userService);
    }
    @Override
    protected void configure(HttpSecurity http) throws Exception {
        ApplicationContext applicationContext =
                        http.getSharedObject(ApplicationContext.class);
        http.apply(new UrlAuthorizationConfigurer<>(applicationContext))
            .withObjectPostProcessor(new
                    ObjectPostProcessor<FilterSecurityInterceptor>() {
                @Override
                public <O extends FilterSecurityInterceptor> O
                                                postProcess(O object) {
            object.setSecurityMetadataSource(customSecurityMetadataSource);
                    return object;
                }
            });
        http.formLogin()
            .and()
            .csrf().disable();
    }
}
```

关于用户的配置无需多说，我们重点来看 configure(HttpSecurity) 方法。

由于访问路径规则和所需要的权限之间的映射关系已经保存在数据库中，所以我们就没有必要在 Java 代码中配置映射关系了，同时这里的权限对比也不会用到权限表达式，所以我们通过 UrlAuthorizationConfigurer 来进行配置。

在配置的过程中，通过 withObjectPostProcessor 方法调用 ObjectPostProcessor 对象后置处理器，在对象后置处理器中，将 FilterSecurityInterceptor 中的 SecurityMetadataSource 对象替换为我们自定义的 customSecurityMetadataSource 对象即可。

测试

接下来创建 HelloController，代码如下：

```java
@RestController
public class HelloController {
    @GetMapping("/admin/hello")
    public String admin() {
        return "hello admin";
    }
    @GetMapping("/user/hello")
    public String user() {
        return "hello user";
    }
    @GetMapping("/guest/hello")
    public String guest() {
        return "hello guest";
    }
    @GetMapping("/hello")
    public String hello() {
        return "hello";
    }
}
```

最后启动项目进行测试。

首先使用 admin/123 进行登录，该用户具备 ROLE_ADMIN 角色，ROLE_ADMIN 可以访问/admin/hello、/user/hello 以及/guest/hello 三个接口。

接下来使用 user/123 进行登录，该用户具备 ROLE_USER 角色，ROLE_USER 可以访问/user/hello 以及/guest/hello 两个接口。

最后使用 javaboy/123 进行登录，该用户具备 ROLE_GUEST 角色，ROLE_GUEST 可以访问/guest/hello 接口。

由于/hello 接口不包含在 URL-权限映射关系中，所以任何用户都可以访问/hello 接口，包括匿名用户。如果希望所有的 URL 地址都必须在数据库中配置 URL-权限映射关系后才能访问，那么可以通过如下配置实现：

```java
http.apply(new UrlAuthorizationConfigurer<>(applicationContext))
        .withObjectPostProcessor(new
                    ObjectPostProcessor<FilterSecurityInterceptor>() {
            @Override
            public <O extends FilterSecurityInterceptor> O
                                        postProcess(O object) {
                object.setSecurityMetadataSource(customSecurityMetadataSource);
```

```
            object.setRejectPublicInvocations(true);
            return object;
        }
    });
```

通过设置 FilterSecurityInterceptor 中的 rejectPublicInvocations 属性为 true，就可以关闭 URL 的公开访问，所有 URL 必须具备对应的权限才能访问。

13.5 基于方法的权限管理

基于方法的权限管理主要是通过 AOP 来实现的，Spring Security 中通过 MethodSecurityInterceptor 来提供相关的实现。不同在于 FilterSecurityInterceptor 只是在请求之前进行前置处理，MethodSecurityInterceptor 除了前置处理之外还可以进行后置处理。前置处理就是在请求之前判断是否具备相应的权限，后置处理则是对方法的执行结果进行二次过滤。前置处理和后置处理分别对应了不同的实现类，我们分别来看。

13.5.1 注解介绍

目前在 Spring Boot 中基于方法的权限管理主要是通过注解来实现，我们需要通过 @EnableGlobalMethodSecurity 注解开启权限注解的使用，用法如下：

```
@Configuration
@EnableGlobalMethodSecurity(prePostEnabled = true,
                            securedEnabled = true, jsr250Enabled = true)
public class SecurityConfig extends WebSecurityConfigurerAdapter {
}
```

这个注解中我们设置了三个属性：

- prePostEnabled：开启 Spring Security 提供的四个权限注解，@PostAuthorize、@PostFilter、@PreAuthorize 以及@PreFilter，这四个注解支持权限表达式，功能比较丰富。
- securedEnabled：开启 Spring Security 提供的@Secured 注解，该注解不支持权限表达式。
- jsr250Enabled：开启 JSR-250 提供的注解，主要包括@DenyAll、@PermitAll 以及 @RolesAllowed 三个注解，这些注解也不支持权限表达式。

这些注解的含义分别如下：

- @PostAuthorize：在目标方法执行之后进行权限校验。
- @PostFilter：在目标方法执行之后对方法的返回结果进行过滤。
- @PreAuthorize：在目标方法执行之前进行权限校验。
- @PreFilter：在目标方法执行之前对方法参数进行过滤。
- @Secured：访问目标方法必须具备相应的角色。

- @DenyAll：拒绝所有访问。
- @PermitAll：允许所有访问。
- @RolesAllowed：访问目标方法必须具备相应的角色。

这些基于方法的权限管理相关的注解，一般来说只要设置 prePostEnabled=true 就够用了。

另外还有一种比较"古老"的方法配置基于方法的权限管理，那就是通过 XML 文件配置方法拦截规则，目前已经很少有用 XML 文件来配置 Spring Security 了，所以对于这种方式我们不做过多介绍。感兴趣的读者可以查看官网的相关介绍：https://docs.spring.io/spring-security/site/docs/5.4.0/reference/html5/#secure-object-impls。

13.5.2　基本用法

接下来我们通过几个简单的案例来学习上面几种不同注解的用法。

首先创建一个 Spring Boot 项目，引入 Web 和 Spring Security 依赖，项目创建完成后，添加如下配置文件：

```
@EnableGlobalMethodSecurity(prePostEnabled = true,
                    securedEnabled = true, jsr250Enabled = true)
public class SecurityConfig{
}
```

为了方便起见，我们将使用单元测试进行验证，所以这里就不进行额外的配置了，通过 @EnableGlobalMethodSecurity 注解开启其他权限注解的使用即可。

接下来创建一个 User 类以备后续使用：

```
public class User {
    private Integer id;
    private String username;

    //省略getter/setter
}
```

准备工作完成后，我们来逐个讲解一下前面注解的用法。

@PreAuthorize

@PreAuthorize 可以在目标方法执行之前对其进行安全校验，在安全校验时，可以直接使用我们在 13.3.7 小节中介绍的权限表达式。例如可以定义如下方法：

```
@Service
public class HelloService {
    @PreAuthorize("hasRole('ADMIN')")
    public String hello() {
        return "hello";
    }
}
```

这里使用了权限表达式 hasRole，表示执行该方法必须具备 ADMIN 角色才可以访问，否则不可以访问。我们在单元测试中来测试该方法：

```
@SpringBootTest
class BasedOnMethodApplicationTests {
    @Autowired
    HelloService helloService;
    @Test
    @WithMockUser(roles = "ADMIN")
    void preauthorizeTest01() {
        String hello = helloService.hello();
        assertNotNull(hello);
        assertEquals("hello", hello);
    }
}
```

通过@WithMockUser(roles = "ADMIN") 注解设定当前执行的用户角色是 ADMIN，然后调用 helloService 中的方法进行测试即可。如果将用户角色设置为其他字符，那单元测试就不会通过。

当然，这里除了 hasRole 表达式之外，也可以使用其他权限表达式，包括在 13.4.3 小节中自定义的表达式也可以使用。也可以同时使用多个权限表达式，如下所示：

```
@Service
public class HelloService {
    @PreAuthorize("hasRole('ADMIN') and authentication.name=='javaboy'")
    public String hello() {
        return "hello";
    }
}
```

表示访问者名称必须是 javaboy，而且还需要同时具备 ADMIN 角色，才可以访问该方法。此时通过如下代码对其进行测试：

```
@SpringBootTest
class BasedOnMethodApplicationTests {
    @Autowired
    HelloService helloService;
    @Test
    @WithMockUser(roles = "ADMIN",username = "javaboy")
    void preauthorizeTest01() {
        String hello = helloService.hello();
        assertNotNull(hello);
        assertEquals("hello", hello);
    }
}
```

在@PreAuthorize 注解中，还可以通过#引用方法的参数，并对其进行校验，例如如下方法表示请求者的用户名必须等于方法参数 name 的值，方法才可以被执行：

```
@PreAuthorize("authentication.name==#name")
public String hello(String name) {
    return "hello:" + name;
}
```

测试方法如下：

```
@Test
@WithMockUser(username = "javaboy")
void preauthorizeTest02() {
    String hello = helloService.hello("javaboy");
    assertNotNull(hello);
    assertEquals("hello:javaboy", hello);
}
```

当模拟的用户名和方法参数相等时，单元测试就可以通过。

@PreFilter

@PreFilter 主要是对方法的请求参数进行过滤，它里边包含了一个内置对象 filterObject 表示要过滤的参数，如果方法只有一个参数，则内置的 filterObject 对象就代表该参数；如果方法有多个参数，则需要通过 filterTarget 来指定 filterObject 到底代表哪个对象：

```
@PreFilter(value = "filterObject.id%2!=0",filterTarget = "users")
public void addUsers(List<User> users, Integer other) {
    System.out.println("users = " + users);
}
```

上面代码表示对方法参数 users 进行过滤，将 id 为奇数的 user 保留。

然后通过单元测试对该方法进行测试：

```
@Test
@WithMockUser(username = "javaboy")
void preFilterTest01() {
    List<User> users = new ArrayList<>();
    for (int i = 0; i < 10; i++) {
        users.add(new User(i, "javaboy:" + i));
    }
    helloService.addUsers(users, 99);
}
```

执行单元测试方法，addUsers 方法中只会打印出 id 为奇数的 user 对象。

@PostAuthorize

@PostAuthorize 是在目标方法执行之后进行权限校验。可能有读者会觉得奇怪，目标方法都执行完了才去做权限校验意义何在？其实这个主要是在 ACL 权限模型中会用到，目标方法执行完毕后，通过@PostAuthorize 注解去校验目标方法的返回值是否满足相应的权限要求。

从技术角度来讲，@PostAuthorize 注解中也可以使用权限表达式，但是在实际开发中权限表达式一般都是结合@PreAuthorize 注解一起使用的。@PostAuthorize 包含一个内置对象

returnObject，表示方法的返回值，开发者可以对返回值进行校验：

```java
@PostAuthorize("returnObject.id==1")
public User getUserById(Integer id) {
    return new User(id, "javaboy");
}
```

这个表示方法返回的 user 对象的 id 必须为 1，调用才会顺利通过，否则就会抛出异常。

然后通过单元测试对该方法进行测试：

```java
@Test
@WithMockUser(username = "javaboy")
void postAuthorizeTest01() {
    User user = helloService.getUserById(1);
    assertNotNull(user);
    assertEquals(1,user.getId());
    assertEquals("javaboy",user.getUsername());
}
```

如果调用时传入的参数为 1，单元测试就会顺利通过。

这里先通过这样一个简单的例子来了解一下@PostAuthorize 注解的基本用法，在第 14 章的 ACL 权限模型讲解中，我们将会再次介绍@PostAuthorize 注解的其他用法。

@PostFilter

@PostFilter 注解在目标方法执行之后，对目标方法的返回结果进行过滤，该注解中包含了一个内置对象 filterObject，表示目标方法返回的集合/数组中的具体元素：

```java
@PostFilter("filterObject.id%2==0")
public List<User> getAll() {
    List<User> users = new ArrayList<>();
    for (int i = 0; i < 10; i++) {
        users.add(new User(i, "javaboy:" + i));
    }
    return users;
}
```

这段代码表示 getAll 方法的返回值 users 集合中 user 对象的 id 必须为偶数。

然后我们通过单元测试对其进行测试，代码如下：

```java
@Test
@WithMockUser(roles = "ADMIN")
void postFilterTest01() {
    List<User> all = helloService.getAll();
    assertNotNull(all);
    assertEquals(5, all.size());
    assertEquals(2,all.get(1).getId());
}
```

@Secured

@Secured 注解也是 Spring Security 提供的权限注解,不同于前面四个注解,该注解不支持权限表达式,只能做一些简单的权限描述。

```
@Secured({"ROLE_ADMIN","ROLE_USER"})
public User getUserByUsername(String username) {
    return new User(99, username);
}
```

这段代码表示用户需要具备 ROLE_ADMIN 或者 ROLE_USER 角色,才能访问 getUserByUsername 方法。

然后我们通过单元测试对其进行测试,代码如下:

```
@Test
@WithMockUser(roles = "ADMIN")
void securedTest01() {
    User user = helloService.getUserByUsername("javaboy");
    assertNotNull(user);
    assertEquals(99,user.getId());
    assertEquals("javaboy", user.getUsername());
}
```

注意,这里不需要给角色添加 ROLE_ 前缀,系统会自动添加。

@DenyAll

@DenyAll 是 JSR-250 提供的方法注解,看名字就知道这是拒绝所有访问:

```
@DenyAll
public String denyAll() {
    return "DenyAll";
}
```

然后我们通过单元测试对其进行测试,代码如下:

```
@Test
@WithMockUser(username = "javaboy")
void denyAllTest01() {
    helloService.denyAll();
}
```

在单元测试过程中,就会抛出异常。

@PermitAll

@PermitAll 也是 JSR-250 提供的方法注解,看名字就知道这是允许所有访问:

```
@PermitAll
public String permitAll() {
    return "PermitAll";
}
```

然后我们通过单元测试对其进行测试，代码如下：

```
@Test
@WithMockUser(username = "javaboy")
void permitAllTest01() {
    String s = helloService.permitAll();
    assertNotNull(s);
    assertEquals("PermitAll", s);
}
```

@RolesAllowed

@RolesAllowed 也是 JSR-250 提供的注解，可以添加在方法上或者类上，当添加在类上时，表示该注解对类中的所有方法生效；如果类上和方法上都有该注解，并且起冲突，则以方法上的注解为准。我们来看一个简单的案例：

```
@RolesAllowed({"ADMIN","USER"})
public String rolesAllowed() {
    return "RolesAllowed";
}
```

这个表示访问 rolesAllowed 方法需要具备 ADMIN 或者 USER 角色，然后我们通过单元测试对其进行测试，代码如下：

```
@Test
@WithMockUser(roles = "ADMIN")
void rolesAllowedTest01() {
    String s = helloService.rolesAllowed();
    assertNotNull(s);
    assertEquals("RolesAllowed", s);
}
```

这就是常见的方法权限注解。

13.5.3 原理剖析

MethodSecurityInterceptor

在 13.4.4 小节中，我们介绍了 AbstractSecurityInterceptor 中的三大方法，当我们基于 URL 请求地址进行权限控制时，使用的 AbstractSecurityInterceptor 实现类是 FilterSecurityInterceptor，而当我们基于方法进行权限控制时，使用的 AbstractSecurityInterceptor 实现类是 MethodSecurityInterceptor。

MethodSecurityInterceptor 提供了基于 AOP Alliance 的方法拦截，该拦截器中所使用的 SecurityMetadataSource 类型为 MethodSecurityMetadataSource。MethodSecurityInterceptor 中最重要的就是 invoke 方法，我们一起来看一下：

```
public Object invoke(MethodInvocation mi) throws Throwable {
    InterceptorStatusToken token = super.beforeInvocation(mi);
```

```
    Object result;
    try {
        result = mi.proceed();
    }
    finally {
        super.finallyInvocation(token);
    }
    return super.afterInvocation(token, result);
}
```

invoke 方法的逻辑非常清晰明了。首先调用父类的 beforeInvocation 方法进行权限校验，校验通过后，调用 mi.proceed()方法继续执行目标方法，然后在 finally 代码块中调用 finallyInvocation 方法完成一些清理工作，最后调用父类的 afterInvocation 方法进行请求结果过滤。

在 13.4.4 小节中，我们介绍了 FilterSecurityInterceptor 是通过 ExpressionUrlAuthorizationConfigurer 或者 UrlAuthorizationConfigurer 进行配置的，那么 MethodSecurityInterceptor 又是通过谁配置的呢？在前面的配置中，我们使用到了@EnableGlobalMethodSecurity 注解，所以就以该注解为线索展开分析。

@EnableGlobalMethodSecurity

@EnableGlobalMethodSecurity 用来开启方法的权限注解，我们来看一下该注解的定义：

```
@Retention(value = java.lang.annotation.RetentionPolicy.RUNTIME)
@Target(value = { java.lang.annotation.ElementType.TYPE })
@Documented
@Import({ GlobalMethodSecuritySelector.class })
@EnableGlobalAuthentication
@Configuration
public @interface EnableGlobalMethodSecurity {
    //省略其他
}
```

从该类的定义上可以看到，它引入了一个配置 GlobalMethodSecuritySelector，该类的作用主要是用来导入外部配置类，我们来看一下该类的定义：

```
final class GlobalMethodSecuritySelector implements ImportSelector {
    public String[] selectImports(AnnotationMetadata importingClassMetadata) {
        Class<EnableGlobalMethodSecurity> annoType =
                                    EnableGlobalMethodSecurity.class;
        Map<String, Object> annotationAttributes = importingClassMetadata
                .getAnnotationAttributes(annoType.getName(), false);
        AnnotationAttributes attributes = AnnotationAttributes
                .fromMap(annotationAttributes);
        Class<?> importingClass = ClassUtils
                .resolveClassName(importingClassMetadata.getClassName(),
                        ClassUtils.getDefaultClassLoader());
        boolean skipMethodSecurityConfiguration =
GlobalMethodSecurityConfiguration.class.isAssignableFrom(importingClass);
```

```
        AdviceMode mode = attributes.getEnum("mode");
        boolean isProxy = AdviceMode.PROXY == mode;
        String autoProxyClassName =
            isProxy ? AutoProxyRegistrar.class.getName() :
            GlobalMethodSecurityAspectJAutoProxyRegistrar.class.getName();
        boolean jsr250Enabled = attributes.getBoolean("jsr250Enabled");
        List<String> classNames = new ArrayList<>(4);
        if (isProxy) {
            classNames
                .add(MethodSecurityMetadataSourceAdvisorRegistrar.class.getName());
        }
        classNames.add(autoProxyClassName);
        if (!skipMethodSecurityConfiguration) {
            classNames.add(GlobalMethodSecurityConfiguration.class.getName());
        }
        if (jsr250Enabled) {
            classNames.add(Jsr250MetadataSourceConfiguration.class.getName());
        }
        return classNames.toArray(new String[0]);
    }
}
```

这里只有一个 selectImports 方法，该方法的参数 importingClassMetadata 中保存了 @Enable GlobalMethodSecurity 注解的元数据，包括各个属性的值、注解是加在哪个配置类上等。

selectImports 方法的逻辑比较简单，要导入的外部配置类有如下几种：

- MethodSecurityMetadataSourceAdvisorRegistrar：如果使用的是 Spring 自带的 AOP，则该配置类会被导入。该类主要用来向 Spring 容器中注册一个 MethodSecurityMetadataSourceAdvisor 对象，这个对象中定义了 AOP 中的 pointcut 和 advice。
- autoProxyClassName：注册自动代理创建者，根据不同的代理模式而定。
- GlobalMethodSecurityConfiguration：这个配置类用来提供 MethodSecurityMetadataSource 和 MethodInterceptor 两个关键对象。如果开发者自定义配置类继承自 GlobalMethodSecurityConfiguration，则这里不会导入这个外部配置类。
- Jsr250MetadataSourceConfiguration：如果开启了 JSR-250 注解，则会导入该配置类。该配置类主要用来提供 JSR-250 注解所需的 Jsr250MethodSecurityMetadataSource 对象。

这四个导入的外部配置类中，MethodSecurityMetadataSourceAdvisorRegistrar 是用来配置 MethodSecurityMetadataSourceAdvisor 的，而 MethodSecurityMetadataSourceAdvisor 则提供了 AOP 所需的 pointcut 和 advice。先来看 MethodSecurityMetadataSourceAdvisorRegistrar：

```
class MethodSecurityMetadataSourceAdvisorRegistrar implements
        ImportBeanDefinitionRegistrar {
    public void registerBeanDefinitions(AnnotationMetadata
            importingClassMetadata,BeanDefinitionRegistry registry) {
        BeanDefinitionBuilder advisor = BeanDefinitionBuilder
            .rootBeanDefinition(MethodSecurityMetadataSourceAdvisor.class);
```

```java
        advisor.setRole(BeanDefinition.ROLE_INFRASTRUCTURE);
        advisor.addConstructorArgValue("methodSecurityInterceptor");
        advisor.addConstructorArgReference("methodSecurityMetadataSource");
        advisor.addConstructorArgValue("methodSecurityMetadataSource");
        MultiValueMap<String, Object> attributes =
    importingClassMetadata
.getAllAnnotationAttributes(EnableGlobalMethodSecurity.class.getName());
        Integer order = (Integer) attributes.getFirst("order");
        if (order != null) {
            advisor.addPropertyValue("order", order);
        }
        registry.registerBeanDefinition("metaDataSourceAdvisor",
            advisor.getBeanDefinition());
    }
}
```

这个类很好理解，在 registerBeanDefinitions 方法中，首先定义 BeanDefinitionBuilder，然后给目标对象 MethodSecurityMetadataSourceAdvisor 的构造方法设置参数，参数一共有三个：第一个是要引用的 MethodInterceptor 对象名；第二是要引用的 MethodSecurityMetadataSource 对象名；第三个参数和第二个一样，只不过一个是引用，一个是字符串。所有属性都配置好之后，将其注册到 Spring 容器中。

我们再来看 MethodSecurityMetadataSourceAdvisor：

```java
public class MethodSecurityMetadataSourceAdvisor extends
                AbstractPointcutAdvisor implements BeanFactoryAware {
    private transient MethodSecurityMetadataSource attributeSource;
    private transient MethodInterceptor interceptor;
    private final Pointcut pointcut =
                            new MethodSecurityMetadataSourcePointcut();
    private BeanFactory beanFactory;
    private final String adviceBeanName;
    private final String metadataSourceBeanName;
    private transient volatile Object adviceMonitor = new Object();
    public MethodSecurityMetadataSourceAdvisor(String adviceBeanName,
            MethodSecurityMetadataSource attributeSource,
          String attributeSourceBeanName) {
        this.adviceBeanName = adviceBeanName;
        this.attributeSource = attributeSource;
        this.metadataSourceBeanName = attributeSourceBeanName;
    }
    public Pointcut getPointcut() {
        return pointcut;
    }
    public Advice getAdvice() {
        synchronized (this.adviceMonitor) {
            if (interceptor == null) {
                interceptor = beanFactory.getBean(this.adviceBeanName,
```

```
                MethodInterceptor.class);
        }
        return interceptor;
    }
}
public void setBeanFactory(BeanFactory beanFactory) throws BeansException {
    this.beanFactory = beanFactory;
}
class MethodSecurityMetadataSourcePointcut extends
            StaticMethodMatcherPointcut    implements Serializable {
    public boolean matches(Method m, Class targetClass) {
        Collection attributes = attributeSource.getAttributes(m, targetClass);
        return attributes != null && !attributes.isEmpty();
    }
}
}
```

MethodSecurityMetadataSourceAdvisor 继承自 AbstractPointcutAdvisor，主要定义了 AOP 的 pointcut 和 advice。MethodSecurityMetadataSourceAdvisor 构造方法所需的三个参数就是前面 MethodSecurityMetadataSourceAdvisorRegistrar 类中提供的三个参数。

pointcut 也就是切点，可以简单理解为方法的拦截规则，即哪些方法需要拦截，哪些方法不需要拦截。不用看代码我们也知道，加了权限注解的方法需要拦截下来，没加权限注解的方法则不需要拦截。

这里的 pointcut 对象就是内部类 MethodSecurityMetadataSourcePointcut，在它的 matches 方法中，定义了具体的拦截规则。通过 attributeSource.getAttributes 方法去查看目标方法上有没有相应的权限注解，如果有，则返回 true，目标方法就被拦截下来；如果没有，则返回 false，目标方法就不会被拦截。这里的 attributeSource 实际上就是 MethodSecurityMetadataSource 对象，也就是我们在 13.3.6 小节中介绍的提供权限元数据的类。

advice 也就是增强/通知，就是将方法拦截下来之后要增强的功能。advice 由 getAdvice() 方法返回，在该方法内部，就是去 Spring 容器中查找一个名为 methodSecurityInterceptor 的 MethodInterceptor 对象，这就是 advice。

此时，读者已经明白了 AOP 的切点和增强/通知是如何定义的了，这里涉及两个关键对象：一个名为 methodSecurityInterceptor 的 MethodInterceptor 对象和一个名为 methodSecurityMetadataSource 的 MethodSecurityMetadataSource 对象。

这两个关键对象在 GlobalMethodSecurityConfiguration 类中定义，相关的方法比较长，我们先来看 MethodSecurityMetadataSource 对象的定义：

```
@Bean
public MethodSecurityMetadataSource methodSecurityMetadataSource() {
    List<MethodSecurityMetadataSource> sources = new ArrayList<>();
    ExpressionBasedAnnotationAttributeFactory attributeFactory =
        new ExpressionBasedAnnotationAttributeFactory(getExpressionHandler());
    MethodSecurityMetadataSource customMethodSecurityMetadataSource =
```

```java
                                customMethodSecurityMetadataSource();
    if (customMethodSecurityMetadataSource != null) {
        sources.add(customMethodSecurityMetadataSource);
    }
    boolean hasCustom = customMethodSecurityMetadataSource != null;
    boolean isPrePostEnabled = prePostEnabled();
    boolean isSecuredEnabled = securedEnabled();
    boolean isJsr250Enabled = jsr250Enabled();
    if (!isPrePostEnabled && !isSecuredEnabled && !isJsr250Enabled &&
                                                        !hasCustom) {
        throw new IllegalStateException("");
    }
    if (isPrePostEnabled) {
        sources.add(new
                PrePostAnnotationSecurityMetadataSource(attributeFactory));
    }
    if (isSecuredEnabled) {
        sources.add(new SecuredAnnotationSecurityMetadataSource());
    }
    if (isJsr250Enabled) {
        GrantedAuthorityDefaults grantedAuthorityDefaults =
                getSingleBeanOrNull(GrantedAuthorityDefaults.class);
    Jsr250MethodSecurityMetadataSource jsr250MethodSecurityMetadataSource =
            this.context.getBean(Jsr250MethodSecurityMetadataSource.class);
        if (grantedAuthorityDefaults != null) {
            jsr250MethodSecurityMetadataSource.setDefaultRolePrefix(
                grantedAuthorityDefaults.getRolePrefix());
        }
        sources.add(jsr250MethodSecurityMetadataSource);
    }
    return new DelegatingMethodSecurityMetadataSource(sources);
}
```

可以看到，这里首先创建了一个 List 集合，用来保存所有的 MethodSecurityMetadataSource 对象，然后调用 customMethodSecurityMetadataSource 方法去获取自定义的 MethodSecurityMetadataSource，默认情况下该方法返回 null，如果项目有需要，开发者可以重写 customMethodSecurityMetadataSource 方法来提供自定义的 MethodSecurityMetadataSource 对象。接下来就是根据注解中配置的属性值，来向 sources 集合中添加相应的 MethodSecurityMetadataSource 对象：

- 如果 @EnableGlobalMethodSecurity 注解配置了 prePostEnabled=true，则加入 PrePostAnnotationSecurityMetadataSource 对象来解析相应的注解。
- 如果 @EnableGlobalMethodSecurity 注解配置了 securedEnabled=true，则加入 SecuredAnnotationSecurityMetadataSource 对象来解析相应的注解。
- 如果 @EnableGlobalMethodSecurity 注解配置了 jsr250Enabled=true，则加入 Jsr250MethodSecurityMetadataSource 对象来解析相应的注解。
- 最后构建一个代理对象 DelegatingMethodSecurityMetadataSource 返回即可。

可以看到，默认提供的 MethodSecurityMetadataSource 对象实际上是一个代理对象，它包含多个不同的 MethodSecurityMetadataSource 实例。

回顾前面所讲的切点定义，在判断一个方法是否需要被拦截下来时，由这些被代理的对象逐个去解析目标方法是否含有相应的注解（例如，PrePostAnnotationSecurityMetadataSource 可以检查出目标方法是否含有@PostAuthorize、@PostFilter、@PreAuthorize 以及@PreFilter 四种注解），如果有，则请求就会被拦截下来。

举个反例，如果开发者在项目中使用了 @Secured 注解，但是却没有在 @EnableGlobalMethodSecurity 注解中配置 securedEnabled=true，那么这里就不会加入 SecuredAnnotationSecurityMetadataSource 对象到代理对象中去，进而导致在切点定义的方法中，SecuredAnnotationSecurityMetadataSource 对象不会参与到目标方法注解的解析中，而其他的 SecurityMetadataSource 又无法解析目标方法上的@Secured，所以最终目标方法就不会被拦截。因此，使用哪个权限注解，一定要先在@EnableGlobalMethodSecurity 中开启对应的配置。

再来看 MethodInterceptor 的定义：

```java
@Bean
public MethodInterceptor methodSecurityInterceptor(
        MethodSecurityMetadataSource methodSecurityMetadataSource) {
    this.methodSecurityInterceptor = isAspectJ()
        ? new AspectJMethodSecurityInterceptor()
        : new MethodSecurityInterceptor();
    methodSecurityInterceptor
                .setAccessDecisionManager(accessDecisionManager());
    methodSecurityInterceptor
                .setAfterInvocationManager(afterInvocationManager());
    methodSecurityInterceptor
            .setSecurityMetadataSource(methodSecurityMetadataSource);
    RunAsManager runAsManager = runAsManager();
    if (runAsManager != null) {
        methodSecurityInterceptor.setRunAsManager(runAsManager);
    }
    return this.methodSecurityInterceptor;
}
protected AccessDecisionManager accessDecisionManager() {
    List<AccessDecisionVoter<?>> decisionVoters = new ArrayList<>();
    if (prePostEnabled()) {
        ExpressionBasedPreInvocationAdvice expressionAdvice =
                        new ExpressionBasedPreInvocationAdvice();
        expressionAdvice.setExpressionHandler(getExpressionHandler());
        decisionVoters
            .add(new PreInvocationAuthorizationAdviceVoter(expressionAdvice));
    }
    if (jsr250Enabled()) {
        decisionVoters.add(new Jsr250Voter());
    }
    RoleVoter roleVoter = new RoleVoter();
```

```java
            GrantedAuthorityDefaults grantedAuthorityDefaults =
                    getSingleBeanOrNull(GrantedAuthorityDefaults.class);
            if (grantedAuthorityDefaults != null) {
                roleVoter.setRolePrefix(grantedAuthorityDefaults.getRolePrefix());
            }
            decisionVoters.add(roleVoter);
            decisionVoters.add(new AuthenticatedVoter());
            return new AffirmativeBased(decisionVoters);
    }
    protected AfterInvocationManager afterInvocationManager() {
        if (prePostEnabled()) {
            AfterInvocationProviderManager invocationProviderManager =
                                    new AfterInvocationProviderManager();
            ExpressionBasedPostInvocationAdvice postAdvice =
            new ExpressionBasedPostInvocationAdvice(    getExpressionHandler());
            PostInvocationAdviceProvider postInvocationAdviceProvider =
                            new PostInvocationAdviceProvider(postAdvice);
            List<AfterInvocationProvider> afterInvocationProviders =
                                                        new ArrayList<>();
            afterInvocationProviders.add(postInvocationAdviceProvider);
            invocationProviderManager.setProviders(afterInvocationProviders);
            return invocationProviderManager;
        }
        return null;
    }
```

MethodInterceptor 的创建，首先看代理的方式，默认使用 Spring 自带的 AOP，所以使用 MethodSecurityInterceptor 来创建对应的 MethodInterceptor 实例。然后给 methodSecurityInterceptor 对象设置 AccessDecisionManager 决策管理器，默认的决策管理器是 AffirmativeBased，根据 @EnableGlobalMethodSecurity 注解的配置，在角色管理器中配置不同的投票器；接下来给 methodSecurityInterceptor 配置后置处理器，如果@EnableGlobalMethodSecurity 注解配置了 prePostEnabled=true，则添加一个后置处理器 PostInvocationAdviceProvider，该类用来处理@PostAuthorize 和@PostFilter 两个注解；最后再把前面创建好的 MethodSecurityMetadataSource 对象配置给 methodSecurityInterceptor。

至于 methodSecurityInterceptor 对象的工作逻辑，我们在本小节一开始就已经介绍了。

13.6 小　结

本章主要介绍了 Spring Security 中权限管理的核心概念、基本用法以及相关原理，特别是对基于 URL 地址的权限管理和基于方法的权限管理都做了详细介绍。万变不离其宗，我们在实际项目中设计出来的各种各样的权限管理系统，最底层使用的技术都不会变。本章的内容，需要读者慢慢理解，反复实践。

第 14 章

权限模型

权限管理是一个非常复杂烦琐的工作,为了开发出高效的、易于维护的权限管理系统,我们需要一个"指导思想",这个"指导思想"就是权限模型。从计算机应用出现就一直伴随着权限管理,权限模型也在这个过程中得到了长足发展。时至今日,有的权限模型已经逐渐被人们遗忘,而有的权限模型则一时"风头无两"。本章将介绍一些常见的权限模型。

本章涉及的主要知识点有:

- 常见的权限模型。
- ACL。
- RBAC。

14.1 常见的权限模型

权限模型有很多种,这里选择几种知名度较高的权限管理模型进行介绍。

DAC

DAC(Discretionary access control)指自主访问控制。它是 Trusted Computer System Evaluation Criteria(TCSEC)定义的一种访问控制模型。在这种访问控制模型中,系统会根据被操作对象的权限控制列表中的信息,来决定当前用户能够对其进行哪些操作,用户可以将其具备的权限直接或者间接授予其他用户,这也是其称为自主访问控制的原因。

自主访问控制经常与强制访问控制(MAC)对比。

MAC

MAC(Mandatory access control)指强制访问控制,也叫非自主访问控制。这种访问控制方式可以限制主体对对象或目标执行某种操作的能力。通过强制访问控制，安全策略由安全策略管理员集中控制，用户无权覆盖策略。

ABAC

ABAC（Attribute-Based Access Control）基于属性的访问控制，有时也被称为 PBAC（Policy-Based Access Control）或者 CBAC（Claims-Based Access Control）。

ABAC 也被称为下一代权限管理模型。

基于属性的访问控制中一般来说包含四类属性：用户属性、环境属性、操作属性以及资源属性，通过动态计算一个或者一组属性是否满足某一条件来进行授权。

当然，上面介绍的这三种权限模型并非本章重点，因为我们在 Java 企业级开发中很少会用到它们。

在企业级开发领域，目前最为流行的权限管理模型当属 RBAC。除了 RBAC 之外，还有一个 ACL 权限模型，Spring Security 中针对 ACL 也提供了相关的依赖，所以本章重点介绍这两种权限模型。

14.2 ACL

14.2.1 ACL 权限模型介绍

ACL（Access Control List，访问控制列表）是一种比较古老的权限控制模型。它是一种面向资源的访问控制模型，在 ACL 中我们所做的所有权限配置都是针对资源的。

在 ACL 权限模型中，会对系统中的每一个资源配置一个访问控制列表，这个列表中记录了用户/角色对于资源的操作权限，当需要访问这些资源时，会首先检查访问控制列表中是否存在当前用户/角色的访问权限，如果存在，则允许相应的操作，否则就拒绝相应的操作。

ACL 的一个核心思路就是：将某个对象的某种权限授予某个用户或某种角色，它们之间的关系是多对多，即一个用户/角色可以具备某个对象的多种权限，某个对象的权限也可以被多个用户/角色所持有。

举个简单例子：现在有一个 User 对象，针对该对象有查询、修改、删除等权限，可以将这些权限赋值给某一个用户 javaboy，也可以将这些权限赋值给某一个角色 ADMIN，那么 javaboy 用户或者具备 ADMIN 角色的其他用户就具有执行相应操作的权限。从这个角度看，ACL 是一种粒度非常细的权限控制，它可以精确到某一个资源的某一个权限。这些权限数据都记录在数据库中，这带来的另外一个问题就是需要维护的权限数据量非常庞大，特别是对于一些大型系统而言，这一问题尤其突出，大量的数据需要维护可能会造成系统性能下降。不过对于一些简单的小型系统而言，使用 ACL 还是可以的，没有任何问题。

ACL 的使用非常简单，在搞明白实现原理的基础上，开发者可以不使用任何权限框架也能快速实现 ACL 权限模型。当然 Spring Security 也为 ACL 提供了相应的依赖 spring-security-acl，如果项目中用到了 ACL，这个依赖需要自己额外添加。

14.2.2　ACL 核心概念介绍

接下来介绍一些 ACL 中的核心概念。

Sid

Sid 代表了用户和角色（注意是和不是或），它有两种：

- GrantedAuthoritySid
- PrincipalSid

前者代表角色，后者代表用户。根据本书前面章节的介绍，大家已经知道，在 Spring Security 中，用户和角色信息都是保存在 Authentication 对象中的，即 ACL 中的 Sid 我们可以从 Authentication 对象中提取出来的，具体的提取方法是 SidRetrievalStrategyImpl#getSids，我们来看一下相关源码：

```
public List<Sid> getSids(Authentication authentication) {
    Collection<? extends GrantedAuthority> authorities = roleHierarchy
        .getReachableGrantedAuthorities(authentication.getAuthorities());
    List<Sid> sids = new ArrayList<>(authorities.size() + 1);
    sids.add(new PrincipalSid(authentication));
    for (GrantedAuthority authority : authorities) {
        sids.add(new GrantedAuthoritySid(authority));
    }
    return sids;
}
```

可以看到，首先从 authentication 对象中提取出当前用户的角色，然后构造代表当前用户的 PrincipalSid 存入 sids 集合中，接下来构造代表用户角色的 GrantedAuthoritySid 存入 sids 集合中，最后将 sids 集合返回。

在 ACL 中，查看一个用户的身份，就是查看他的 Sid。

ObjectIdentity

ObjectIdentity 是一个域对象，这是官方的说法，有点拗口。实际上这就是我们要操作的对象。

例如有一个 User 对象，如果直接去记录能够对 User 对象执行哪些操作，这就会导致高耦合。所以我们需要对其解耦，将所有需要操作的对象通过 ObjectIdentity 抽象出来，用 ObjectIdentity 代表所有需要控制的对象，这样就能实现权限管理系统和具体业务解耦。

ObjectIdentity 中有两个关键方法，getType 和 getIdentifier。一般来说，getType 方法返回资源类的全路径，例如 org.javaboy.acl.model.User；getIdentifier 方法则返回资源的 id。通过这两个方法，就能够锁定一个具体的对象。

Acl（Acl 对象）

看名字就知道，这是整个系统的核心调度部分。

一个 Acl 对象关联一个 ObjectIdentity 和多个 ACE，同时一个 Acl 对象还拥有一个 Sid，这个 Sid 表示这个 Acl 对象的所有者，Acl 对象的所有者可以修改甚至删除该 Acl 对象。

AccessControlEntry

AccessControlEntry 简写为 ACE，一个 AccessControlEntry 对象代表一条权限记录，每一个 AccessControlEntry 中包含一个 Sid 和一个 Permission 对象，表示某个 Sid 具备某种权限。同时每一个 AccessControlEntry 对象都对应了一个 Acl 对象，一个 Acl 对象则可以对应多个 AccessControlEntry。

由于 Acl 对象还关联了一个 ObjectIdentity 对象，有了这层对应关系，就可以知道某个 Sid 具备某个 ObjectIdentity 的某种 Permission，如图 14-1 所示。

图 14-1　ObjectIdentity、Acl 以及 ACE 的关联关系

Permission

Permission 就是具体的权限对象，在 Spring Security 的默认实现中，最多支持 32 种权限。Spring Security 中默认定义了五种权限：

```java
public class BasePermission extends AbstractPermission {
    public static final Permission READ = new BasePermission(1 << 0, 'R');
    public static final Permission WRITE = new BasePermission(1 << 1, 'W');
    public static final Permission CREATE = new BasePermission(1 << 2, 'C');
    public static final Permission DELETE = new BasePermission(1 << 3, 'D');
    public static final Permission ADMINISTRATION =
                                    new BasePermission(1 << 4, 'A');
    protected BasePermission(int mask) {
        super(mask);
    }
    protected BasePermission(int mask, char code) {
        super(mask, code);
    }
}
```

如果默认的五种权限不能满足需求，开发者也可以自定义类继承自 BasePermission，从而扩展权限的定义。

AclService

AclService 接口中主要定义了一些解析 Acl 对象的方法，例如通过 ObjectIdentity 对象查询出其对应的 Acl 对象。

AclService 主要有两类实现接口：

- JdbcAclService
- JdbcMutableAclService

前者主要是针对 Acl 的查询操作，后者支持 Acl 的添加、更新以及删除等操作。一般情况下我们使用 JdbcMutableAclService。

至此，ACL 中一些核心概念就介绍完了。

14.2.3　ACL 数据库分析

在 ACL 中，所有和权限相关的数据都保存在数据库中，在 spring-security-acl 依赖中，为数据库提供了相应的脚本，如图 14-2 所示。针对不同的数据库提供了不同的脚本，开发者根据自己的实际情况选择合适的数据库脚本执行即可。

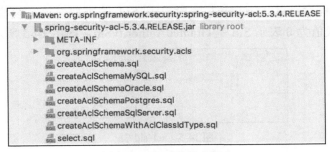

图 14-2　ACL 数据库脚本

脚本中主要涉及四张表，如图 14-3 所示。

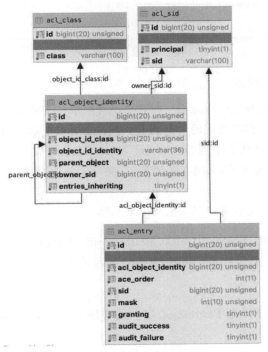

图 14-3　Acl 中的四张表

接下来逐一分析这四张表的作用。

acl_class

acl_class 表用来保存资源类的全路径，它有两个字段，其中 id 是自增长的主键，class 字段中保存资源类的全路径名，如图 14-4 所示。

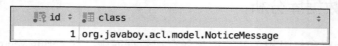

图 14-4　acl_class 表

acl_sid

acl_sid 表用来保存 Sid，principal 字段表示 Sid 的类型，该字段值为 1 表示 Sid 是 PrincipalSid，该字段值为 0 表示 Sid 是 GrantedAuthoritySid，如图 14-5 所示。

id	principal	sid
2	1	hr
13	1	javaboy
1	1	manager
3	0	ROLE_EDITOR

图 14-5　acl_sid 表

acl_object_identity

acl_object_identity 表用来保存需要进行访问控制的对象信息，如图 14-6 所示。

id	object_id_class	object_id_identity	parent_object	owner_sid	entries_inheriting
14	1	1	<null>	13	1
18	1	99	<null>	13	1

图 14-6　acl_object_identity 表

这个表涉及的字段比较多，含义如下：

- object_id_class：关联 acl_class.id，即需要进行权限控制的类信息。
- object_id_identity：需要控制的对象的 id，和 object_id_class 字段共同锁定一个具体对象。
- parent_object：父对象 ID，关联一条 acl_object_identity 记录。
- owner_sid：这个 ACL 记录拥有者的 Sid，拥有者可以对该 ACL 记录执行更新、删除等操作。
- entries_inheriting：是否需要继承父对象的权限。

简单来说，这个表中的 object_id_class 和 object_id_identity 字段锁定了要进行权限控制的对象，具体如何控制呢？则要看 acl_entry 表中的关联关系了。

acl_entry

acl_entry 表用来保存 Sid 和 Permission 之间的对应关系，如图 14-7 所示。

id	acl_object_identity	ace_order	sid	mask	granting	audit_success	audit_failure
15	14	0	2	1	1	0	0
16	14	1	2	2	1	0	0
22	18	0	1	4	1	0	0

图 14-7　acl_entry 表

这个表涉及的字段比较多，含义如下：

- acl_object_identity：关联 acl_object_identity.id。
- ace_order：权限顺序。acl_object_identity 和 ace_order 的组合要唯一。
- sid：关联 acl_sid.id，该字段关联一个具体的用户/角色。
- mask：权限掩码。
- granting：表示当前记录是否生效。
- audit_success/audit_failure：审计信息。

简单来说，acl_entry 中的一条记录，关联了一个要操作的对象（acl_object_identity 和 ace_order 字段），关联了 Sid（sid 字段），也描述了权限（mask），将权限涉及的东西都在该字段中整合起来了。

至此，ACL 中涉及的相关表就介绍完了。

接下来我们通过一个案例，来学习 ACL 的具体用法。

14.2.4 实战

14.2.4.1 准备工作

首先创建一个 Spring Boot 项目，由于这里涉及数据库操作，所以除了 Spring Security 依赖之外，还需要加入数据库驱动以及 MyBatis 依赖。

另外由于没有 ACL 相关的 starter，所以需要我们手动添加 ACL 依赖，由于 ACL 还依赖于 Ehcache 缓存，所以还需要加上缓存依赖。

最终的 pom.xml 文件如下：

```xml
<dependency>
    <groupId>org.springframework.boot</groupId>
    <artifactId>spring-boot-starter-security</artifactId>
</dependency>
<dependency>
    <groupId>org.springframework.boot</groupId>
    <artifactId>spring-boot-starter-web</artifactId>
</dependency>
<dependency>
    <groupId>org.springframework.security</groupId>
    <artifactId>spring-security-acl</artifactId>
    <version>5.3.4.RELEASE</version>
</dependency>
<dependency>
```

```xml
    <groupId>org.mybatis.spring.boot</groupId>
    <artifactId>mybatis-spring-boot-starter</artifactId>
    <version>2.1.3</version>
</dependency>
<dependency>
    <groupId>mysql</groupId>
    <artifactId>mysql-connector-java</artifactId>
</dependency>
<dependency>
    <groupId>net.sf.ehcache</groupId>
    <artifactId>ehcache</artifactId>
    <version>2.10.4</version>
</dependency>
<dependency>
    <groupId>com.alibaba</groupId>
    <artifactId>druid-spring-boot-starter</artifactId>
    <version>1.1.23</version>
</dependency>
<dependency>
    <groupId>org.springframework</groupId>
    <artifactId>spring-context-support</artifactId>
</dependency>
```

项目创建成功之后，我们在 ACL 的 jar 包（spring-security-acl-5.3.4.RELEASE.jar）中可以找到数据库脚本文件，选择适合自己数据库的脚本文件复制到数据库中执行，执行后生成四张表，这四张表前面已经介绍过了，这里不再赘述。

最后，再在项目的 application.properties 文件中配置数据库连接信息，代码如下：

```
spring.datasource.url=jdbc:mysql:///acls?useUnicode=true&characterEncoding=UTF-8&serverTimezone=Asia/Shanghai
spring.datasource.username=root
spring.datasource.password=123
spring.datasource.driver-class-name=com.mysql.cj.jdbc.Driver
```

至此，准备工作就算完成了。接下来我们来看配置。

14.2.4.2　Acl 配置

创建 AclConfig 类，完成 ACL 相关配置，代码如下：

```java
@Configuration
@EnableGlobalMethodSecurity(prePostEnabled = true, securedEnabled = true)
public class AclConfig {
    @Autowired
    DataSource dataSource;
    @Bean
    public AclAuthorizationStrategy aclAuthorizationStrategy() {
        return new AclAuthorizationStrategyImpl(
                        new SimpleGrantedAuthority("ROLE_ADMIN"));
```

```java
}
@Bean
public PermissionGrantingStrategy permissionGrantingStrategy() {
    return new DefaultPermissionGrantingStrategy(new ConsoleAuditLogger());
}
@Bean
public AclCache aclCache() {
    return new EhCacheBasedAclCache(
            aclEhCacheFactoryBean().getObject(),
            permissionGrantingStrategy(), aclAuthorizationStrategy());
}
@Bean
public EhCacheFactoryBean aclEhCacheFactoryBean() {
    EhCacheFactoryBean ehCacheFactoryBean = new EhCacheFactoryBean();
    ehCacheFactoryBean.setCacheManager(aclCacheManager().getObject());
    ehCacheFactoryBean.setCacheName("aclCache");
    return ehCacheFactoryBean;
}
@Bean
public EhCacheManagerFactoryBean aclCacheManager() {
    return new EhCacheManagerFactoryBean();
}
@Bean
public LookupStrategy lookupStrategy() {
    return new BasicLookupStrategy(dataSource, aclCache(),
            aclAuthorizationStrategy(), new ConsoleAuditLogger());
}
@Bean
public AclService aclService() {
    return new JdbcMutableAclService(dataSource, lookupStrategy(),
                                                aclCache());
}
@Bean
PermissionEvaluator permissionEvaluator() {
    AclPermissionEvaluator permissionEvaluator =
                        new AclPermissionEvaluator(aclService());
    return permissionEvaluator;
}
}
```

这里的配置实例比较多，我们来逐个解释：

- @EnableGlobalMethodSecurity 注解表示开启项目中@PreAuthorize、@PostAuthorize 以及 @Secured 注解的使用，我们将通过这些注解配置访问权限。
- 由于引入了数据库的一整套依赖，并且配置了数据库连接信息，所以这里可以直接注入 DataSource 实例以备后续使用。
- AclAuthorizationStrategy 实例用来判断当前的认证主体是否有修改 ACL 的权限，准确来说是三种权限：修改 ACL 的 owner、修改 ACL 的审计信息、修改 ACE 本身。这个接口只有一个实现类就是 AclAuthorizationStrategyImpl，我们在创建实例时，可以传入三个参数，分别对应了这三种权限，也可以传入一个参数，表示这一个角色可以做三件事情。

- PermissionGrantingStrategy 接口提供了一个 isGranted 方法，这个方法就是最终真正进行权限比对的方法，该接口只有一个实现类 DefaultPermissionGrantingStrategy，直接新建就行了。
- 在 ACL 中，由于权限比对总是要查询数据库，造成了性能问题，因此引入了 Ehcache 做缓存。AclCache 共有两个实现类：SpringCacheBasedAclCache 和 EhCacheBasedAclCache。我们前面已经引入了 Ehcache 实例，所以这里配置 EhCacheBasedAclCache 实例即可（如果使用 SpringCacheBasedAclCache，则不需要引入 Ehcache 依赖）。
- LookupStrategy 可以通过 ObjectIdentity 解析出对应的 ACL。LookupStrategy 只有一个实现类就是 BasicLookupStrategy，直接新建即可。
- AclService 已经在上文介绍过了，这里不再赘述。
- PermissionEvaluator 是为表达式 hasPermission 提供支持的。由于本案例后面使用 @PreAuthorize("hasPermission(#noticeMessage, 'WRITE')") 这样的注解进行权限控制，因此这里需要配置一个 PermissionEvaluator 实例，当进行权限校验时，就会调用到 AclPermissionEvaluator#hasPermission 方法。

以上就是 Acl 的配置类。

14.2.4.3 业务配置

假设我们现在有一个通知消息类 NoticeMessage，代码如下：

```java
public class NoticeMessage {
    private Integer id;
    private String content;
    @Override
    public String toString() {
        return "NoticeMessage{" +
                "id=" + id +
                ", content='" + content + '\'' +
                '}';
    }
    public Integer getId() {
        return id;
    }
    public void setId(Integer id) {
        this.id = id;
    }
    public String getContent() {
        return content;
    }
    public void setContent(String content) {
        this.content = content;
```

```
    }
}
```

然后根据该类创建一张数据表 system_message，SQL 脚本如下：

```sql
CREATE TABLE `system_message` (
  `id` int(11) unsigned NOT NULL AUTO_INCREMENT,
  `content` varchar(255) COLLATE utf8mb4_unicode_ci DEFAULT NULL,
  PRIMARY KEY (`id`)
) ENGINE=InnoDB DEFAULT CHARSET=utf8mb4 COLLATE=utf8mb4_unicode_ci;
```

接下来的权限控制测试都将针对 NoticeMessage 来进行。

创建 NoticeMessageMapper，并添加几个测试方法：

```java
@Mapper
public interface NoticeMessageMapper {
    List<NoticeMessage> findAll();
    NoticeMessage findById(Integer id);
    void save(NoticeMessage noticeMessage);
    void update(NoticeMessage noticeMessage);
}
```

创建对应的 NoticeMessageMapper.xml，代码如下：

```xml
<!DOCTYPE mapper
        PUBLIC "-//mybatis.org//DTD Mapper 3.0//EN"
        "http://mybatis.org/dtd/mybatis-3-mapper.dtd">
<mapper namespace="org.javaboy.acl.mapper.NoticeMessageMapper">
    <select id="findAll" resultType="org.javaboy.acl.model.NoticeMessage">
        select * from system_message;
    </select>
    <select id="findById" resultType="org.javaboy.acl.model.NoticeMessage">
        select * from system_message where id=#{id};
    </select>
    <insert id="save" parameterType="org.javaboy.acl.model.NoticeMessage">
        insert into system_message (id,content) values (#{id},#{content});
    </insert>
    <update id="update" parameterType="org.javaboy.acl.model.NoticeMessage">
        update system_message set content = #{content} where id=#{id};
    </update>
</mapper>
```

这是 MyBatis 的基本操作，这里不再赘述。

接下来创建 NoticeMessageService，代码如下：

```java
@Service
public class NoticeMessageService {
    @Autowired
    NoticeMessageMapper noticeMessageMapper;
    @PostFilter("hasPermission(filterObject, 'READ')")
    public List<NoticeMessage> findAll() {
```

```java
        List<NoticeMessage> all = noticeMessageMapper.findAll();
        return all;
    }
    @PostAuthorize("hasPermission(returnObject, 'READ')")
    public NoticeMessage findById(Integer id) {
        return noticeMessageMapper.findById(id);
    }
    @PreAuthorize("hasPermission(#noticeMessage, 'CREATE')")
    public NoticeMessage save(NoticeMessage noticeMessage) {
        noticeMessageMapper.save(noticeMessage);
        return noticeMessage;
    }
    @PreAuthorize("hasPermission(#noticeMessage, 'WRITE')")
    public void update(NoticeMessage noticeMessage) {
        noticeMessageMapper.update(noticeMessage);
    }
}
```

通过在 Service 类上添加注解来实现权限控制，几个权限注解我们在本书第 13 章中都有介绍，不同的是这里注解中的表达式变成了 hasPermission：

- @PostFilter("hasPermission(filterObject, 'READ')")：在方法执行完成后，过滤返回的集合或数组，筛选出当前用户/角色具有 READ 权限的数据。filterObject 表示方法的返回的集合/数组中的元素。
- @PostAuthorize("hasPermission(returnObject, 'READ')")：在方法执行完成后，进行权限校验，如果表达式计算结果为 false，即当前用户/角色不具备返回对象的 READ 权限，将抛出异常。
- @PreAuthorize("hasPermission(#noticeMessage, 'CREATE')")：在方法调用之前，进行权限校验，判断当前用户/角色是否具备 noticeMessage 对象的 CREATE 权限。#noticeMessage 表示对方法参数 noticeMessage 的引用。
- @PreAuthorize("hasPermission(#noticeMessage, 'WRITE')")：在方法调用之前，进行权限校验，判断当前用户/角色是否具备 noticeMessage 对象的 WRITE 权限。

ACL 作为一种权限模型，底层实现原理还是本书第 13 章中所讲的权限实现原理，ACL 只是对这些原理的一个具体的应用，NoticeMessageService 中几个权限注解的原理这里就不再赘述了，hasPermission 表达式的具体实现则在 AclPermissionEvaluator#hasPermission 方法中。

配置完成，接下来我们进行测试。

14.2.4.4 测试

为了方便测试，我们首先准备几条测试数据，SQL 脚本如下：

```sql
INSERT INTO `acl_class` (`id`, `class`)
VALUES
    (1,'org.javaboy.acls.model.NoticeMessage');
INSERT INTO `acl_sid` (`id`, `principal`, `sid`)
```

```
VALUES
    (2,1,'hr'),
    (1,1,'manager'),
    (3,0,'ROLE_EDITOR');
INSERT INTO `system_message` (`id`, `content`)
VALUES
    (1,'111'),
    (2,'222'),
    (3,'333');
```

在测试数据中,首先向 acl_class 表中添加了一条类记录;然后添加了三个 Sid,两个是用户,一个是角色;最后添加了三个 NoticeMessage 实例。

目前没有任何用户/角色能够访问到 system_message 中的三条数据。例如,执行如下代码获取到的集合为空:

```
@Autowired
NoticeMessageService noticeMessageService;
@Test
@WithMockUser(username = "manager")
public void test01() {
    List<NoticeMessage> all = noticeMessageService.findAll();
    Assertions.assertEquals(0,all.size());
}
```

> **提　示**
>
> @WithMockUser(username = "manager") 表示以 manager 用户身份进行访问,这是为了方便使用单元测试。读者可以自己给 Spring Security 配置用户,然后在 Controller 中添加测试接口进行测试。

现在我们对其进行权限配置。

首先我们想设置让 hr 这个用户可以读取 system_message 表中 id 为 1 的记录,方式如下:

```
@Autowired
JdbcMutableAclService jdbcMutableAclService;
@Test
@WithMockUser(username = "javaboy")
@Transactional
@Rollback(value = false)
public void test02() {
    ObjectIdentity objectIdentity =
                    new ObjectIdentityImpl(NoticeMessage.class, 1);
    Permission p = BasePermission.READ;
    MutableAcl acl = jdbcMutableAclService.createAcl(objectIdentity);
    acl.insertAce(acl.getEntries().size(), p, new PrincipalSid("hr"), true);
    jdbcMutableAclService.updateAcl(acl);
}
```

这里配置的 mock user 是 javaboy，也就是这个 Acl 对象创建好之后，它的 owner 是 javaboy，但是我们前面 acl_sid 表的预设数据中没有 javaboy，所以上面这段代码执行后，会自动向 acl_sid 表中添加一条记录，值为 javaboy。

在方法执行过程中，会分别向 acl_entry、acl_object_identity 以及 acl_sid 三张表中添加记录，因此需要添加事务。因为我们是在单元测试中执行，事务会自动回滚，为了确保能够看到数据库中数据的变化，所以需要添加 @Rollback(value = false) 注解，让事务不要自动回滚。

在方法内部，首先分别创建 ObjectIdentity 和 Permission 对象，然后创建一个 Acl 对象，接下来调用 acl.insertAce 方法，将 ACE 保存到一个集合中，最后调用 updateAcl 方法去更新 Acl 对象。acl_object_identity 和 acl_entry 两张表的更新操作也将在 updateAcl 这个方法中完成。

该方法执行完成后，数据库中就会有相应的记录了。

接下来，使用 hr 这个用户就可以读取到 id 为 1 的记录了，代码如下：

```java
@Autowired
NoticeMessageService noticeMessageService;
@Test
@WithMockUser(username = "hr")
public void test03() {
    List<NoticeMessage> all = noticeMessageService.findAll();
    assertNotNull(all);
    assertEquals(1, all.size());
    assertEquals(1, all.get(0).getId());
    NoticeMessage byId = noticeMessageService.findById(1);
    assertNotNull(byId);
    assertEquals(1, byId.getId());
}
```

这里的演示用了两个方法。首先我们调用了 findAll，这个方法会查询出所有的数据，然后返回结果会被自动过滤，只剩下 hr 用户具有读取权限的数据，即 id 为 1 的数据；另一个调用的就是 findById 方法，传入参数为 1，这个好理解。

此时的 hr 用户并不具备修改对象的权限，我们可以继续使用上面的代码，让 hr 这个用户可以修改 id 为 1 的记录，代码如下：

```java
@Autowired
JdbcMutableAclService jdbcMutableAclService;
@Test
@WithMockUser(username = "javaboy")
@Transactional
@Rollback(value = false)
public void test04() {
    ObjectIdentity objectIdentity =
                    new ObjectIdentityImpl(NoticeMessage.class, 1);
    Permission p = BasePermission.WRITE;
    MutableAcl acl =
            (MutableAcl) jdbcMutableAclService.readAclById(objectIdentity);
    acl.insertAce(acl.getEntries().size(), p, new PrincipalSid("hr"), true);
```

```
    jdbcMutableAclService.updateAcl(acl);
}
```

和前面的 test02 方法相比,这里主要有两个变化:

- 定义的 Permission 对象是 WRITE。
- 调用 readAclById 方法来获取一个 Acl 对象。在 test02 方法中,由于当时创建的 objectIdentity 对象没有关联任何 Acl 对象,所以我们需要调用 createAcl 方法去创建一个全新的 Acl 对象,而经过前面的调用,现在我们创建的 objectIdentity 对象已经关联了相应的 Acl 对象了,所以这里调用 readAclById 方法去获取已经存在的 Acl 对象即可。

方法执行完毕后,我们再进行 hr 用户写权限的测试:

```
@Autowired
NoticeMessageService noticeMessageService;
@Test
@WithMockUser(username = "hr")
public void test05() {
    NoticeMessage msg = noticeMessageService.findById(1);
    assertNotNull(msg);
    assertEquals(1, msg.getId());
    msg.setContent("javaboy-1111");
    noticeMessageService.update(msg);
    msg = noticeMessageService.findById(1);
    assertNotNull(msg);
    assertEquals("javaboy-1111", msg.getContent());
}
```

假设现在想让 manager 这个用户去创建一个 id 为 99 的 NoticeMessage,默认情况下,manager 是没有这个权限的。我们现在可以为其赋权:

```
@Autowired
JdbcMutableAclService jdbcMutableAclService;
@Test
@WithMockUser(username = "javaboy")
@Transactional
@Rollback(value = false)
public void test06() {
    ObjectIdentity objectIdentity =
                    new ObjectIdentityImpl(NoticeMessage.class, 99);
    Permission p = BasePermission.CREATE;
    MutableAcl acl = jdbcMutableAclService.createAcl(objectIdentity);
    acl.insertAce(acl.getEntries().size(), p,
                            new PrincipalSid("manager"), true);
    jdbcMutableAclService.updateAcl(acl);
}
```

和 test02 相比,这里构造的 objectIdentity 对象 id 是 99,权限是 CREATE。该方法执行完毕后,接下来就可以使用 manager 用户来添加一条 id 为 99 的数据了:

```
@Autowired
NoticeMessageService noticeMessageService;
@Test
@WithMockUser(username = "manager")
public void test07() {
    NoticeMessage noticeMessage = new NoticeMessage();
    noticeMessage.setId(99);
    noticeMessage.setContent("999");
    noticeMessageService.save(noticeMessage);
}
```

执行该方法，就可以将 id 为 99 的数据添加到数据库中。添加成功后，manager 这个用户没有 id 为 99 的数据读取权限，可以参考前面案例自行添加。

14.2.4.5　Acl 小结

从上面的案例中可以感受到 ACL 是一种面向资源的访问控制模型，我们所做的所有的权限配置都是针对资源的。权限控制粒度也非常细，细到每一个资源的 CURD。当然，这么细的权限粒度也带来了权限数据量庞大的问题，不仅维护麻烦，而且扩展性也较弱。

14.3　RBAC

14.3.1　RBAC 权限模型介绍

RBAC（Role-based access control，基于角色的访问控制）是一种以角色为基础的访问控制，它是一种较新且广为使用的权限控制机制，这种机制不是直接给用户赋予权限，而是将权限赋予角色。

RBAC 权限模型将用户按角色进行归类，通过用户的角色来确定用户对某项资源是否具备操作权限。RBAC 简化了用户与权限的管理，它将用户与角色关联、角色与权限关联、权限与资源关联，这种模式使得用户的授权管理变得非常简单和易于维护。

RBAC 权限模型有三原则：

（1）最小权限：给角色配置的权限是其完成任务所需要的最小权限集合。

（2）职责分离：通过相互独立互斥的角色来共同完成任务。

（3）数据抽象：通过权限的抽象来体现，RBAC 支持的数据抽象程度与 RBAC 的实现细节有关。

RBAC 的应用非常广泛，除了常规的企业级应用，RBAC 也广泛应用在医疗、国防等领域。

关于 RBAC 的更多介绍，读者可以参考这个文档：https://csrc.nist.gov/projects/Role-Based-Access-Control。

14.3.2 RBAC 权限模型分类

RBAC 权限模型有四种不同的分类，我们分别来看一下。

RBAC0

RBAC0 是最简单的用户、角色、权限模型，也是 RBAC 权限模型中最核心的一部分，后面其他模型都是在此基础上建立的。在 RBAC0 中，一个用户可以具备多个角色，一个角色可以具备多个权限，最终用户所具备的权限是用户所具备的角色的权限并集，如图 14-8 所示。

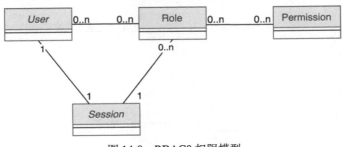

图 14-8 RBAC0 权限模型

RBAC1

RBAC1 则是在 RABC0 的基础上引入了角色继承，让角色有了上下级关系，如图 14-9 所示。

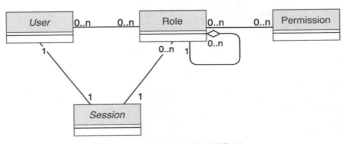

图 14-9 RBAC1 权限模型

角色继承实际上也非常容易，在本书第 13.3.2 小节中，我们已经介绍过角色继承的实现方式，这里不再赘述。

RBAC2

RBAC2 也是在 RBAC0 的基础上进行扩展，引入了静态职责分离和动态职责分离。要理解职责分离，得先理解角色互斥。在实际项目中，有一些角色是互斥的、对立的，例如，财务这个角色一般是不能和其他角色兼任的，否则自己报账自己审批。通过职责分离可以解决这一问题，职责分离有两种，静态职责分离和动态职责分离：

- 静态职责分离（Static Separation of Duty，SSD）：在权限配置阶段就做限制。比如，同一用户不能授予互斥的角色，用户只能有有限个角色，用户获得高级权限之前要有低级权限等。

- 动态职责分离（Dynamic Separation of Duty，DSD）：在运行阶段进行限制。比如，运行时同一用户下 5 个角色中只能同时有 2 个角色激活等。

RBAC2 权限模型如图 14-10 所示。

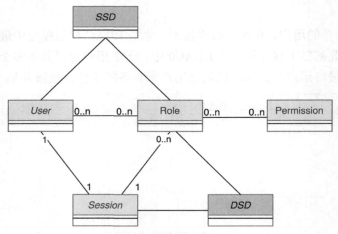

图 14-10　RBAC2 权限模型

RBAC3

RBAC3 是 RBAC1 和 RBAC2 的合体，如图 14-11 所示。

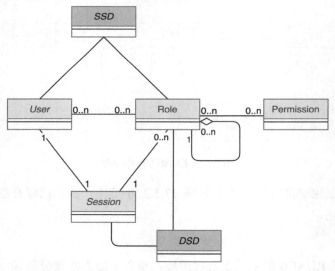

图 14-11　RBAC3 权限模型

这就是 RBAC 权限模型的四种不同分类。如果这四种模型依然不能满足我们的项目需求，开发者也可以在此基础上进行扩展，例如用户分组、角色分组等。

14.3.3 RBAC 小结

不同于 Acl 在使用时需要额外引入 spring-security-acl 依赖，RBAC 权限模型通过本书第 13 章中介绍的权限管理知识就可以快速实现。从技术角度来说，RBAC 用到的都是非常基础的权限管理知识，这也是 RBAC 比较受欢迎的原因之一，因为不需要再去额外学习新技术，利用已有的权限管理技术，结合 RBAC 这样一种"指导思想"，就能很快实现 RBAC 权限模型。所以这里就不给出案例了。有需要的读者可以关注本书技术支持微信公众号"江南一点雨"，回复 RBAC，查询笔者对于一些开源 RBAC 项目的分析文章。

14.4 小　结

权限模型主要是从权限设计层面去指导我们如何设计一个权限管理系统，至于技术点则没有新意，所以本章用到的技术，基本上还是第 13 章中介绍的。另外，不同的项目都会有不同的权限管理需求，在实际开发中要结合实际情况选择合适的权限模型，并对其进行适当的改造，不可拘泥于理论。

第 15 章

OAuth2

在当今的互联网应用中，OAuth2 是一个非常重要的认证协议，在很多场景下都会用到它，Spring Security 对 OAuth2 协议提供了相应的支持，开发者在 Spring Security 中可以非常方便地使用 OAuth2 协议，这也是 Spring Security 的魅力之一。

本章涉及的主要知识点有：

- OAuth2 简介。
- 四种授权模式。
- GitHub 授权登录。
- 授权服务器与资源服务器。
- 使用 JWT。

15.1 OAuth2 简介

OAuth 是一个开放标准，该标准允许用户让第三方应用访问该用户在某一网站上存储的私密资源（如头像、照片、视频等），并且在这个过程中无须将用户名和密码提供给第三方应用。通过令牌（token）可以实现这一功能，每一个令牌授权一个特定的网站在特定的时段内允许访问特定的资源。

OAuth 让用户可以授权第三方网站灵活访问它们存储在另外一些资源服务器上的特定信息，而非所有内容。对于用户而言，我们在互联网应用中最常见的 OAuth 应用就是各种第三方登录，例如 QQ 授权登录、微信授权登录、微博授权登录、GitHub 授权登录等。

例如用户想通过 QQ 登录今日头条，这时今日头条就是一个第三方应用，今日头条需要访问用户存储在 QQ 服务器上的一些基本信息，就需要得到用户的授权。如果用户把自己的 QQ 用户名/密码告诉今日头条，那么今日头条就能访问用户存储在 QQ 服务器上的所有数据，并且用户只有修改密码才能收回授权，这种授权方式安全隐患很大，如果使用 OAuth 协议就能很好地解决这一问题。

OAuth2 是 OAuth 协议的下一版本，但不兼容 OAuth 1.0。OAuth2 关注客户端开发者的简易性，同时为 Web 应用、桌面应用、移动设备、IoT 设备提供专门的认证流程。

15.2　OAuth2 四种授权模式

OAuth2 协议一共支持四种不同的授权模式：

（1）授权码模式：常见的第三方平台登录功能基本都是使用这种模式。

（2）简化模式：简化模式是不需要第三方服务端参与，直接在浏览器中向授权服务器申请令牌（token），如果网站是纯静态页面，则可以采用这种方式。

（3）密码模式：密码模式是用户把用户名/密码直接告诉客户端，客户端使用这些信息向授权服务器申请令牌（token）。这需要用户对客户端高度信任，例如客户端应用和服务提供商就是同一家公司。

（4）客户端模式：客户端模式是指客户端使用自己的名义而不是用户的名义向服务提供者申请授权。严格来说，客户端模式并不能算作 OAuth 协议解决问题的一种解决方案，但是，对于开发者而言，在一些为移动端提供的授权服务器上使用这种模式还是非常方便的。

如图 15-1 所示，无论哪种授权模式，其授权流程都是相似的，只不过在个别步骤上有所差异而已。

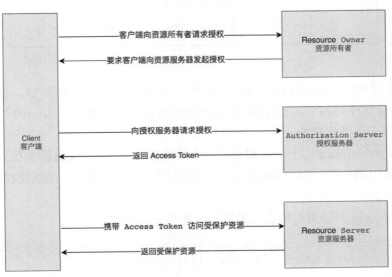

图 15-1　OAuth2 授权流程图

从图中我们可以看出 OAuth2 中包含四种不同的角色：
- Client：第三方应用。
- Resource Owner：资源所有者。
- Authorization Server：授权服务器。
- Resource Server：资源服务器。

接下来我们对四种授权模式进行详细介绍。

15.2.1 授权码模式

授权码模式（Authorization Code）是最安全并且使用最广泛的一种 OAuth2 授权模式，同时也是最复杂的一种授权模式，其具体的授权流程如图 15-2 所示（图片来自 RFC6749 文档，https://tools.ietf.org/html/rfc6749）。

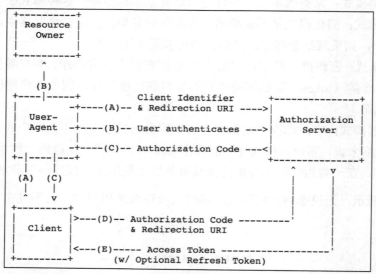

图 15-2 授权码模式流程图

这个流程图初看有点复杂，这里结合一个具体的案例来讲解：假设现在想给 www.javaboy.org 这个网站引入 GitHub 第三方登录功能，如果使用了 OAuth2 协议中的授权码模式，那么流程应该是这样的：

首先 www.javaboy.org 这个网站相当于一个第三方应用，在该网站的首页上放一个登录超链接，用户（服务方的用户，例如 GitHub 用户）单击这个超链接，就会去请求授权服务器（GitHub 的授权服务器）。

www.javaboy.org 网站首页的登录超链接可能是下面这样的：

```
https://github.com/oauth/authorize?response_type=code&client_id=javaboy&redirect_uri=www.javaboy.org&scope=all&state=123
```

这个请求链接中的参数比较多，我们解释一下：

- response_type：该参数表示授权类型，使用授权码模式的时候这里固定为 code，表示要求返回授权码，拿到授权码之后，再根据授权码去获取 Access Token。
- client_id：该参数表示客户端 id，也就是第三方应用的 id。有的读者可能对这个参数不理解，这里解释一下：如果想让 www.javaboy.org 接入 GitHub 第三方登录功能，开发者首先要去 GitHub 开放平台注册，去填入第三方应用的基本信息，信息填完之后，会获取到一个 APPID，在第三方应用发起授权请求时需要携带上该 APPID，也就是这里超链接中的 client_id 参数。从这里我们也可以看出，授权服务器在校验的时候，会做两件事：① 校验客户端的身份；② 校验用户身份。
- redirect_uri：该参数表示在登录校验成功/失败后，跳转的地址，即校验成功后，跳转到 www.javaboy.org 中的哪个页面。跳转的时候，还会携带上一个授权码参数，开发者再根据这个授权码获取 Access Token。
- scope：该参数表示授权范围，即 www.javaboy.org 这个第三方应用拿着获取到的 Access Token 能干什么。
- state：授权服务器会原封不动地返回该参数，通过对该参数的校验，可以防止 CSRF 攻击。

当用户单击登录超链接时，系统会将用户导入授权服务器的登录页面，这对应了图 15-2 中所示的步骤 A；用户选择是否给予授权，这对应了图 15-2 中所示的步骤 B；如果用户同意授权，则授权服务器会将页面重定向到 redirect_uri 指定的地址，同时携带一个授权码参数（如果一开始的链接提供了 state 参数，这里也会将 state 参数原封不动返回），这对应了图 15-2 中所示的步骤 C；根据步骤 C 中获取到的授权码，再结合自有的 client_id 和 grant_type、redirect_uri 等参数，向授权服务器请求令牌，这一步是在客户端的后端进行的，对用户不可见，这对应了图 15-2 中所示的步骤 D；授权服务器对参数进行校验之后，会返回 Access Token 和 Refresh Token，这个过程对应了图 15-2 中所示的步骤 E。

这就是授权码模式的工作流程。一般认为授权码模式是四种模式中最安全的一种模式，因为这种模式的 Access Token 不用经过浏览器或者移动端 App，是直接从项目的后端获取，并从后端发送到资源服务器上，这样就很大程度上减少了 Access Token 泄漏的风险。

15.2.2 简化模式

现在程序员搭建博客网站都流行静态网站，例如 Hexo、Jekyll 等，这类技术栈有一个共同的特点就是没有后端，开发者不需要将精力放在维护后端网站上，只需要专注于内容创作即可。

对于这种没有后端，只有页面的网站，如果想接入 GitHub 第三方登录功能，该怎么办呢？这就是本小节要介绍的简化模式（Implicit），图 15-3 所示表示简化模式的工作流程（图片来自 RFC6749 文档）。

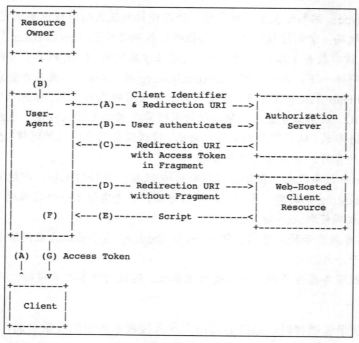

图 15-3　简化模式流程图

它的工作流程为：

首先在 www.javaboy.org 网站上有一个 GitHub 第三方登录的超链接，这个超链接如下所示：

```
https://github.com/oauth/authorize?response_type=token&client_id=javaboy&redirect_uri=www.javaboy.org&scope=all&state=123
```

这里的参数和前面授权码模式的参数基本相同，只有 response_type 的参数值不一样，这里是 token，表示要求授权服务器直接返回 Access Token，和授权码模式相比，这里省略了获取授权码那一步，所以叫作简化模式。

当用户单击登录超链接，系统会将用户导入授权服务器的登录页面，这对应了图 15-3 中所示的步骤 A；用户选择是否给予授权，这对应了图 15-3 中所示的步骤 B；如果用户同意授权，则授权服务器会将页面重定向到 redirect_uri 指定的地址，同时在 URL 中添加锚点参数携带 Access Token，如下所示：

```
http://localhost:8082/index.html#access_token=9fda1800-3b57-4d32-ad01-05ff700d44cc&token_type=bearer&expires_in=1940&state=123
```

这里采用锚点参数而不是 URL 地址参数，主要是为了避免中间人攻击，其中 token_type 表示令牌类型，expires_in 表示令牌过期时间，这对应了图 15-3 中所示的步骤 C。

接下来客户端向资源服务器发送请求，这个请求不需要携带令牌，这对应了图 15-3 中所示的步骤 D；资源服务器返回一段 JS 脚本，这对应了图 15-3 中所示的步骤 E；客户端执行 JS 脚本，提取出令牌 Access Token，这对应了图 15-3 中所示的步骤 F、步骤 G。

这就是简化模式的工作流程，它的弊端很明显，因为没有后端，所以非常不安全，除非对安全性要求不高，否则不建议使用。

15.2.3 密码模式

使用密码模式（Password）有一个前提就是高度信任第三方应用，只有高度信任第三方应用，才可以让用户在第三方应用的页面上输入登录凭证。我们来看一下密码模式的流程，如图 15-4 所示（图片来自 RFC6749 文档）。

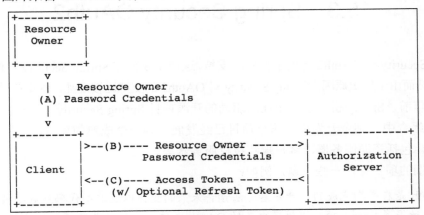

图 15-4　密码模式工作流程图

密码模式的流程比较简单：

首先用户在第三方应用的登录页面输入登录凭证（用户名/密码），这对应了图 15-4 中所示的步骤 A；第三方应用将用户的登录信息发送给授权服务器，获取到令牌 Access Token，这对应了图 15-4 中所示的步骤 B；授权服务器检查用户的登录信息，如果没有问题，则返回 Access Token 和 Refresh Token，这对应了图 15-4 中所示的步骤 C。

15.2.4 客户端模式

有的应用可能没有前端页面，只有一个后台，这种时候如果要用 OAuth2，就可以考虑客户端模式（Client Credentials），我们来看一个客户端模式的流程，如图 15-5 所示。

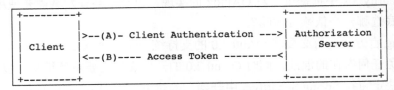

图 15-5　客户端模式流程图

客户端模式的流程很简单，只有两步：

（1）客户端发送一个请求到授权服务器，对应图 15-5 中所示的步骤 A。

（2）授权服务器验证通过后，会直接返回 Access Token 给客户端，对应图 15-5 中所示的步骤 B。

由于客户端模式是以客户端的名义向授权服务器请求令牌，所以授权服务器的令牌 Access Token 颁发给客户端而非用户。

以上就是 OAuth2 中的四种不同授权模式。

15.3　Spring Security OAuth2

Spring Security 对 OAuth2 提供了很好的支持，这使得我们在 Spring Security 中使用 OAuth2 非常方便。然而由于历史原因，Spring Security 对 OAuth2 的支持比较混乱，这里简单梳理一下。

大约十年前，Spring 引入了一个社区驱动的开源项目 Spring Security OAuth，并将其纳入 Spring 项目组合中。到今天为止，这个项目已经发展成为一个成熟的项目，可以支持大部分 OAuth 规范，包括资源服务器、客户端和授权服务器等。

然而早期的项目存在一些问题，例如：

- OAuth 是在早期完成的，开发者无法预料未来的变化以及这些代码到底要被怎么使用，这导致很多 Spring 项目提供了自己的 OAuth 支持，也就带来了 OAuth 支持的碎片化。
- 最早的 OAuth 项目同时支持 OAuth1.0 和 OAuth2.0，而现在 OAuth1.0 早已经不再使用，可以放弃了。
- 现在我们有更多的库可以选择，可以在这些库的基础上去开发，以便更好地支持 JWT 等新技术。

基于以上这些原因，官方决定重写 Spring Security OAuth，以便更好地协调 Spring 和 OAuth，并简化代码库，使 Spring 的 OAuth 支持更加灵活。

然而，在重写的过程中，发生了不少波折。

2018 年 1 月 30 日，Spring 官方发了一个通知，表示要逐渐停止现有的 OAuth2 支持，然后在 Spring Security 5 中构建下一代 OAuth2.0 支持。这么做的原因是因为当时 OAuth2 的落地方案比较混乱，在 Spring Security OAuth、Spring Cloud Security、Spring Boot 1.5.x 以及当时最新的 Spring Security 5.x 中都提供了对 OAuth2 的实现。以至于当开发者需要使用 OAuth2 时，不得不问，到底选哪一个依赖合适呢？

所以 Spring 官方决定有必要将 OAuth2.0 的支持统一到一个项目中，以便为用户提供明确的选择，并避免任何潜在的混乱，同时 OAuth2.0 的开发文档也要重新编写，以方便开发人员学习。所有的决定将在 Spring Security 5 中开始，构建下一代 OAuth2.0 的支持。

从那个时候起，Spring Security OAuth 项目就正式处于维护模式。官方将提供至少一年的错误/安全修复程序，并且会考虑添加次要功能，但不会添加主要功能。同时将 Spring Security OAuth 中的所有功能重构到 Spring Security 5.x 中。

到了 2019 年 11 月 14 日，Spring 官方又发布一个通知，这次的通知首先表示 Spring Security OAuth 在迁往 Spring Security 5.x 的过程非常顺利，大部分迁移工作已经完成了，剩下的将在 5.3 版本中完成迁移，在迁移的过程中还添加了许多新功能，包括对 OpenID Connect1.0 的支持。同时还宣布将不再支持授权服务器，不支持的原因有两个：

（1）在 2019 年，已经有大量的商业和开源授权服务器可用。

（2）授权服务器是使用一个库来构建产品，而 Spring Security 作为框架，并不适合做这件事情。

一石激起千层浪，许多开发者表示对此难以接受。这件事也在 Spring 社区引发了激烈的讨论，好在 Spring 官方愿意倾听来自社区的声音。

到了 2020 年 4 月 15 日，Spring 官方宣布启动 Spring Authorization Server 项目。这是一个由 Spring Security 团队领导的社区驱动的项目，致力于向 Spring 社区提供 Authorization Server 支持，也就是说，Spring 又重新支持授权服务器了。

2020 年 8 月 21 日，Spring Authorization Server 0.0.1 正式发布！

现在读者了解了 OAuth2 在 Spring 家族中的发展历程了。本章后面的案例中，客户端和资源服务器都将采用最新的方式来构建，授权服务器依然采用旧的方式来构建，因为目前的 Spring Authorization Server 0.0.1 功能较少且 BUG 较多。

一般来说，当我们在项目中使用 OAuth2 时，都是开发客户端，授权服务器和资源服务器都是由外部提供。例如我们想在 www.javaboy.org 这个网站上集成 GitHub 第三方登录，只需要开发自己的客户端即可，认证服务器和授权服务器都是由 GitHub 提供的。

15.4　GitHub 授权登录

我们通过一个 GitHub 授权登录来体验下 OAuth2 认证流程。

15.4.1　准备工作

首先我们需要将第三方应用的信息注册到 GitHub 上，打开 https://github.com/settings/developers 链接，单击 New OAuth App，注册一个新的应用，如图 15-6 所示。

图 15-6　单击 New OAuth App 注册一个新的应用

在打开的页面中,填入应用的基本信息:

- Application name:应用名称。
- Homepage URL:项目主页面。
- Application description:项目描述信息(可选)。
- Authorization callback URL:认证成功后的回调页面,默认的回调 URI 地址模板为 {baseUrl}/login/oauth2/code/{registrationId},其中 registrationId 是 ClientRegistration 的唯一标识符,如果这里使用了默认的回调地址,则在接下来的 Spring Boot 项目中就不必提供回调接口了。

如图 15-7 所示,信息填完之后,单击下方的 Register application 按钮完成注册。

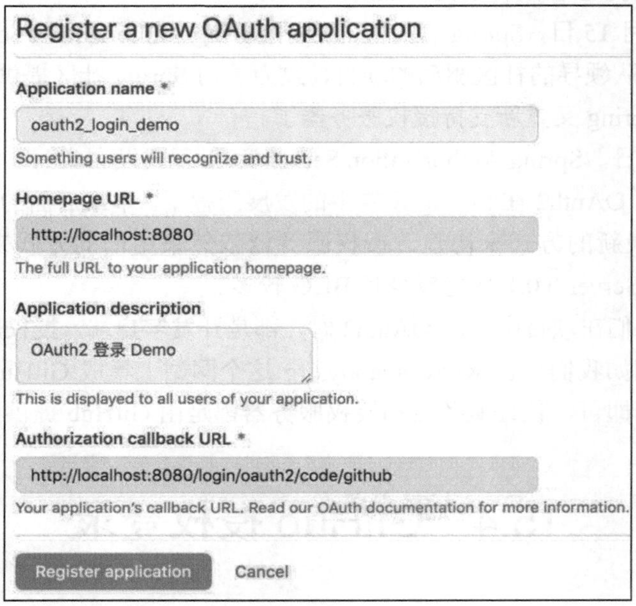

图 15-7 注册一个应用

注册成功后,会获取到一个 Client ID 和一个 Client Secret,如图 15-8 所示。

图 15-8 Client ID 和 Client Secret

保存好 Client ID 和 Client Secret，在接下来的项目中我们会用到这两个参数。

准备工作就算完成了。

15.4.2 项目开发

创建一个 Spring Boot 项目，引入 Web、Spring Security 以及 OAuth2 Client 依赖，如图 15-9 所示。

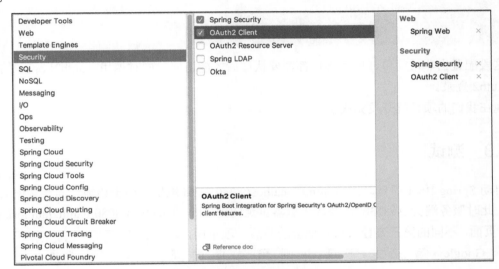

图 15-9　创建项目时添加三个依赖

项目创建成功后，在 application.properties 文件中配置刚刚申请到的 Client ID 和 Client Secret，代码如下：

```
spring.security.oauth2.client.registration.github.client-id=aa9e79846df9
cbc6201f
spring.security.oauth2.client.registration.github.client-secret=c324b934
43594fe84d106bb32c904799e1839e6a
```

接下来提供一个测试接口：

```
@RestController
public class HelloController {
    @GetMapping("/hello")
    public DefaultOAuth2User hello() {
        return ((DefaultOAuth2User) SecurityContextHolder.getContext()
                            .getAuthentication().getPrincipal());
    }
}
```

在测试接口中获取当前登录用户信息。注意，此时的用户对象是 DefaultOAuth2User，将获取到的当前登录用户对象返回。

最后我们再来简单配置一下 Spring Security：

```
@Configuration
public class SecurityConfig extends WebSecurityConfigurerAdapter {
    @Override
    protected void configure(HttpSecurity http) throws Exception {
        http.authorizeRequests()
                .anyRequest().authenticated()
                .and()
                .oauth2Login();
    }
}
```

这段配置也非常简单，所有接口都需要认证后才能访问，同时调用 oauth2Login()方法开启 OAuth2 登录。

现在我们的项目就开发完成了。

15.4.3 测试

启动 Spring Boot 项目，访问 http://localhost:8080/hello 地址，由于该地址需要认证后才能访问，此时服务端会返回 302，要求浏览器重定向到 http://localhost:8080/oauth2/authorization/github 页面。不同的第三方登录只是地址的最后一项不同，如果是 Google 第三方登录，则登录页面是 /oauth2/authorization/google。

当浏览器去请求该页面时，服务端检测到这是一个授权请求，于是再次返回 302，要求浏览器重定向到 GitHub 授权页面 https://github.com/login/oauth/authorize?response_type=code&client_id=aa9e79846df9cbc6201f&scope=read:user&state=vY8CJuRg2WlROVyo4mBZUn__1ksl6ieBjkmOQyGYA0A%3D&redirect_uri=http://localhost:8080/login/oauth2/code/github，这个 URL 地址的参数比较多，但都是我们前面 15.2.1 小节中介绍的授权码模式中的参数，因此这里不做过多解释。

接下来 GitHub 的授权服务器还会再次要求重定向，但是这就和我们这里的 OAuth2 没有关系了，最终来到 GitHub 认证页面，如图 15-10 所示。

图 15-10　GitHub 认证页面

输入用户名/密码，完成认证后，GitHub 授权服务器又会要求浏览器重定向到提前配置好的 Authorization callback URL 地址上，同时还会携带一个授权码参数 http://localhost:8080/login/oauth2/code/github?code=b15a5ed5198650e47e6a&state=cRYcu1Xg3E0sdlJAlbbi1CNeJT5-PdBOYVEaUkxcF8g%3D。

当浏览器请求该地址时，客户端会根据这里的授权码 code，向 GitHub 授权服务器的 https://github.com/login/oauth/access_token 接口去请求 Access Token，拿到 Access Token 之后，再向 https://api.github.com/user 地址发送请求，获取用户信息。

前面几步都是浏览器中可见的，最后获取令牌 Access Token 和获取用户信息的过程，是在后端完成的，浏览器将不可见。

所有工作都完成后，最终会自动跳转回 http://localhost:8080/hello 页面，在该页面可以看到用户登录成功后的信息，如图 15-11 所示。

图 15-11　认证成功后获取到的用户信息

15.4.4　原理分析

可以看到，接入 GitHub 第三方登录整个过程非常顺畅，开发者几乎不需要做什么事情，GitHub 上注册应用，项目中配置一下 Client ID 和 Client Secret，然后再开启一下 OAuth2 登录就可以了。

那么 Spring Security 如何得知 GitHub 授权地址、用户接口、令牌接口等信息？

由于用户接口、令牌接口、授权地址等信息一般不会轻易变化，所以 Spring Security 将一些常用的第三方登录如 Google、GitHub、Facebook、Okta 的信息收集起来，保存在一个枚举类 CommonOAuth2Provider 中，当我们在 application.properties 中配置 GitHub 时，就会自动选择枚举类中的 GITHUB。我们来看一下 CommonOAuth2Provider 中关于 GITHUB 信息的定义：

```
GITHUB {
    @Override
    public Builder getBuilder(String registrationId) {
        ClientRegistration.Builder builder = getBuilder(registrationId,
                ClientAuthenticationMethod.BASIC, DEFAULT_REDIRECT_URL);
        builder.scope("read:user");
        builder.authorizationUri("https://github.com/login/oauth/authorize");
        builder.tokenUri("https://github.com/login/oauth/access_token");
        builder.userInfoUri("https://api.github.com/user");
        builder.userNameAttributeName("id");
        builder.clientName("GitHub");
        return builder;
    }
},
```

可以看到，需要用到的地址都提前定义好了。

当我们开启 OAuth2 自动登录之后，在 Spring Security 过滤器链中多了两个过滤器：

（1）OAuth2AuthorizationRequestRedirectFilter

（2）OAuth2LoginAuthenticationFilter

回顾一下 15.2.1 小节所讲的授权码模式工作流程，当用户在没有登录时就去访问 http://localhost:8080/hello 地址，会被自动导入到 GitHub 授权页面，这个过程是由 OAuth2AuthorizationRequestRedirectFilter 过滤器完成的。

接下来用户进行 GitHub 登录，登录成功后，GitHub 授权服务器会调用回调地址，同时返回一个授权码，客户端再根据授权码去 GitHub 授权服务器上获取 Access Token，有了 Access Token 就可以获取用户信息了，这个过程是由 OAuth2LoginAuthenticationFilter 过滤器来完成的。

接下来我们对这里涉及的几个关键类进行简单分析。

OAuth2ClientRegistrationRepositoryConfiguration

OAuth2ClientRegistrationRepositoryConfiguration 是一个配置类，当项目启动时，该类会自动加载，并向 Spring 容器中注册一个 InMemoryClientRegistrationRepository 实例，该实例保存了客户端注册表信息，代码如下：

```
@Configuration(proxyBeanMethods = false)
@EnableConfigurationProperties(OAuth2ClientProperties.class)
@Conditional(ClientsConfiguredCondition.class)
class OAuth2ClientRegistrationRepositoryConfiguration {
    @Bean
    @ConditionalOnMissingBean(ClientRegistrationRepository.class)
    InMemoryClientRegistrationRepository
        clientRegistrationRepository(OAuth2ClientProperties properties) {
```

```
            List<ClientRegistration> registrations = new ArrayList<>(
                    OAuth2ClientPropertiesRegistrationAdapter
                            .getClientRegistrations(properties).values());
            return new InMemoryClientRegistrationRepository(registrations);
    }
}
```

可以看到，clientRegistrationRepository 方法的参数实际上就是我们在 application.properties 中配置的 GitHub 的 Client ID 和 Client Secret。接下来调用 getClientRegistrations 方法，会将 CommonOAuth2Provider 枚举类中预设的 GitHub 信息和用户配置的 GitHub 信息合并然后返回。如果 application.properties 中只是配置了 GitHub 信息，则这里的 registrations 集合中就只有一项；如果 application.properties 中还配置了 Facebook、Google 等信息，则 registrations 集合中就包含多项。

OAuth2AuthorizationRequestRedirectFilter

OAuth2AuthorizationRequestRedirectFilter 过滤器主要是判断当前请求是否是授权请求，如果是授权请求，则进行重定向到 GitHub 授权页面，否则执行下一个过滤器。

我们来看一下该过滤器的 doFilterInternal 方法：

```
@Override
protected void doFilterInternal(HttpServletRequest request,
                                HttpServletResponse response,
        FilterChain filterChain) throws ServletException, IOException {
    try {
        OAuth2AuthorizationRequest authorizationRequest =
                    this.authorizationRequestResolver.resolve(request);
        if (authorizationRequest != null) {
            this.sendRedirectForAuthorization(request, response,
                                            authorizationRequest);
            return;
        }
    } catch (Exception failed) {
        //省略其他
    }
    try {
        filterChain.doFilter(request, response);
    } catch (IOException ex) {
    //省略其他
    }
}
```

首先调用 authorizationRequestResolver.resolve 方法将当前请求解析为一个 OAuth2AuthorizationRequest 对象：如果当前请求是授权请求（如 http://localhost:8080/oauth2/authorization/github），则根据 InMemoryClientRegistrationRepository 中保存的客户端注册表信息，构造一个 OAuth2AuthorizationRequest 对象并返回；如果当前请求不是授权请求，而是一

个普通请求，则这里返回的 OAuth2AuthorizationRequest 对象为 null。

如果获取到的 authorizationRequest 对象不为 null，即当前请求是授权请求，则调用 sendRedirectForAuthorization 方法进行重定向，重定向的地址就是 GitHub 的授权地址（即枚举类 CommonOAuth2Provider 中 authorizationUri 方法所配置的地址）。当然这里的地址会在该地址上再自动加上 response_type、client_id、scope、state 以及 redirect_uri 参数（这些参数都可以从枚举类中获取）。另外，在重定向之前，还会将当前授权请求保存到一个 Map 集合中，并将 Map 集合保存到 HttpSession 中，以备后续使用。

OAuth2LoginAuthenticationFilter

通过前面的讲解，可能有读者会疑惑，GitHub 授权服务器登录成功后的回调地址是 http://localhost:8080/login/oauth2/code/github，但是我们的项目中并没有定义这样一个接口，为什么还能调用成功呢？这就是 OAuth2LoginAuthenticationFilter 过滤器所起的作用了！

OAuth2LoginAuthenticationFilter 继承自 AbstractAuthenticationProcessingFilter，它目前的角色相当于我们之前所讲的 UsernamePasswordAuthenticationFilter 过滤器的角色。在 AbstractAuthenticationProcessingFilter 过滤器中会拦截下认证请求进行处理。我们来看一下 AbstractAuthenticationProcessingFilter#doFilter 方法：

```
public void doFilter(ServletRequest req, ServletResponse res,
            FilterChain chain)throws IOException, ServletException {
//省略
    if (!requiresAuthentication(request, response)) {
        chain.doFilter(request, response);
        return;
    }
//省略
}
```

如果使用了 OAuth2 登录，这里的逻辑就是判断当前请求接口是否是 /login/oauth2/code/* 格式，如果是，说明这是一个认证请求，将该请求拦截下来交给 OAuth2LoginAuthenticationFilter #attemptAuthentication 方法去处理；如果不是，则继续执行下一个过滤器。

我们来看一下 OAuth2LoginAuthenticationFilter#attemptAuthentication 方法：

```
public Authentication attemptAuthentication(HttpServletRequest request,
        HttpServletResponse response)throws AuthenticationException {
//1
MultiValueMap<String, String> params =
    OAuth2AuthorizationResponseUtils.toMultiMap(request.getParameterMap());
    if (!OAuth2AuthorizationResponseUtils.isAuthorizationResponse(params)) {
        //省略
    }
//2
    OAuth2AuthorizationRequest authorizationRequest =
                    this.authorizationRequestRepository
                        .removeAuthorizationRequest(request, response);
```

```java
    if (authorizationRequest == null) {
        //省略
    }
//3
    String registrationId =
      authorizationRequest.getAttribute(OAuth2ParameterNames.REGISTRATION_ID);
    ClientRegistration clientRegistration =
      this.clientRegistrationRepository.findByRegistrationId(registrationId);
    if (clientRegistration == null) {
        //省略
    }
//4
    String redirectUri =
      UriComponentsBuilder.fromHttpUrl(UrlUtils.buildFullRequestUrl(request))
        .replaceQuery(null)
        .build()
        .toUriString();
    OAuth2AuthorizationResponse authorizationResponse =
            OAuth2AuthorizationResponseUtils.convert(params, redirectUri);
    Object authenticationDetails =
                this.authenticationDetailsSource.buildDetails(request);
    OAuth2LoginAuthenticationToken authenticationRequest = new
                OAuth2LoginAuthenticationToken(clientRegistration,
                new OAuth2AuthorizationExchange(authorizationRequest,
                                            authorizationResponse));
    authenticationRequest.setDetails(authenticationDetails);
    OAuth2LoginAuthenticationToken authenticationResult =
        (OAuth2LoginAuthenticationToken)
            this.getAuthenticationManager().authenticate(authenticationRequest);
    OAuth2AuthenticationToken oauth2Authentication =
        new OAuth2AuthenticationToken(
            authenticationResult.getPrincipal(),
            authenticationResult.getAuthorities(),
            authenticationResult.getClientRegistration().getRegistrationId());
    oauth2Authentication.setDetails(authenticationDetails);
    OAuth2AuthorizedClient authorizedClient = new OAuth2AuthorizedClient(
        authenticationResult.getClientRegistration(),
        oauth2Authentication.getName(),
        authenticationResult.getAccessToken(),
        authenticationResult.getRefreshToken());
    this.authorizedClientRepository.saveAuthorizedClient(authorizedClient,
                            oauth2Authentication, request, response);
    return oauth2Authentication;
}
```

注释 1 是对请求参数进行校验，请求必须包含授权码 code 和 state 两个参数，否则会抛出异常。

注释 2 是从 HttpSession 中获取在 OAuth2AuthorizationRequestRedirectFilter 过滤器中保存

的授权请求，如果获取到的对象为 null，则抛出异常。

注释 3 是检查当前注册应用中是否有授权请求时的应用，如果没有，则抛出异常。

注释 4 是构造一个未经认证的 OAuth2LoginAuthenticationToken 对象，并调用 authenticate 方法进行认证。认证成功后，最终封装成一个 OAuth2AuthenticationToken 对象并返回。这一块读者可以回顾本书 3.1.4 小节的分析，流程都是一样的，只是实现细节有所差异而已。需要注意的是，这里认证时调用的 AuthenticationProvider 是 OAuth2LoginAuthenticationProvider。

OAuth2LoginAuthenticationProvider

OAuth2LoginAuthenticationProvider 负责最终的校验工作，作用类似 3.1.2 小节所讲的 DaoAuthenticationProvider，我们来看一下它的 authenticate 方法：

```java
public Authentication authenticate(Authentication authentication)
                            throws AuthenticationException {
    OAuth2LoginAuthenticationToken loginAuthenticationToken =
                    (OAuth2LoginAuthenticationToken) authentication;
//1
    if (loginAuthenticationToken.getAuthorizationExchange()
        .getAuthorizationRequest().getScopes().contains("openid")) {
        return null;
    }
//2
    OAuth2AuthorizationCodeAuthenticationToken
                            authorizationCodeAuthenticationToken;
    try {
        authorizationCodeAuthenticationToken =
            (OAuth2AuthorizationCodeAuthenticationToken)
            this.authorizationCodeAuthenticationProvider
            .authenticate(new OAuth2AuthorizationCodeAuthenticationToken(
                loginAuthenticationToken.getClientRegistration(),
                loginAuthenticationToken.getAuthorizationExchange()));
    } catch (OAuth2AuthorizationException ex) {
    }
//3
    OAuth2AccessToken accessToken =
                authorizationCodeAuthenticationToken.getAccessToken();
    Map<String, Object> additionalParameters =
        authorizationCodeAuthenticationToken.getAdditionalParameters();
//4
OAuth2User oauth2User = this.userService.loadUser(new OAuth2UserRequest(
                loginAuthenticationToken.getClientRegistration(),
                    accessToken, additionalParameters));
    Collection<? extends GrantedAuthority> mappedAuthorities =
        this.authoritiesMapper.mapAuthorities(oauth2User.getAuthorities());
//5
OAuth2LoginAuthenticationToken authenticationResult =
    new OAuth2LoginAuthenticationToken(
```

```
            loginAuthenticationToken.getClientRegistration(),
            loginAuthenticationToken.getAuthorizationExchange(),
            oauth2User,
            mappedAuthorities,
            accessToken,
            authorizationCodeAuthenticationToken.getRefreshToken());
    authenticationResult.setDetails(loginAuthenticationToken.getDetails());
    return authenticationResult;
}
```

注释 1 是判断是否为 OpenID Connect 认证，如果是，则返回 null，请求交给 OidcAuthorizationCodeAuthenticationProvider 去处理。

注释 2 是根据授权码 code 去请求 https://github.com/login/oauth/access_token 接口获取 Access Token，这一步调用到了 OAuth2AuthorizationCodeAuthenticationProvider#authenticate 方法，并在该方法中调用 DefaultAuthorizationCodeTokenResponseClient#getTokenResponse 方法发起网络请求，底层使用的网络请求工具是 RestTemplate。

注释 3 是根据上一步的结果，提取出 accessToken 对象。

注释 4 是根据获取到的 accessToken 对象，向 https://api.github.com/user 地址发起请求，获取用户信息，并最终封装为一个 OAuth2User 对象。

注释 5 是构造一个 OAuth2LoginAuthenticationToken 对象并返回。

我们在 GitHub 上配置的 http://localhost:8080/login/oauth2/code/github 地址其实类似于登录请求，当 GitHub 授权服务器重定向到该地址时，重定向请求携带了授权码参数，客户端根据授权码获取 Access Token，再根据 Access Token 加载到用户对象，最终构建 OAuth2LoginAuthenticationToken 并返回。

至此，整个 GitHub 授权登录就分析完了，我们再结合 15.2.1 小节中介绍的授权码模式的工作流程，应该就很好理解了。

15.4.5 自定义配置

15.4.5.1 自定义 ClientRegistrationRepository

完全使用自动化配置虽然方便，但是灵活性却降低了。假如我们在 GitHub 上注册 App 时，填写的回调地址不是 http://localhost:8080/login/oauth2/code/github，而是其他地址，此时就需要我们手动配置了。

举个简单例子，假设我们在 GitHub 上注册 App 时填写的回调地址是 http://localhost:8080/authorization_code，那么可以通过如下方式配置客户端。

在 application.properties 文件中修改重定向地址：

```
spring.security.oauth2.client.registration.github.client-id=aa9e79846df9
cbc6201f
spring.security.oauth2.client.registration.github.client-secret=c324b934
43594fe84d106bb32c904799e1839e6a
```

```
spring.security.oauth2.client.registration.github.redirect-uri=http://lo
calhost:8080/authorization_code
```

然后修改认证请求处理地址:

```
@Configuration
public class SecurityConfig extends WebSecurityConfigurerAdapter {
    @Override
    protected void configure(HttpSecurity http) throws Exception {
        http.authorizeRequests()
                .anyRequest().authenticated()
                .and()
                .oauth2Login()
                .loginProcessingUrl("/authorization_code");
    }
}
```

前面我们已经分析过，GitHub 配置的回调地址相当于登录请求链接，默认的 loginProcessingUrl 是 /login/oauth2/code/github，所以默认情况下，当重定向到 http://localhost:8080/authorization_code 地址时，该请求会被当成一个普通请求，无法在 OAuth2LoginAuthenticationFilter 过滤器中进行登录处理。在只有修改 loginProcessingUrl 地址才能确保当重定向到 http://localhost:8080/authorization_code 地址时，该请求会在 AbstractAuthenticationProcessingFilter#doFilter 方法中被认定为一个登录请求，进而将请求交给 OAuth2LoginAuthenticationFilter#attemptAuthentication 方法去处理，以完成登录操作。

这是一种自定义配置的方式。

我们也可以使用 Java 代码，完成更加丰富、更加灵活的配置。

使用 Java 代码配置时，可以删除 application.properties 中的所有配置，然后修改配置类，代码如下:

```
@Configuration
public class SecurityConfig extends WebSecurityConfigurerAdapter {
    @Override
    protected void configure(HttpSecurity http) throws Exception {
        http.authorizeRequests()
                .anyRequest().authenticated()
                .and()
                .oauth2Login()
                .loginProcessingUrl("/authorization_code");
    }
    @Bean
    public ClientRegistrationRepository clientRegistrationRepository() {
        return new
InMemoryClientRegistrationRepository(githubClientRegistration());
    }
    private ClientRegistration githubClientRegistration() {
        return ClientRegistration.withRegistrationId("github")
                .clientId("aa9e79846df9cbc6201f")
```

```
                .clientSecret("c324b93443594fe84d106bb32c904799e1839e6a")
                .clientAuthenticationMethod(ClientAuthenticationMethod.BASIC)
                .userNameAttributeName("id")
                .authorizationGrantType(AuthorizationGrantType.AUTHORIZATION_CODE)
                .redirectUriTemplate("http://localhost:8080/authorization_code")
                .scope("read:user")
                .authorizationUri("https://github.com/login/oauth/authorize")
                .tokenUri("https://github.com/login/oauth/access_token")
                .userInfoUri("https://api.github.com/user")
                .clientName("GitHub")
                .build();
    }
}
```

我们只需要向 Spring 容器中注册一个 ClientRegistrationRepository 实例,然后在该实例中提供 GitHub 的配置信息即可,此时 OAuth2ClientRegistrationRepositoryConfiguration 配置类自动配置的 ClientRegistrationRepository 实例就会失效。

这种配置方式非常直观也非常灵活,所有需要的配置信息现在都摆出来了,需要修改哪个直接修改即可。

15.4.5.2 自定义用户

默认情况下,GitHub 返回的用户信息被包装成一个 DefaultOAuth2User 对象,但是 DefaultOAuth2User 是通过一个 Map 集合来保存 GitHub 用户信息,这样解析起来并不方便,因此我们也可以自定义用户对象。

自定义用户对象实现 OAuth2User 接口即可,代码如下:

```java
public class GitHubOAuth2User implements OAuth2User {
    private List<GrantedAuthority> authorities =
            AuthorityUtils.createAuthorityList("ROLE_USER");
    private Map<String, Object> attributes;
    private String id;
    private String name;
    private String login;
    private String email;
    @Override
    public Collection<? extends GrantedAuthority> getAuthorities() {
        return this.authorities;
    }
    @Override
    public Map<String, Object> getAttributes() {
        if (this.attributes == null) {
            this.attributes = new HashMap<>();
            this.attributes.put("id", this.getId());
            this.attributes.put("name", this.getName());
            this.attributes.put("login", this.getLogin());
            this.attributes.put("email", this.getEmail());
```

```
        }
        return attributes;
    }
    public String getId() {
        return this.id;
    }
    public void setId(String id) {
        this.id = id;
    }
    @Override
    public String getName() {
        return this.name;
    }
    public void setName(String name) {
        this.name = name;
    }
    public String getLogin() {
        return this.login;
    }
    public void setLogin(String login) {
        this.login = login;
    }
    public String getEmail() {
        return this.email;
    }
    public void setEmail(String email) {
        this.email = email;
    }
}
```

这里定义的 id、name、login 以及 email 属性，都是 GitHub 返回的用户信息中所包含的，如果还想映射其他用户信息，则继续定义相应的属性即可。最后在配置类中使用该自定义用户对象：

```
protected void configure(HttpSecurity http) throws Exception {
    http.authorizeRequests()
            .anyRequest().authenticated()
            .and()
            .oauth2Login()
            .userInfoEndpoint()
            .customUserType(GitHubOAuth2User.class,"github")
            .and()
            .loginProcessingUrl("/authorization_code");
}
```

配置完成后，在通过 Access Token 去加载用户信息这一环节中，将不再使用 DefaultOAuth2UserService 类去完成加载，而是使用 CustomUserTypesOAuth2UserService，该类支持自定义用户对象。

配置完成后，重启项目完成认证，此时再从 SecurityContextHolder 中提取出来的用户对象就不再是 DefaultOAuth2User，而是 GitHubOAuth2User 了。

15.5　授权服务器与资源服务器

前面的 GitHub 授权登录主要向大家展示了 OAuth2 中客户端的工作模式。对于大部分的开发者而言，日常接触到的 OAuth2 都是开发客户端，例如接入 QQ 登录、接入微信登录等。不过也有少量场景，可能需要开发者提供授权服务器与资源服务器，接下来我们就通过一个完整的案例向大家演示如何搭建授权服务器与资源服务器。

搭建授权服务器，我们可以选择一些现成的开源项目，直接运行即可，例如：

- Keycloak：RedHat 公司提供的开源工具，提供了很多实用功能，例如单点登录、支持 OpenID、可视化后台管理等。
- Apache Oltu：Apache 上的开源项目，最近几年没怎么维护了。

这里随便举出两例，类似的开源项目很多（这也是 Spring Security 官方一开始说不提供授权服务器的原因之一）。企业应用中，建议使用成熟的开源项目搭建授权服务器。

当然我们也可以使用 Spring Security 最新发布的 Spring Authorization Server 来搭建授权服务器。截至本书写作时，spring-authorization-server 发布了 0.0.1 版，但是这个版本功能较少而且问题较多，因此这里依然使用较早的 spring-security-oauth2 来搭建授权服务器，可能有一些类过期了，不过这不影响大家理解授权服务器的功能。

接下来我们将搭建一个包含授权服务器、资源服务器以及客户端在内的 OAuth2 案例。

15.5.1　项目规划

首先把项目分为三部分：

- 授权服务器：采用较早的 spring-security-oauth2 来搭建授权服务器。
- 资源服务器：采用最新的 Spring Security 5.x 搭建资源服务器。
- 客户端：采用最新的 Spring Security 5.x 搭建客户端。

同时为了避免测试时互相影响，我们需要修改电脑的 hosts 文件，在 hosts 文件中增加如下解析规则：

```
127.0.0.1       auth.javaboy.org
127.0.0.1       res.javaboy.org
127.0.0.1       client.javaboy.org
```

- auth.javaboy.org：表示授权服务器域名。
- res.javaboy.org：表示资源服务器域名。

- client.javaboy.org：表示客户端域名。

为了完整地演示 OAuth2 案例，我们一共需要三个项目，其中：
- 授权服务器端口为 8881。
- 资源服务器端口为 8882。
- 客户端端口为 8883。

项目规划完成。

15.5.2 项目搭建

15.5.2.1 授权服务器搭建

创建一个名为 auth_server 的 Spring Boot 项目，引入 Web 依赖和 spring-security-oauth2 依赖，代码如下：

```xml
<dependency>
    <groupId>org.springframework.boot</groupId>
    <artifactId>spring-boot-starter-web</artifactId>
</dependency>
<dependency>
    <groupId>org.springframework.security.oauth</groupId>
    <artifactId>spring-security-oauth2</artifactId>
    <version>2.5.0.RELEASE</version>
</dependency>
```

接下来提供一个 Spring Security 的基本配置：

```java
@Configuration
public class SecurityConfig extends WebSecurityConfigurerAdapter {
    @Bean
    PasswordEncoder passwordEncoder() {
        return new BCryptPasswordEncoder();
    }
    @Override
    protected void configure(AuthenticationManagerBuilder auth)
                                                    throws Exception {
        auth.inMemoryAuthentication()
            .withUser("javaboy")
            .password(passwordEncoder().encode("123"))
            .roles("ADMIN")
            .and()
            .withUser("sang")
            .password(passwordEncoder().encode("123"))
            .roles("USER");
    }
    @Override
    @Bean
```

```
    public AuthenticationManager authenticationManagerBean()
                                                    throws Exception {
        return super.authenticationManagerBean();
    }
    @Override
    protected void configure(HttpSecurity http) throws Exception {
        http.csrf().disable().formLogin();
    }
}
```

为了方便起见，这里的用户直接创建在内存中，一共两个用户 javaboy/123 和 sang/123，角色分别是 ADMIN 和 USER。这里配置的用户就是我们项目的用户，例如用 GitHub 登录第三方网站，在这个过程中，需要先从 GitHub 获取授权，登录 GitHub 需要用户名/密码信息，这里配置的用户相当于 GitHub 的用户。

另外，由于我们希望让这个授权服务器同时支持授权码模式、简化模式、密码模式以及客户端模式，在支持密码模式时，需要用到 AuthenticationManager 实例，所以在这里暴露出一个 AuthenticationManager 实例。

基本的用户信息配置完成后，接下来我们来配置授权服务器：

```
//1
@Configuration
public class AccessTokenConfig {
    @Bean
    TokenStore tokenStore() {
        return new InMemoryTokenStore();
    }
}
//2
@EnableAuthorizationServer
@Configuration
public class AuthorizationServer extends AuthorizationServerConfigurerAdapter {
    @Autowired
    TokenStore tokenStore;
    @Autowired
    ClientDetailsService clientDetailsService;
    @Autowired
    AuthenticationManager authenticationManager;
    @Autowired
    PasswordEncoder passwordEncoder;
    //3
    @Bean
    AuthorizationServerTokenServices tokenServices() {
        DefaultTokenServices services = new DefaultTokenServices();
        services.setClientDetailsService(clientDetailsService);
        services.setSupportRefreshToken(true);
        services.setTokenStore(tokenStore);
        services.setAccessTokenValiditySeconds(60 * 60 * 2);
```

```java
            services.setRefreshTokenValiditySeconds(60 * 60 * 24 * 3);
            return services;
        }
        //4
        @Override
        public void configure(AuthorizationServerSecurityConfigurer security)
                                                        throws Exception {
            security.checkTokenAccess("permitAll()")
                    .allowFormAuthenticationForClients();
        }
        //5
        @Override
        public void configure(ClientDetailsServiceConfigurer clients)
                                                        throws Exception {
            clients.inMemory()
                    .withClient("my_client")
                    .secret(passwordEncoder.encode("123"))
                    .authorizedGrantTypes("authorization_code","refresh_token",
                              "implicit","password","client_credentials")
                    .scopes("read:user","read:msg")
                    .redirectUris("http://client.javaboy.org:8883/login/oauth2/code/javaboy");
        }
        //6
        @Override
        public void configure(AuthorizationServerEndpointsConfigurer endpoints)
                                                        throws Exception {
            endpoints
                    .authenticationManager(authenticationManager)
                    .authorizationCodeServices(authorizationCodeServices())
                    .tokenServices(tokenServices());
        }
        @Bean
        AuthorizationCodeServices authorizationCodeServices() {
            return new InMemoryAuthorizationCodeServices();
        }
    }
}
```

这段配置比较长，我们来逐个解释一下：

- 注释 1 中配置了一个 TokenStore 的实例，这是配置生成的 Access Token 要保存到哪里，可以存在内存中，也可以存在 Redis 中，如果用到了 JWT，就不需要保存了。这里我们配置的实例是 InMemoryTokenStore，即生成的令牌存在内存中。
- 注释 2 创建了一个 AuthorizationServer 类继承自 AuthorizationServerConfigurerAdapter，用来对授权服务器做进一步的详细配置，配置类上通过 @EnableAuthorizationServer 注解开启授权服务器的自动化配置。在该配置类中，主要重写三个 configure 方法。
- 注释 3 配置向 Spring 容器中注册了一个 AuthorizationServerTokenServices 实例，该实例

主要配置了生成的 Access Token 的一些基本信息：例如 Access Token 是否支持刷新、Access Token 的存储位置、Access Token 的有效期以及 Refresh Token 的有效期等。
- 注释 4 配置了令牌端点的安全约束，这里设置了 checkTokenAccess 端点可以自由访问。该端点的作用是当资源服务器收到 Access Token 之后，需要去授权服务器校验 Access Token 的合法性，就会访问这个端点。
- 注释 5 配置客户端的详细信息，需要提前在这里配置好客户端信息，这就类似于 GitHub 第三方登录时，我们需要提前在 GitHub 上注册我们的应用信息。客户端的信息可以存在数据库中，也可以存在内存中，这里保存在内存中，分别配置了客户端的 id、secret、授权类型、授权范围以及重定向 uri。OAuth2 四种授权类型不包含 refresh_token 这种类型，但是在 Spring Security 实现中，refresh_token 也被算作一种。
- 注释 6 配置授权码服务和令牌服务。authorizationCodeServices 用来配置授权码（code）的存储，tokenServices 用来配置令牌的存储。

最后将该项目的端口修改为 8881。至此，我们的授权服务器就算搭建成功了。

15.5.2.2　资源服务器搭建

在资源服务器搭建之前，我们需要了解 Access Token 令牌，它可以分为两种：
- 透明令牌，如 JWT。
- 不透明令牌。

透明令牌是指令牌本身就携带了用户信息，不透明则是指令牌本身是一个无意义的字符串。如果是透明令牌，如 JWT，那么资源服务器在收到令牌之后，可以自行解析并校验；如果是不透明令牌，那么资源服务器在收到令牌之后，就只能调用授权服务器的端口去校验令牌是否合法。由于前面搭建的授权服务器使用的是不透明令牌，所以这里资源服务器中对令牌的处理也按不透明令牌来处理。

接下来开始搭建资源服务器，我们采用目前最新的方案来搭建。

首先创建一个名为 res_server 的项目，添加 Web、Spring Security 以及 OAuth2 Resource 依赖，最终的 pom.xml 文件内容如下：

```xml
<dependency>
    <groupId>org.springframework.boot</groupId>
    <artifactId>spring-boot-starter-oauth2-resource-server</artifactId>
</dependency>
<dependency>
    <groupId>org.springframework.boot</groupId>
    <artifactId>spring-boot-starter-security</artifactId>
</dependency>
<dependency>
    <groupId>org.springframework.boot</groupId>
    <artifactId>spring-boot-starter-web</artifactId>
</dependency>
<dependency>
    <groupId>com.nimbusds</groupId>
```

```xml
        <artifactId>oauth2-oidc-sdk</artifactId>
        <version>6.23</version>
</dependency>
```

项目创建成功后,在 application.yml 文件中配置令牌解析路径以及客户端 id 和 secret:

```yaml
spring:
  security:
    oauth2:
      resourceserver:
        opaque:
          introspection-uri: http://auth.javaboy.org:8881/oauth/check_token
          introspection-client-id: my_client
          introspection-client-secret: 123
server:
  port: 8882
```

introspection-uri 属性配置的就是令牌校验地址,客户端从授权服务器上申请到令牌之后,拿着令牌来资源服务器读取数据,资源服务器收到令牌后,调用该地址去校验令牌是否合法。

接下来配置资源服务器:

```java
@Configuration
public class OAuth2ResourceServerSecurityConfiguration extends
                                            WebSecurityConfigurerAdapter {
@Value("${spring.security.oauth2.resourceserver.opaque.introspection-uri}")
    String introspectionUri;
@Value("${spring.security.oauth2.resourceserver.opaque.introspection-client-id}")
    String clientId;
@Value("${spring.security.oauth2.resourceserver.opaque.introspection-client-secret}")
    String clientSecret;
    @Override
    protected void configure(HttpSecurity http) throws Exception {
        http.authorizeRequests()
                .anyRequest().authenticated()
                .and()
                .oauth2ResourceServer()
                .opaqueToken().introspectionUri(introspectionUri)
                .introspectionClientCredentials(clientId, clientSecret);
    }
}
```

将 application.yml 中配置的三个属性注入进来,然后在 configure(HttpSecurity)方法中开启不透明令牌的配置,传入三个相关的参数即可。

最后再定义一个测试接口,代码如下:

```java
@RestController
public class HelloController {
```

```
    @GetMapping("/hello")
    public String hello() {
        return "hello res server";
    }
}
```

至此，我们的资源服务器就算配置成功了。

15.5.2.3 客户端应用搭建

客户端应用搭建和我们前面 GitHub 授权登录比较像。

创建一个名为 client01 的 Spring Boot 项目，引入 Web、Spring Security、Thymeleaf 以及 OAuth2 Client 依赖，在之前旧的 OAuth2 Client 中，负责发送网络请求的是 OAuth2RestTemplate，但是在目前的最新方案中，OAuth2RestTemplate 被 WebClient 所替代，所以我们还需要在项目中引入 WebFlux，最终的 pom.xml 文件内容如下：

```xml
<dependency>
    <groupId>org.springframework.boot</groupId>
    <artifactId>spring-boot-starter-oauth2-client</artifactId>
</dependency>
<dependency>
    <groupId>org.springframework.boot</groupId>
    <artifactId>spring-boot-starter-security</artifactId>
</dependency>
<dependency>
    <groupId>org.springframework.boot</groupId>
    <artifactId>spring-boot-starter-web</artifactId>
</dependency>
<dependency>
    <groupId>org.springframework</groupId>
    <artifactId>spring-webflux</artifactId>
</dependency>
<dependency>
    <groupId>io.projectreactor.netty</groupId>
    <artifactId>reactor-netty</artifactId>
</dependency>
<dependency>
    <groupId>org.springframework.boot</groupId>
    <artifactId>spring-boot-starter-thymeleaf</artifactId>
</dependency>
<dependency>
    <groupId>org.thymeleaf.extras</groupId>
    <artifactId>thymeleaf-extras-springsecurity5</artifactId>
</dependency>
```

由于我们要在这个项目中同时演示授权码模式、客户端模式以及密码模式，所以接下来在 application.yml 中对三种授权模式所需要的参数分别进行配置：

```
server:
```

```yaml
      port: 8883
spring:
  security:
    oauth2:
      client:
        registration:
          auth-code:
            provider: javaboy
            client-id: my_client
            client-secret: 123
            authorization-grant-type: authorization_code
            redirect-uri:http://client.javaboy.org:8883/login/oauth2/code/javaboy
            scope: read:msg
          client-creds:
            provider: javaboy
            client-id: my_client
            client-secret: 123
            authorization-grant-type: client_credentials
            scope: read:msg
          password:
            provider: javaboy
            client-id: my_client
            client-secret: 123
            authorization-grant-type: password
            scope: read:msg
        provider:
          javaboy:
            authorization-uri: http://auth.javaboy.org:8881/oauth/authorize
            token-uri: http://auth.javaboy.org:8881/oauth/token
```

这里提供了三个客户端，名字分别是 auth-code（授权码模式）、client-creds（客户端模式）以及 password（密码模式），三个客户端中指定了各自的参数，大家可以对照 15.2 节去理解这些参数，这里不再赘述。

另外还提供了一个名为 javaboy 的 provider，并配置了授权服务器的认证地址以及令牌获取地址。

接下来我们需要提供一个 WebClient 实例，利用 WebClient 可以方便地发起认证请求，这也是最新的 OAuth2 Client 推荐的方式。如果不用 WebClient，那在发起请求时需要开发者自己去拼接各种参数，比较麻烦。

```java
@Configuration
public class WebClientConfig {
    @Bean
    WebClient webClient(OAuth2AuthorizedClientManager
                                                authorizedClientManager) {
        ServletOAuth2AuthorizedClientExchangeFilterFunction oauth2Client =new
        ServletOAuth2AuthorizedClientExchangeFilterFunction(authorizedClientManager);
        return WebClient.builder()
```

```java
                .apply(oauth2Client.oauth2Configuration())
                .build();
    }
    @Bean
    OAuth2AuthorizedClientManager
                authorizedClientManager(ClientRegistrationRepository
                clientRegistrationRepository,
            OAuth2AuthorizedClientRepository authorizedClientRepository) {
        OAuth2AuthorizedClientProvider authorizedClientProvider =
            OAuth2AuthorizedClientProviderBuilder.builder()
                    .authorizationCode()
                    .refreshToken()
                    .clientCredentials()
                    .password()
                    .build();
        DefaultOAuth2AuthorizedClientManager authorizedClientManager =
        new DefaultOAuth2AuthorizedClientManager(
            clientRegistrationRepository, authorizedClientRepository);
        authorizedClientManager
        .setAuthorizedClientProvider(authorizedClientProvider);
        authorizedClientManager.setContextAttributesMapper(contextAttributesMapper());
        return authorizedClientManager;
    }
    private Function<OAuth2AuthorizeRequest, Map<String, Object>>
                                                    contextAttributesMapper() {
        return authorizeRequest -> {
            Map<String, Object> contextAttributes = Collections.emptyMap();
            HttpServletRequest servletRequest =
            authorizeRequest.getAttribute(HttpServletRequest.class.getName());
            String username =
                servletRequest.getParameter(OAuth2ParameterNames.USERNAME);
            String password =
                servletRequest.getParameter(OAuth2ParameterNames.PASSWORD);
            if (StringUtils.hasText(username) && StringUtils.hasText(password)) {
                contextAttributes = new HashMap<>();
            contextAttributes
                .put(OAuth2AuthorizationContext.USERNAME_ATTRIBUTE_NAME, username);
            contextAttributes
                .put(OAuth2AuthorizationContext.PASSWORD_ATTRIBUTE_NAME, password);
            }
            return contextAttributes;
        };
    }
}
```

这里主要提供了两个 Bean：WebClient 和 OAuth2AuthorizedClientManager。前者用来发起网络请求，在 WebClient 配置时，需要用到 OAuth2AuthorizedClientManager 实例。

OAuth2AuthorizedClientManager 主要用来管理授权的客户端，它的职责是通过

OAuth2AuthorizedClientProvider 对不同的客户端进行授权。不同的授权模式会对应不同的 OAuth2AuthorizedClientProvider 实例，例如：

- 授权码模式对应 AuthorizationCodeOAuth2AuthorizedClientProvider。
- 密码模式对应 PasswordOAuth2AuthorizedClientProvider。
- 客户端模式对应 ClientCredentialsOAuth2AuthorizedClientProvider。
- 刷新令牌对象 RefreshTokenOAuth2AuthorizedClientProvider。

通过 OAuth2AuthorizedClientProviderBuilder 来构建所需要的 OAuth2AuthorizedClientProvider 实例，并添加到 OAuth2AuthorizedClientManager 对象中。

另外，由于密码模式还需要用到用户输入的用户名/密码，所以这里通过 contextAttributesMapper 将请求中的用户名/密码提取出来存入 contextAttributes 中。

接下来提供 SecurityConfig，代码如下：

```java
@Configuration
public class SecurityConfig extends WebSecurityConfigurerAdapter {
    @Override
    protected void configure(HttpSecurity http) throws Exception {
        http
                .authorizeRequests()
                .anyRequest().authenticated()
                .and()
                .formLogin()
                .loginPage("/login.html")
                .defaultSuccessUrl("/index")
                .permitAll()
                .and()
                .csrf().disable()
                .oauth2Client();
    }
    @Bean
    public UserDetailsService users() {
        return new InMemoryUserDetailsManager(User
                .withUsername("javaboy")
                .password("{noop}123").roles("USER").build());
    }
}
```

这段配置大家应该都很熟悉了，这里不再赘述。登录页面比较简单，这里给出登录表单，代码如下：

```html
<form id="login-form" class="form" action="/login.html" method="post">
    <h3 class="text-center text-info">登录</h3>
    <div class="form-group">
        <label for="username" class="text-info">用户名:</label><br>
        <input type="text" name="username" id="username" value="javaboy"
                                           class="form-control">
```

```html
    </div>
    <div class="form-group">
        <label for="password" class="text-info">密码:</label><br>
        <input type="text" name="password" id="password" value="123"
                                                    class="form-control">
    </div>
    <div class="form-group">
        <input type="submit" name="submit" class="btn btn-info btn-md"
                                                    value="登录">
    </div>
</form>
```

登录成功后跳转到 index 页面，该页面是由 Thymeleaf 渲染的，代码如下：

```html
<div class="panel panel-default">
    <div class="panel-heading">
        <h3 class="panel-title">请选择一种授权模式</h3>
    </div>
    <ul class="list-group">
        <li class="list-group-item">
            //1
            <a th:href="@{/authorize?grant_type=authorization_code}">
                    <span style="font-size:medium">授权码模式</span></a>
        </li>
        <li class="list-group-item">
         //2
            <a th:href="@{/authorize?grant_type=client_credentials}">
                    <span style="font-size:medium">客户端模式</span></a>
        </li>
        <li class="list-group-item">
            <div>
            //3
                <p style="font-size:medium">密码模式</p>
                <form th:action="@{/authorize}" method="post">
                    <div class="form-row">
                        <div class="form-group">
                            <label for="username"
                                    style="font-size:small">用户名</label>
                            <input type="text" id="username" name="username"
                                value="javaboy" class="form-control input-sm">
                        </div>
                    </div>
                    <div class="form-row">
                        <div class="form-group">
                            <label for="password"
                                    style="font-size:small">密码</label>
                            <input type="password" id="password"
                        name="password" value="123" class="form-control input-sm">
                        </div>
```

```
                </div>
                <input type="hidden" id="grant_type" name="grant_type"
                                                    value="password">
                <button type="submit"
                            class="btn btn-primary btn-sm">授权</button>
            </form>
        </div>
    </li>
</ul>
//4
<div th:if="${msg}" class="panel-footer">
    <div th:text="${msg}"></div>
</div>
</div>
```

- 注释 1，这是一个授权码模式的超链接，单击该超链接会触发授权码模式进行校验。
- 注释 2，这是一个客户端模式的超链接，单击该超链接会触发客户端模式进行校验。
- 注释 3，这是密码模式，这种模式需要用到用户名/密码，在下方的输入框中输入用户名/密码，然后单击"授权"按钮，触发密码模式进行校验。
- 注释 4，在授权成功后，会返回一个 msg，在当前页面渲染出来。

无论是哪种授权模式，都调用了 /authorize 接口，只是参数不同而已，因此我们还需要提供 /authorize 接口，代码如下：

```
@Controller
public class HelloController {
    @Autowired
    WebClient webClient;
    private String helloUri="http://res.javaboy.org:8882/hello";
    //1
    @GetMapping(value = "/authorize",params = "grant_type=authorization_code")
    public String authorization_code_grant(Model model) {
        String msg = retrieveMessages("auth-code");
        model.addAttribute("msg", msg);
        return "index";
    }
    //2
    @GetMapping(value = "/authorize",params = "grant_type=client_credentials")
    public String client_credentials_grant(Model model) {
        String msg = retrieveMessages("client-creds");
        model.addAttribute("msg", msg);
        return "index";
    }
    //3
    @PostMapping(value = "/authorize", params = "grant_type=password")
    public String password_grant(Model model) {
        String msg = retrieveMessages("password");
        model.addAttribute("msg", msg);
```

```
        return "index";
    }
    //4
    private String retrieveMessages(String clientRegistrationId) {
        return webClient
                .get()
                .uri(helloUri)
                .attributes(clientRegistrationId(clientRegistrationId))
                .retrieve()
                .bodyToMono(String.class)
                .block();
    }
    //5
    @GetMapping("/")
    public String root() {
        return "redirect:/index";
    }
    @GetMapping("/index")
    public String index() {
        return "index";
    }
}
```

- 注释 1，处理授权码模式的接口，调用 retrieveMessages 方法去请求 helloUri 地址，将获取结果存入 Model 中。
- 注释 2 与注释 3，功能类似，不再赘述。
- 注释 4，通过 WebClient 发起请求。
- 注释 5，两个页面映射。

最后将项目端口改为 8883。

至此，我们的整个工程就搭建完成了。

15.5.3 测试

首先在浏览器中访问 http://client.javaboy.org:8883/index 地址，由于用户未登录，所以会重定向到 http://client.javaboy.org:8883/login.html 页面进行登录，如图 15-12 所示。

图 15-12 客户端登录页面

用户在该页面上完成登录，注意这个时候是登录客户端而不是授权服务器，客户端登录成功后，就可以看到项目首页了，如图 15-13 所示。

图 15-13　客户端首页

在客户端首页，用户可以选择任何一种授权模式。例如单击授权码模式，此时就会访问到客户端的 /authorize 接口，然后 WebClient 向 http://res.javaboy.org:8882/hello 接口发起请求，但是由于当前客户端还没有在授权服务器上进行认证，所以又会跳转到授权服务器的登录页面，如图 15-14 所示。

图 15-14　授权服务器的登录页面

在该页面输入用户名/密码进行登录（注意此时是登录授权服务器），登录成功后，就会看到一个授权页面，如图 15-15 所示。

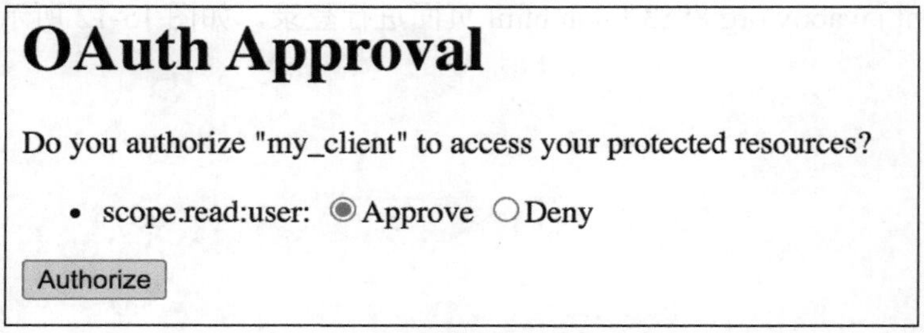

图 15-15　授权页面

选择 Approve 然后单击下方的"授权"按钮，表示批准本次授权。这里也可以自动批准，在授权服务器中配置客户端信息时，通过调用.autoApprove(true)方法可以设置自动批准，这样在登录完成后就不会看到该页面了。

至此，整个授权过程完成，再次来到 http://client.javaboy.org:8883/authorize 地址，此时页面底部就可以看到接口数据了，如图 15-16 所示。

图 15-16　授权成功后获取到接口数据

客户端模式和密码模式测试方式类似，这里不再赘述。

15.5.4　原理分析

先来看资源服务器。

当我们在资源服务器中配置了 .oauth2ResourceServer().opaqueToken() 之后，实际上向 Spring Security 过滤器链中添加了一个 BearerTokenAuthenticationFilter 过滤器，在该过滤器中将完成令牌的解析与校验，我们来看一下它的 doFilterInternal 方法：

```
protected void doFilterInternal(HttpServletRequest request,
                HttpServletResponse response, FilterChain filterChain)
                    throws ServletException, IOException {
    String token;
    try {
        //1
        token = this.bearerTokenResolver.resolve(request);
    } catch ( OAuth2AuthenticationException invalid ) {
        this.authenticationEntryPoint.commence(request, response, invalid);
        return;
    }
```

```
    //2
    if (token == null) {
        filterChain.doFilter(request, response);
        return;
    }
//3
    BearerTokenAuthenticationToken authenticationRequest =
                        new BearerTokenAuthenticationToken(token);
    authenticationRequest
        .setDetails(this.authenticationDetailsSource.buildDetails(request));
    try {
        AuthenticationManager authenticationManager =
                this.authenticationManagerResolver.resolve(request);
        Authentication authenticationResult =
                authenticationManager.authenticate(authenticationRequest);
        //4
        SecurityContext context = SecurityContextHolder.createEmptyContext();
        context.setAuthentication(authenticationResult);
        SecurityContextHolder.setContext(context);
        filterChain.doFilter(request, response);
    } catch (AuthenticationException failed) {
        SecurityContextHolder.clearContext();
        this.authenticationFailureHandler
                .onAuthenticationFailure(request, response, failed);
    }
}
```

- 注释1，从当前请求中解析出 Access Token，Access Token 默认放在请求头中，从请求头中取出 Authorization 字段即可。
- 注释2，如果第一步获取到的令牌为空，就说明用户没有传递 Access Token，此时继续执行后面的过滤器，当前请求将在最后的 FilterSecurityInterceptor 中被检查出权限不足而抛出异常。
- 注释3，构造一个 BearerTokenAuthenticationToken 对象并传入 Access Token 令牌，然后调用 authenticate 方法完成令牌校验。最终承担校验任务的是 OpaqueTokenAuthenticationProvider#authenticate 方法，在该方法中会调用授权服务器的/oauth/check_token 接口，令牌校验成功后，该接口会返回该令牌对应的用户信息。
- 注释4，将获取到的 authenticationResult 对象存入 SecurityContext 中，完成登录。

这就是资源服务器的大致工作流程，还是比较简单的。

从这个流程中大家也可以看到，客户端每次从资源服务器中请求数据时，资源服务器都会调用授权服务器的接口去校验令牌的合法性，这无形中增大了授权服务器的压力，后面我们将通过 JWT 来解决这一问题。

客户端的工作原理就要简单很多了。

客户端的请求是由 WebClient 发起的，底层真正发起 HTTP 请求的依然是 RestTemplate。

这里涉及的核心类就是 DefaultOAuth2AuthorizedClientManager 和 OAuth2AuthorizedClient Provider，这两者的关系类似于我们之前讲的 ProviderManager 和 AuthenticationProvider 的关系。

DefaultOAuth2AuthorizedClientManager 类的核心方法是 authorize，由 WebClient 发起的请求会在这里被拦截下来，我们来看一下该方法：

```java
public OAuth2AuthorizedClient authorize(OAuth2AuthorizeRequest
                                        authorizeRequest) {
//1
   String clientRegistrationId = authorizeRequest.getClientRegistrationId();
   OAuth2AuthorizedClient authorizedClient =
                              authorizeRequest.getAuthorizedClient();
   Authentication principal = authorizeRequest.getPrincipal();
   HttpServletRequest servletRequest =
      getHttpServletRequestOrDefault(authorizeRequest.getAttributes());
   HttpServletResponse servletResponse =
      getHttpServletResponseOrDefault(authorizeRequest.getAttributes());
//2
   OAuth2AuthorizationContext.Builder contextBuilder;
   if (authorizedClient != null) {
      contextBuilder =
      OAuth2AuthorizationContext.withAuthorizedClient(authorizedClient);
   } else {
      authorizedClient = this.authorizedClientRepository
       .loadAuthorizedClient(clientRegistrationId, principal, servletRequest);
      if (authorizedClient != null) {
         contextBuilder =
      OAuth2AuthorizationContext.withAuthorizedClient(authorizedClient);
      } else {
         ClientRegistration clientRegistration =
        this.clientRegistrationRepository.findByRegistrationId(clientRegistrationId);
         contextBuilder =
       OAuth2AuthorizationContext.withClientRegistration(clientRegistration);
      }
   }
   OAuth2AuthorizationContext authorizationContext = contextBuilder
       .principal(principal)
       .attributes(attributes -> {
          Map<String, Object> contextAttributes =
             this.contextAttributesMapper.apply(authorizeRequest);
          if (!CollectionUtils.isEmpty(contextAttributes)) {
             attributes.putAll(contextAttributes);
          }
       })
       .build();
//3
   try {
      authorizedClient =
         this.authorizedClientProvider.authorize(authorizationContext);
```

```
        } catch (OAuth2AuthorizationException ex) {
        }
//4
        if (authorizedClient != null) {
            this.authorizationSuccessHandler.onAuthorizationSuccess(
                    authorizedClient, principal, createAttributes(servletRequest,
                                                servletResponse));
        } else {
            if (authorizationContext.getAuthorizedClient() != null) {
                return authorizationContext.getAuthorizedClient();
            }
        }
        return authorizedClient;
}
```

- 注释 1，authorizeRequest 是一个包含 clientRegistrationId 标志的客户端请求，从中可以提取出 clientRegistrationId、客户端对象、已经登录用户信息以及原始的 HttpServletRequest 与 HttpServletResponse。
- 注释 2，构造 OAuth2AuthorizationContext 对象，用来保存授权请求时所需要的一些必要信息，构造该对象需要用到客户端对象。对于已经认证过的 authorizedClient 可以从 authorizedClientRepository 中直接获取，而没有认证过的 authorizedClient 则只能从 clientRegistrationRepository 中获取客户端信息，然后构造 OAuth2AuthorizationContext 并配置已登录用户对象以及额外附加的用户信息，额外附加的用户信息主要是指密码模式中跟随原始请求一起传来的用户名/密码。
- 注释 3，通过 authorizedClientProvider.authorize 方法进行授权，最终会调用到不同的 OAuth2AuthorizedClientProvider 实例。如果客户端同时支持多种不同的授权模式，则多个 OAuth2AuthorizedClientProvider 实例会被 DelegatingOAuth2AuthorizedClientProvider 对象代理，在代理对象中再去遍历不同的 OAuth2AuthorizedClientProvider 实例，选择合适的 OAuth2AuthorizedClientProvider 实例进行处理，最终调用的请求发送工具依然是 RestTemplate。
- 注释 4，在 OAuth2AuthorizedClientProvider 中认证成功后，会返回认证成功后的 authorizedClient 对象，该对象中就包含了 Access Token 和 Refresh Token。如果该客户端已经认证过了，并且 Access Token 还没有过期，则返回的 authorizedClient 为 null，此时直接从 authorizationContext 取出旧的 authorizedClient 返回即可。

这就是客户端的一个大致工作流程，OAuth2AuthorizedClientProvider 中的实现细节都比较容易，我们就不一一讲解了。

15.5.5 自定义请求

前面的案例中，我们通过 WebClient 来发送认证请求，整个授权过程包括参数的拼接都是由框架帮我们完成的。

有时候，我们可能需要拿到令牌 Access Token，然后自己调用资源服务器的接口来获取数据，这在前后端分离中非常有用。在客户端中获取令牌 Access Token 的方式很简单，我们在 client01 项目的 HelloController 中添加如下接口：

```
@GetMapping("/token")
@ResponseBody
public String token(@RegisteredOAuth2AuthorizedClient
                                OAuth2AuthorizedClient authorizedClient) {
   OAuth2AccessToken accessToken = authorizedClient.getAccessToken();
   return accessToken.getTokenValue();
}
```

注入 OAuth2AuthorizedClient 对象，然后就可以从该对象中提取出 Access Token 以及 Refresh Token。如果当前客户端只支持一种授权模式，则直接按照上面的写法来；如果当前客户端支持多种授权模式，则需要在 @RegisteredOAuth2AuthorizedClient 注解中指明 registrationId，代码如下：

```
@GetMapping("/token")
@ResponseBody
public String token(@RegisteredOAuth2AuthorizedClient("auth-code")
                        OAuth2AuthorizedClient authorizedClient) {
   OAuth2AccessToken accessToken = authorizedClient.getAccessToken();
   return accessToken.getTokenValue();
}
```

有了 Access Token，开发者就可以利用 Access Token 来请求资源服务器的其他接口了。HTTP 请求工具可以利用 Spring 提供的 RestTemplate，也可以使用自己擅长的其他 HTTP 请求工具，如 HttpClient、OkHttp 等，这个过程就比较简单了，这里不再赘述。

15.6 使用 Redis

在 15.5.2.1 小节搭建授权服务器时，我们配置将生成的令牌 Access Token 存在内存中，当时提供的 TokenStore 实例是 InMemoryTokenStore：

```
@Configuration
public class AccessTokenConfig {
   @Bean
   TokenStore tokenStore() {
      return new InMemoryTokenStore();
   }
}
```

事实上，TokenStore 接口有多种不同的实现类，如图 15-17 所示。

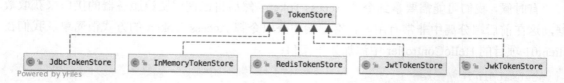

图 15-17　TokenStore 实现类

可以看到，我们有多种方式来存储 Access Token：

- InMemoryTokenStore：这是我们之前使用的方式，即将 Access Token 存到内存中，单机使用这个没有问题，但是在分布式环境下不推荐使用。
- JdbcTokenStore：将 Access Token 保存到数据中，方便和其他应用共享令牌信息。
- JwtTokenStore：这个其实不算是存储，因为使用了 JWT 之后，在生成的 JWT 中就有用户的所有信息，服务端不需要保存。
- RedisTokenStore：将 Access Token 存到 Redis 中。
- JwkTokenStore：这个只在资源服务器上使用，主要作用是解码 JWT 并使用相应的 JWK 验证其签名（JWS）。

虽然这里支持的方案比较多，但是我们常用的方案实际上主要是这两个：RedisTokenStore 和 JwtTokenStore，JwtTokenStore 后面会做介绍，这里我们先来看一下 RedisTokenStore。

首先我们启动一个 Redis。

接下来在 15.5.2.1 小节搭建的授权服务器基础上，添加 Redis 依赖，代码如下：

```
<dependency>
    <groupId>org.springframework.boot</groupId>
    <artifactId>spring-boot-starter-data-redis</artifactId>
</dependency>
```

然后在 application.properties 中添加 Redis 配置：

```
spring.redis.host=127.0.0.1
spring.redis.port=6379
spring.redis.password=123
```

配置完成后，我们修改 TokenStore 的实例，代码如下：

```
@Configuration
public class AccessTokenConfig {
    @Autowired
    RedisConnectionFactory redisConnectionFactory;
    @Bean
    TokenStore tokenStore() {
        return new RedisTokenStore(redisConnectionFactory);
    }
}
```

只需要修改一下 TokenStore 实例即可。

配置完成后，再去启动 auth_server，此时授权服务器生成的 Access Token 令牌就会保存到 Redis 中。Access Token 在 Redis 中的有效期就是令牌的有效期，也正是因为 Redis 中的这种过期机制，让它在存储 Access Token 时具有天然的优势。

15.7 客户端信息存入数据库

在 15.5.2.1 小节搭建的授权服务器中，客户端信息是直接存储在内存中的。然而在实际项目中，这种方式并不可取，一方面客户端信息无法实现动态添加与删除，另一方面硬编码的客户端信息也不好维护，所以我们需要将客户端信息存入数据库中。

涉及客户端信息保存的接口主要是 ClientDetailsService，这个接口主要有两个实现类，如图 15-18 所示。

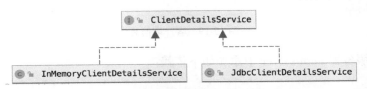

图 15-18 ClientDetailsService 的实现类

InMemoryClientDetailsService 就是将客户端信息存入内存中，也就是我们之前案例所采用的存储方式；JdbcClientDetailsService 则是将客户端信息存入数据库中。

由于官方没有给出使用 JdbcClientDetailsService 存储客户端信息时的数据库脚本，所以我们可以根据 JdbcClientDetailsService 中定义的 SQL 来分析出数据库表结构。JdbcClientDetailsService 部分源码：

```
private static final String CLIENT_FIELDS_FOR_UPDATE =
"resource_ids, scope, "   + "authorized_grant_types, web_server_redirect_uri,
    authorities, access_token_validity, "+"refresh_token_validity,
                                   additional_information, autoapprove";
private static final String CLIENT_FIELDS = "client_secret, " +
                                                CLIENT_FIELDS_FOR_UPDATE;
```

根据这两个属性就能确定数据库字段名，最终分析出来的 SQL 如下：

```
DROP TABLE IF EXISTS `oauth_client_details`;
CREATE TABLE `oauth_client_details` (
  `client_id` varchar(48) NOT NULL,
  `resource_ids` varchar(256) DEFAULT NULL,
  `client_secret` varchar(256) DEFAULT NULL,
  `scope` varchar(256) DEFAULT NULL,
  `authorized_grant_types` varchar(256) DEFAULT NULL,
  `web_server_redirect_uri` varchar(256) DEFAULT NULL,
  `authorities` varchar(256) DEFAULT NULL,
```

```
  `access_token_validity` int(11) DEFAULT NULL,
  `refresh_token_validity` int(11) DEFAULT NULL,
  `additional_information` varchar(4096) DEFAULT NULL,
  `autoapprove` varchar(256) DEFAULT NULL,
  PRIMARY KEY (`client_id`)
) ENGINE=InnoDB DEFAULT CHARSET=utf8;
```

在数据库中执行该段 SQL 脚本,并将一开始配置在代码中的客户端信息录入数据库中,如图 15-19 所示。

client_id	my_client
resource_ids	<null>
client_secret	$2a$10$oE39aG10kB/rFu2vQeCJTu/V/v4n6DRR0f8WyXRiAYvBpmadoOBE.
scope	read:user,read:msg
authorized_grant_types	authorization_code,refresh_token,implicit,password,client_credentials
web_server_redirect_uri	http://client.javaboy.org:8883/login/oauth2/code/javaboy
authorities	<null>
access_token_validity	7200
refresh_token_validity	7200
additional_information	<null>
autoapprove	true

图 15-19　客户端信息

然后在授权服务器 auth-server 中添加如下依赖:

```xml
<dependency>
    <groupId>org.springframework.boot</groupId>
    <artifactId>spring-boot-starter-jdbc</artifactId>
</dependency>
<dependency>
    <groupId>mysql</groupId>
    <artifactId>mysql-connector-java</artifactId>
</dependency>
```

在 application.properties 中配置一下数据库连接信息:

```
spring.datasource.url=jdbc:mysql:///security15?useUnicode=true&characterEncoding=UTF-8&serverTimezone=Asia/Shanghai
spring.datasource.password=123
spring.datasource.username=root
spring.main.allow-bean-definition-overriding=true
```

最后一条配置是允许 Bean 的覆盖,否则我们自己创建的 ClientDetailsService 将会和系统创建的 ClientDetailsService 实例相冲突。

接下来配置 ClientDetailsService 实例,代码如下:

```java
@EnableAuthorizationServer
@Configuration
public class AuthorizationServer extends AuthorizationServerConfigurerAdapter {
    @Autowired
    DataSource dataSource;
    @Bean
    ClientDetailsService clientDetailsService() {
```

```
        return new JdbcClientDetailsService(dataSource);
    }
    @Bean
    AuthorizationServerTokenServices tokenServices() {
        DefaultTokenServices services = new DefaultTokenServices();
        services.setClientDetailsService(clientDetailsService());
        services.setSupportRefreshToken(true);
        services.setTokenStore(tokenStore);
        return services;
    }
    @Override
    public void configure(ClientDetailsServiceConfigurer clients)
                                                        throws Exception {
        clients.withClientDetails(clientDetailsService());
    }
    //省略其他
}
```

和 15.5.2.1 小节中的案例相比，这里的变化主要在四个方面：

（1）注入 DataSource 实例。

（2）向 Spring 容器注册一个 JdbcClientDetailsService 实例。

（3）在 AuthorizationServerTokenServices 实例中除去令牌有效期设置，令牌有效期将从数据库中加载。

（4）在 configure(ClientDetailsServiceConfigurer)方法中直接配置 JdbcClientDetailsService 实例即可，项目启动后会自动从数据库中加载客户端信息。

配置完成后，重启授权服务器再去进行授权测试即可。

15.8 使用 JWT

在前面的案例中，我们一直都是使用的不透明令牌（Opaque Token），在实际开发中，JWT 令牌目前使用较多，因此本节我们来看一下如何在 OAuth2 中使用 JWT。

15.8.1 JWT

JWT 全称为 Json Web Token，它是一种 JSON 风格的轻量级授权和身份认证规范，可实现无状态、分布式的 Web 应用授权。

JWT 作为一种规范，并没有和某一种语言绑定在一起，开发者可以使用任何语言来实现 JWT。Java 中 JWT 相关的开源库也比较多，例如 jjwt、nimbus-jose-jwt 等。

15.8.2 JWT 数据格式

JWT 包含三部分数据：Header、Payload 与 Signature。

Header

头部，通常头部有两部分信息：

- 声明类型，这里是 JWT。
- 加密算法，自定义。

我们会对头部进行 Base64Url 编码（可解码），得到第一部分数据。

Payload

载荷，就是有效数据，在官方文档中（RFC7519）给了 7 个示例信息：

- iss (issuer)：签发人。
- exp (expiration time)：过期时间。
- sub (subject)：主题。
- aud (audience)：受众。
- nbf (Not Before)：生效时间。
- iat (Issued At)：签发时间。
- jti (JWT ID)：编号。

这部分也会采用 Base64Url 编码，得到第二部分数据。

Signature

签名，是整个数据的认证信息。一般根据前两步的数据，再加上服务的密钥 secret（密钥保存在服务端，不能泄漏给客户端）。这个密钥通过 Header 中配置的加密算法生成，用于验证整个数据完整性和可靠性。

生成的数据格式如图 15-20 所示。

```
eyJhbGciOiJIUzI1NiIsInR5cCI6IkpXVCJ9.
eyJzdWIiOiIxMjM0NTY3ODkwIiwibmFtZSI6IkpvaG4
gRG91IiwiaXNTb2NpYWwiOnRydWV9.
4pcPyMD09olPSyXnrXCjTwXyr4BsezdI1AVTmud2fU4
```

图 15-20　JWT 数据格式

> **注　意**
>
> 这里的数据使用"."隔开成了三部分，分别对应前面提到的三部分。另外，这里数据是不换行的，图片换行只是为了展示方便而已。

15.8.3　OAuth2 中使用 JWT

Spring Security 官方推荐使用 nimbus-jose-jwt 来生成和解析 JWT 令牌，该库同时支持对称加密和非对称加密两种方式处理 JWT，本小节使用目前通用的非对称加密（RSA）来处理 JWT。

非对称加密有两种使用场景：

- 加密场景：公钥负责加密，私钥负责解密。
- 签名场景：私钥负责签名，公钥负责验证。

我们在 JWT 中使用的非对称加密属于签名场景。如果要使用 JWT，我们首先需要创建一个证书文件，这里使用 Java 自带的 keytool 工具来生成 jks 证书文件，该工具在 JDK 的 bin 目录下，生成过程如图 15-21 所示。

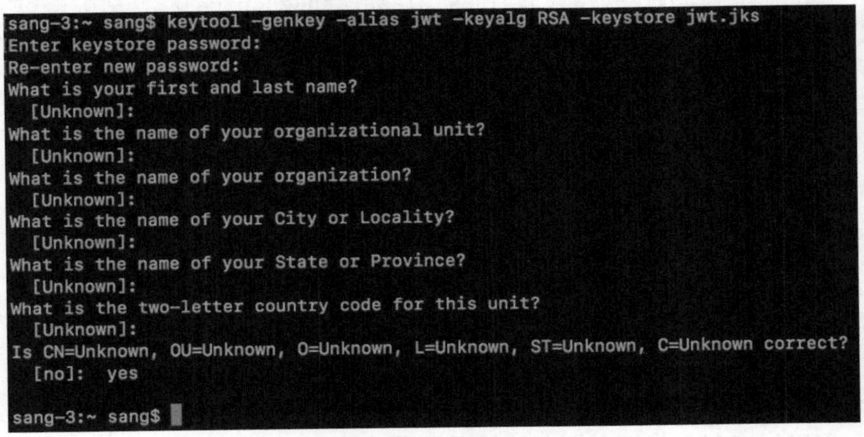

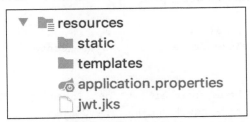

图 15-21　证书生成过程

生成证书命令中，我们设置了生成证书的别名是 jwt，生成的证书文件是 jwt.jks。接下来输入密码以及其他信息即可，命令执行完成后，会在当前目录下生成一个 jwt.jks 文件，将该文件拷贝到 auth_server 项目的 resources 目录下，如图 15-22 所示。

图 15-22　将生成的 jwt.jks 文件复制到 resources 目录下

接下来在 auth_server 项目中添加 JWT 依赖，代码如下（nimbus-jose-jwt 在资源服务器中有提供，所以只需要在授权服务器中添加即可）：

```
<dependency>
    <groupId>org.springframework.security</groupId>
```

```
        <artifactId>spring-security-jwt</artifactId>
        <version>1.1.0.RELEASE</version>
</dependency>
<dependency>
        <groupId>com.nimbusds</groupId>
        <artifactId>nimbus-jose-jwt</artifactId>
</dependency>
```

接下来进行 JWT 配置，首先对密钥进行配置，代码如下：

```
class KeyConfig {
    private static final String KEY_STORE_FILE = "jwt.jks";
    private static final String KEY_STORE_PASSWORD = "123456";
    private static final String KEY_ALIAS = "jwt";
    private static KeyStoreKeyFactory KEY_STORE_KEY_FACTORY =
            new KeyStoreKeyFactory(new ClassPathResource(KEY_STORE_FILE),
                                    KEY_STORE_PASSWORD.toCharArray());
    static RSAPublicKey getVerifierKey() {
        return (RSAPublicKey) getKeyPair().getPublic();
    }
    static RSAPrivateKey getSignerKey() {
        return (RSAPrivateKey) getKeyPair().getPrivate();
    }
    private static KeyPair getKeyPair() {
        return KEY_STORE_KEY_FACTORY.getKeyPair(KEY_ALIAS);
    }
}
```

KEY_STORE_FILE 就是生成的证书文件名，KEY_STORE_PASSWORD 则是生成证书时输入的密码，KEY_ALIAS 指证书别名，然后再通过 getVerifierKey 和 getSignerKey 两个方法分别返回公钥和私钥。

接下来配置 TokenStore，代码如下：

```
@Configuration
public class AccessTokenConfig {
    @Bean
    TokenStore tokenStore() {
        return new JwtTokenStore(jwtAccessTokenConverter());
    }
    @Bean
    public JwtAccessTokenConverter jwtAccessTokenConverter() {
        RsaSigner signer = new RsaSigner(KeyConfig.getSignerKey());
        JwtAccessTokenConverter converter = new JwtAccessTokenConverter();
        converter.setSigner(signer);
        converter.setVerifier(new RsaVerifier(KeyConfig.getVerifierKey()));
        return converter;
    }
    @Bean
    public JWKSet jwkSet() {
```

```
            RSAKey.Builder builder = new RSAKey
                    .Builder(KeyConfig.getVerifierKey())
                    .keyUse(KeyUse.SIGNATURE)
                    .algorithm(JWSAlgorithm.RS256);
            return new JWKSet(builder.build());
    }
}
```

此时提供的 TokenStore 实例是 JwtTokenStore,创建该实例时需要一个 JwtAccessToken Converter 对象,该对象是一个令牌生成工具。JwtAccessTokenConverter 对象在创建时,配置一下签名以及验证者即可。最后还需要提供一个包含公钥的 JWKSet 对象,该对象接下来要暴露给资源服务器。

接下来配置 AuthorizationServer,主要在 AuthorizationServerTokenServices 实例中进行配置,代码如下:

```
@Autowired
TokenStore tokenStore;
@Autowired
JwtAccessTokenConverter jwtAccessTokenConverter;
@Bean
AuthorizationServerTokenServices tokenServices() {
    DefaultTokenServices services = new DefaultTokenServices();
    services.setClientDetailsService(clientDetailsService);
    services.setSupportRefreshToken(true);
    services.setTokenStore(tokenStore);
    TokenEnhancerChain tokenEnhancerChain = new TokenEnhancerChain();
    tokenEnhancerChain
            .setTokenEnhancers(Arrays.asList(jwtAccessTokenConverter));
    services.setTokenEnhancer(tokenEnhancerChain);
    return services;
}
//省略其他
```

主要是在 DefaultTokenServices 中配置 TokenEnhancer,将之前的 JwtAccessTokenConverter 注入进来即可。

最后我们还需要提供一个公钥接口,资源服务器将从该接口中获取到公钥,进而完成对 JWT 的校验:

```
@GetMapping(value = "/oauth2/keys")
public String keys() {
    return jwkSet.toString();
}
```

至此,我们的 auth_server 就改造完成了,接下来对 res_server 进行改造。

当采用 JWT 之后,资源服务器就不需要每次拿到令牌后都去调用授权服务器校验令牌,资源服务器只需要调用授权服务器接口获取到公钥即可。有了公钥,资源服务器就可以自己校验 JWT 令牌了。所以，对资源服务器的改动很简单，代码如下：

```
@Configuration
public class OAuth2ResourceServerSecurityConfiguration extends
                                            WebSecurityConfigurerAdapter {
    @Override
    protected void configure(HttpSecurity http) throws Exception {
        http.authorizeRequests()
                .anyRequest().authenticated()
                .and()
                .oauth2ResourceServer().jwt()
                .jwkSetUri("http://auth.javaboy.org:8881/oauth2/keys");
    }
}
```

开启 JWT 并设置获取 JwkSet 的地址即可。

配置完成后，分别启动 auth_server 和 res_server，测试客户端依然使用 15.5.2.3 小节搭建的客户端，具体的测试过程这里就不再赘述了。

15.9 小 结

本章主要介绍了 OAuth2 的四种授权模式以及 OAuth2 协议在 Spring Security 框架中的落地情况。同时也介绍了 GitHub 授权登录、授权服务器与资源服务器的搭建、OAuth2 客户端的搭建、JWT 的使用，并结合了几个具体的案例向读者演示了如何在 Spring Security 中使用 OAuth2。在微服务日益流行的今天，掌握好 OAuth2 不仅仅可以完成第三方授权登录，也能很好地处理微服务中的鉴权问题。